Solutions Manual

Stephen Hake
John Saxon

SAXON
PUBLISHERS

Saxon Publishers gratefully acknowledges the contributions of the following individuals in the completion of this project:

Authors: Stephen Hake, John Saxon

Editorial: Chris Braun, Matt Maloney, Dana Nixon, Brian E. Rice

Editorial Support Services: Christopher Davey, Jay Allman, Shelley Turner, Jean Van Vleck, Darlene Terry

Production: Alicia Britt, Karen Hammond, Donna Jarrel, Brenda Lopez, Adriana Maxwell, Cristi D. Whiddon

Project Management: Angela Johnson, Becky Cavnar

Printed in the United States of America

ISBN: 1-59141-325-7

Manufacturing Code: 02S0804

Solutions for

Lessons and Investigations

LESSON 1, WARM-UP

a. 30

b. 44

c. 63

d. 15

e. 35

f. 18

Patterns

Final digits: **0, 2, 4, 6, 8**
Not final digits: **1, 3, 5, 7, 9**

LESSON 1, LESSON PRACTICE

a. $5 + 6 =$ **11**

b. $6 + 5 =$ **11**

c. $8 + 0 =$ **8**

d. $4 + 8 + 6 =$ **18**

e. $4 + 5 + 6 =$ **15**

f. Pattern: Some + some more = total
Problem: 5 laps + 8 laps = **13 laps**

g. **2 + 4 = 6**
4 + 2 = 6

h. **1 + 3 + 5 = 9,**
1 + 5 + 3 = 9,
3 + 1 + 5 = 9,
3 + 5 + 1 = 9,
5 + 1 + 3 = 9,
5 + 3 + 1 = 9

i. Since $7 + 3 = 10$, $N =$ **3**

j. Since $4 + 8 = 12$, $A =$ **4**

LESSON 1, MIXED PRACTICE

1. Pattern: Some
$\underline{+ \text{ Some more}}$
Total

Problem: 5 singers
$\underline{+ \ 7 \text{ singers}}$
12 singers

2. Pattern: Some + some more = total
Problem: 6 coins + 3 coins = **9 coins**

3. $9 + 4 =$ **13**

4. $8 + 2 =$ **10**

5. $\quad 4$
$\underline{+ \ 5}$
$\quad 9$
$N =$ **5**

6. $\quad 3$
$\underline{+ \ 5}$
$\quad 8$
$W =$ **3**

7. $\quad 6$
$\underline{+ \ 2}$
$\quad 8$
$P =$ **2**

8. $\quad 0$
$\underline{+ \ 8}$
$\quad 8$
$Q =$ **0**

9. $3 + 4 + 5 =$ **12**

10. $4 + 4 + 4 =$ **12**

11. $6 + 4 = 10$
$R =$ **4**

12. $1 + 5 = 6$
$X =$ **1**

13. $\quad 5$
$\quad 5$
$\underline{+ \ 5}$
$\quad 15$

14.
$$
\begin{array}{r}
8 \\
0 \\
+\ 7 \\
\hline
15
\end{array}
$$

15.
$$
\begin{array}{r}
6 \\
5 \\
+\ 4 \\
\hline
15
\end{array}
$$

16.
$$
\begin{array}{r}
9 \\
9 \\
+\ 9 \\
\hline
27
\end{array}
$$

17.
$$
\begin{array}{r}
1 \\
+\ 9 \\
\hline
10
\end{array}
$$
$M = \mathbf{1}$

18.
$$
\begin{array}{r}
9 \\
+\ 3 \\
\hline
12
\end{array}
$$
$F = \mathbf{3}$

19.
$$
\begin{array}{r}
5 \\
+\ 5 \\
\hline
10
\end{array}
$$
$Z = \mathbf{5}$

20.
$$
\begin{array}{r}
0 \\
+\ 3 \\
\hline
3
\end{array}
$$
$N = \mathbf{3}$

21. $3 + 2 + 5 + 4 + 6 = \mathbf{20}$

22. $2 + 2 + 2 + 2 + 2 + 2 + 2 = \mathbf{14}$

23. $\mathbf{6 + 3 = 9}$ or $\mathbf{3 + 6 = 9}$

24. **One possibility: $4 + 5 + 2 = 11$**

25. $\mathbf{2 + 3 + 4 = 9,}$
    $\mathbf{2 + 4 + 3 = 9,}$
    $\mathbf{3 + 2 + 4 = 9,}$
    $\mathbf{3 + 4 + 2 = 9,}$
    $\mathbf{4 + 2 + 3 = 9,}$
    $\mathbf{4 + 3 + 2 = 9}$

26. **B. 7**

LESSON 2, WARM-UP

a. **50**

b. **36**

c. **49**

d. **17**

e. **19**

f. **73**

Patterns
Final digits: **0, 5**
Numbers in both lists: **10, 20, 30, 40, 50, 60, 70, 80, 90, 100**

LESSON 2, LESSON PRACTICE

a. $10 + A = 17$
$10 + 7 = 17$
$A = \mathbf{7}$

b. $B + 11 = 12$
$1 + 11 = 12$
$B = \mathbf{1}$

c. $14 + C = 20$
$14 + 6 = 20$
$C = \mathbf{6}$

LESSON 2, MIXED PRACTICE

1. Pattern: Some + some more = total
Problem: 5 carrots + 6 carrots = **11 carrots**

2. Pattern: Some + some more = total
Problem: 7 miles + 4 miles = **11 miles**

3. $9 + 4 = 13$
$N = \mathbf{4}$

4. $7 + 8 = \mathbf{15}$

5.
$$\begin{array}{r} 7 \\ + \ 6 \\ \hline 13 \end{array}$$
$$P = 7$$

6.
$$\begin{array}{r} 7 \\ + \ 5 \\ \hline 12 \end{array}$$
$$W = 5$$

7.
$$\begin{array}{r} 4 \\ 8 \\ + \ 5 \\ \hline 17 \end{array}$$

8.
$$\begin{array}{r} 9 \\ 3 \\ + \ 7 \\ \hline 19 \end{array}$$

9.
$$\begin{array}{r} 11 \\ + \ 5 \\ \hline 16 \end{array}$$
$$B = 5$$

10.
$$\begin{array}{r} 9 \\ 7 \\ + \ 3 \\ \hline 19 \end{array}$$

11.
$$\begin{array}{r} 2 \\ 6 \\ + \ 9 \\ \hline 17 \end{array}$$

12.
$$\begin{array}{r} 3 \\ 8 \\ + \ 2 \\ \hline 13 \end{array}$$

13.
$$\begin{array}{r} 9 \\ 5 \\ + \ 3 \\ \hline 17 \end{array}$$

14.
$$\begin{array}{r} 6 \\ + \ 3 \\ \hline 9 \end{array}$$
$$M = 3$$

15.
$$\begin{array}{r} 8 \\ + \ 1 \\ \hline 9 \end{array}$$
$$Q = 1$$

16.
$$\begin{array}{r} 5 \\ + \ 2 \\ \hline 7 \end{array}$$
$$R = 2$$

17.
$$\begin{array}{r} 8 \\ + \ 2 \\ \hline 10 \end{array}$$
$$T = 2$$

18.
$$\begin{array}{r} 8 \\ 4 \\ + \ 6 \\ \hline 18 \end{array}$$

19.
$$\begin{array}{r} 9 \\ + \ 2 \\ \hline 11 \end{array}$$
$$X = 2$$

20.
$$\begin{array}{r} 5 \\ 2 \\ + \ 6 \\ \hline 13 \end{array}$$

21. $20 + 3 = 23$
$$X = 3$$

22. $4 + 5 + 6 = 15,$
$4 + 6 + 5 = 15,$
$5 + 4 + 6 = 15,$
$5 + 6 + 4 = 15,$
$6 + 4 + 5 = 15,$
$6 + 5 + 4 = 15$

23. $4 + 3 = 7$ or $3 + 4 = 7$

24. $5 + 2 = 7$ or $2 + 5 = 7$

25. One possibility: $4 + 2 + 5 = 11$

26. A. 4

LESSON 3, WARM-UP

a. 40

b. 43

c. 53

d. 54

e. 80

f. 75

g. 35; 42; 64

Vocabulary

$$\boxed{\text{addend}} + \boxed{\text{addend}} = \boxed{\text{sum}}$$

$$
\begin{array}{r}
\boxed{\text{addend}} \\
+ \boxed{\text{addend}} \\
\hline
\boxed{\text{sum}}
\end{array}
$$

LESSON 3, LESSON PRACTICE

a. The rule is **"Count down by ones."**
    6, 5, 4

b. The rule is **"Count up by threes."**
    15, 18, 21

c. The rule is "Count down by tens."
    60

d. The rule is "Count up by fours."
    12

e. **2 digits**

f. **4 digits**

g. **10 digits**

h. **9**

i. **1**

j. **0**

LESSON 3, MIXED PRACTICE

1. Pattern:
    Some
    Some more
    + Some more
    _____
    Total

    Problem: 5 dollars
    6 dollars
    + 7 dollars
    _____
    18 dollars

2. Pattern: Some + some more = total
    Problem: 9 songs + 8 songs = **17 songs**

3. (a) **3 digits**

    (b) **3 digits**

    (c) **9 digits**

4. (a) **7**

    (b) **0**

    (c) **9**

5. $9 + 3 = 12$
    $M = 3$

6. $10 + 6 = 16$
    $W = 6$

7. The rule is "Count up by tens."
    40

8. The rule is "Count down by ones."
    19

9. The rule is "Count down by fives."
    20

10. The rule is "Count up by tens."
    100

11. The rule is **"Count up by sixes."**
    24, 30, 36

12. The rule is **"Count up by threes."**
    12, 15, 18

13. The rule is **"Count up by fours."**
    16, 20, 24

14. The rule is **"Count down by nines."**
18, 9, 0

15. The rule is "Count up by fours."
16

16. The rule is "Count up by sixes."
24

17. The rule is "Count down by fives."
20

18. The rule is "Count up by threes."
12

19. 2, 4, 6, 8, 10, 12, 14, 16
16 small rectangles

20. 4, 8, 12, 16, 20, 24
24 X's

21. One possibility: 3 + 6 + 4 = 13

22.   4
    8
    7
$\underline{+\ 5}$
   24

23.   9
    5
    7
$\underline{+\ 8}$
   29

24.   8
    4
    7
$\underline{+\ 2}$
   21

25.   2
    9
    7
$\underline{+\ 5}$
   23

26. D. 7

LESSON 4, WARM-UP

a. **76**

b. **49**

c. **86**

d. **68**

e. **26**

f. **70**

g. **75; 48; 67**

Problem Solving

Number of Coins

Left	Right
0	9
1	8
2	7
3	6
4	5
5	4
6	3
7	2
8	1
9	0

LESSON 4, LESSON PRACTICE

a.
| 100 | 10 | 1 |
2 hundreds 3 tens   1 one

b.
| 100 | 10 | 1 |
2 hundreds   1 ten   3 ones

$213 is less than $231

c. (a) **Ones**

   (b) **Tens**

   (c) **Hundreds**

d. **523**

LESSON 4, MIXED PRACTICE

1. Pattern:
Some
Some more
Some more
+ Some more
Total

Problem:
3 cards
4 cards
5 cards
+ 1 cards
13 cards

2. 6 + 6 = **12**

3. 5¢, 10¢, 15¢, 20¢
20 cents

4.
$$\begin{array}{r} 4 \\ + \ 8 \\ \hline 12 \end{array}$$
$N = $ **8**

5.
$$\begin{array}{r} 4 \\ 5 \\ + \ 3 \\ \hline 12 \end{array}$$

6.
$$\begin{array}{r} 13 \\ + \ 6 \\ \hline 19 \end{array}$$
$Y = $ **6**

7.
$$\begin{array}{r} 7 \\ + \ 7 \\ \hline 14 \end{array}$$
$S = $ **7**

8. 9 + 3 = 12
$N = $ **3**

9. 3 + 5 = 8
$N = $ **3**

10. The rule is **"Count up by threes."**
18, 21, 24

11. The rule is **"Count down by sixes."**
12, 6, 0

12. The rule is **"Count up by fours."**
24, 28, 32

13. The rule is **"Count down by sevens."**
14, 7, 0

14. (a) **5 digits**

(b) **7 digits**

(c) **6 digits**

15. (a) **4**

(b) **7**

(c) **3**

16.

3 hundreds   4 tens   2 ones

17. 4 hundreds, 3 tens, 4 ones
$434

18. The rule is "Count up by sixes."
30

19. The rule is "Count down by fours."
28

20. 2, 4, 6, 8, 10, 12, 14, 16, 18, 20
20 ears

21. **Tens**

22. **5 + 6 = 11 or 6 + 5 = 11**

23. 16 + 4 = 20
$N = $ **4**

24. 19 + 6 = 25
$B = $ **6**

25. **6 + 7 + 8 = 21,**
6 + 8 + 7 = 21,
7 + 6 + 8 = 21,
7 + 8 + 6 = 21,
8 + 6 + 7 = 21,
8 + 7 + 6 = 21

26. **A. 1**

LESSON 5, WARM-UP

a. 84

b. 46

c. 92

d. 63

e. 90

f. 27; 86; 53

Patterns

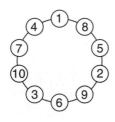

LESSON 5, LESSON PRACTICE

a. Kiyoko   Kayla

third   eighth
4 people

b. Sample answer: 5/12/1993

c. 7/4/(year)

LESSON 5, MIXED PRACTICE

1. Pattern: Some
             Some more
      + Some more
             Total

     Problem: 5 people
              6 people
        + 4 people
          15 people

2. $$\begin{array}{r} 8 \\ + 7 \\ \hline 15 \end{array}$$
$X = 7$

3. $$\begin{array}{r} 8 \\ + 6 \\ \hline 14 \end{array}$$
$Y = 6$

4. $$\begin{array}{r} 8 \\ + 4 \\ \hline 12 \end{array}$$
$Z = 4$

5. $$\begin{array}{r} 7 \\ + 6 \\ \hline 13 \end{array}$$
$N = 6$

6. $$\begin{array}{r} 7 \\ + 3 \\ \hline 10 \end{array}$$
$W = 3$

7. $$\begin{array}{r} 2 \\ + 5 \\ \hline 7 \end{array}$$
$A = 5$

8. $$\begin{array}{r} 6 \\ + 5 \\ \hline 11 \end{array}$$
$R = 6$

9. $$\begin{array}{r} 3 \\ + 2 \\ \hline 5 \end{array}$$
$T = 2$

10. August 15, 1993

11. The rule is **"Count up by threes."**
21, 24, 27

12. The rule is **"Count up by fours."**
28, 32, 36

13. The rule is **"Count up by sevens."**
49, 56, 63

14. The rule is "Count up by sixes."
36

15. The rule is "Count up by fives."
35

SOLUTIONS

16.

4 hundreds   3 tens   2 ones

17. $5 + 5 + 5 = 15$

18. Hundreds

19. 235

20. The rule is "Count up by threes."
3, 6, 9, **12**

21. 2, 4, 6, 8, 10, 12, 14
14 eyes

22.
$$\begin{array}{r} 5 \\ 8 \\ 4 \\ 7 \\ 4 \\ + \ 3 \\ \hline \mathbf{31} \end{array}$$

23.
$$\begin{array}{r} 5 \\ 7 \\ 3 \\ 8 \\ 4 \\ + \ 2 \\ \hline \mathbf{29} \end{array}$$

24.
$$\begin{array}{r} 9 \\ 7 \\ 6 \\ 5 \\ 4 \\ + \ 2 \\ \hline \mathbf{33} \end{array}$$

25.
$$\begin{array}{r} 8 \\ 7 \\ 3 \\ 5 \\ 4 \\ + \ 9 \\ \hline \mathbf{36} \end{array}$$

26.

Jenny   Jessica

third   seventh

A. 3

LESSON 6, WARM-UP

a. 43

b. 42

c. 56

d. 55

e. 75

f. 74

Problem Solving

Number of Coins

Left	Right
2	7
3	6
4	5
5	4
6	3
7	2

LESSON 6, LESSON PRACTICE

a.
$$\begin{array}{r} 14 \\ - \ 8 \\ \hline \mathbf{6} \end{array}$$
check: $6 + 8 = 14$

b.
$$\begin{array}{r} 9 \\ - \ 3 \\ \hline \mathbf{6} \end{array}$$
check: $6 + 3 = 9$

c.
$$\begin{array}{r} 15 \\ - \ 7 \\ \hline \mathbf{8} \end{array}$$
check: $8 + 7 = 15$

d.
$$\begin{array}{r} 11 \\ - \ 4 \\ \hline \mathbf{7} \end{array}$$
check: $7 + 4 = 11$

e.
$$\begin{array}{r} 12 \\ - \ 5 \\ \hline \mathbf{7} \end{array}$$
check: $7 + 5 = 12$

f. $5 + 6 = \mathbf{11}$
$6 + 5 = \mathbf{11}$
$11 - 6 = \mathbf{5}$
$11 - 5 = \mathbf{6}$

g. **Sample answer: We can check a subtraction answer by adding the difference to the number subtracted. For example, we can check** $7 - 3 = 4$ **by adding** $4 + 3 = 7.$

LESSON 6, MIXED PRACTICE

1.
$$
\begin{array}{r}
14 \\
- 5 \\
\hline
\mathbf{9}
\end{array}
$$

2.
$$
\begin{array}{r}
15 \\
- 8 \\
\hline
\mathbf{7}
\end{array}
$$

3.
$$
\begin{array}{r}
9 \\
- 4 \\
\hline
\mathbf{5}
\end{array}
$$

4.
$$
\begin{array}{r}
11 \\
- 7 \\
\hline
\mathbf{4}
\end{array}
$$

5.
$$
\begin{array}{r}
12 \\
- 8 \\
\hline
\mathbf{4}
\end{array}
$$

6.
$$
\begin{array}{r}
11 \\
- 6 \\
\hline
\mathbf{5}
\end{array}
$$

7.
$$
\begin{array}{r}
15 \\
- 7 \\
\hline
\mathbf{8}
\end{array}
$$

8.
$$
\begin{array}{r}
9 \\
- 6 \\
\hline
\mathbf{3}
\end{array}
$$

9.
$$
\begin{array}{r}
13 \\
- 5 \\
\hline
\mathbf{8}
\end{array}
$$

10.
$$
\begin{array}{r}
12 \\
- 6 \\
\hline
\mathbf{6}
\end{array}
$$

11.
$$
\begin{array}{r}
8 \\
+ 9 \\
\hline
17
\end{array}
$$
$N = \mathbf{9}$

12.
$$
\begin{array}{r}
6 \\
+ 8 \\
\hline
14
\end{array}
$$
$A = \mathbf{6}$

13. $3 + 8 = 11$
$W = \mathbf{8}$

14. $5 + 8 = 13$
$M = \mathbf{8}$

15. $4 + 6 = 10$
$6 + 4 = 10$
$10 - 4 = 6$
$10 - 6 = 4$

16. The rule is **"Count up by twos."**
22, 24, 26

17. The rule is **"Count up by sevens."**
42, 49, 56

18. The rule is **"Count up by fours."**
32, 36, 40

19. The tenth month of the year is October. October has **31 days.**

20.

100	10	1
3 hundreds	**2 tens**	**6 ones**

21. **Ones**

22. $6 + 7 = 13$
$N = \mathbf{7}$

23. $8 + 8 = 16$
$A = \mathbf{8}$

24. $17 + 3 = 20$
$M = \mathbf{3}$

25. $3 + 4 + 5 = \mathbf{12,}$
$3 + 5 + 4 = \mathbf{12,}$
$4 + 3 + 5 = \mathbf{12,}$
$4 + 5 + 3 = \mathbf{12,}$
$5 + 3 + 4 = \mathbf{12,}$
$5 + 4 + 3 = \mathbf{12}$

26. Pattern: Some
                   + Some more
                   ‾‾‾‾‾‾‾‾‾‾‾‾
                     Total

   Problem: 2 children
              + 3 children
              ‾‾‾‾‾‾‾‾‾‾‾‾
                5 children

   C. 2 + 3 = 5

LESSON 7, WARM-UP

a. 37

b. 53

c. 96

d. 83

e. 96

f. 68

Patterns

(a) Since 24 months after January is January, 25 months after January is **February.**

(b) Twenty-four months before Valentine's Day is in February. Twenty-two months before Valentine's Day is two months after that, which is **April.**

LESSON 7, LESSON PRACTICE

a. **Zero**

b. **Eighty-one**

c. **Ninety-nine**

d. **Five hundred fifteen**

e. **Four hundred forty-four**

f. **Nine hundred nine**

g. **19**

h. **91**

i. **524**

j. **860**

k. One hundred, three tens, and two ones is 132.
One hundred thirty-two

LESSON 7, MIXED PRACTICE

1. Pattern: Some   + some more = total
   Problem: 8 dollars + 6 dollars = **14 dollars**

2. Pattern: Some   + some more = total
   Problem: 8 ounces + 8 ounces = **16 ounces**

3. $7 + 4 = 11$
   $N = $ **4**

4. $8 + 7 = 15$
   $N = $ **7**

5.   13
        − 5
        ‾‾‾‾
          8

6.   16
        − 8
        ‾‾‾‾
          8

7.   13
        − 7
        ‾‾‾‾
          6

8.   12
        − 8
        ‾‾‾‾
          4

9. **214**

10. **532**

11. **Three hundred one**

12. **Three hundred twenty**

13. Three hundreds, one ten, and two ones is 312.
Three hundred twelve

14. $3 + 5 = 8$ or $5 + 3 = 8$

15. The rule is **"Count up by sixes."**
30, 36, 42

16. The rule is **"Count up by threes."**
24, 27, 30

17. The rule is "Count up by sevens."
49

18. The rule is "Count up by eights."
48

19. 3 hundreds, 0 tens, 3 ones
$303

20. $7 + 8 = 15$
   $8 + 7 = 15$
   $15 - 7 = 8$
   $15 - 8 = 7$

21.

Brad's sister   Brad

sixth   twelfth

5 people

22. 5¢, 10¢, 15¢, 20¢, 25¢, 30¢
30 cents

23. $4 + 7 + 8 + 5 + 4 = $ **28**

24. $2 + 3 + 5 + 8 + 5 = $ **23**

25. $5 + 8 + 6 + 4 + 3 + 7 + 2 = $ **35**

26. $12 - 5 = 7$ is a subtraction fact for the fact
family 5, 7, and 12.
$7 + 5 = 12$ is an addition fact for the fact
family 5, 7, and 12.
A. $7 + 5 = 12$

LESSON 8, WARM-UP

a. **65**

b. **72**

c. **57**

d. **94**

e. **90**

f. **89**

Problem Solving

Number of Coins

Left	Right
2	7
3	6
4	5

LESSON 8, LESSON PRACTICE

a. $53
      $+ \$6$
      $59

b. $14
      $+ \$75$
      $89

c. $36
      $+ \$42$
      $78

d. $27
      $+ \$51$
      $78

e. $15
      $+ \$21$
      $36

f. $32
      $+ \$6$
      $38

LESSON 8, MIXED PRACTICE

1. **343**

2. **307**

3.   **Five hundred ninety-two**

4. $\begin{array}{r} 6 \\ +\ 6 \\ \hline 12 \end{array}$

$N = \mathbf{6}$

5. $\begin{array}{r} 7 \\ +\ 3 \\ \hline 10 \end{array}$

$R = \mathbf{3}$

6. $\begin{array}{r} 8 \\ +\ 6 \\ \hline 14 \end{array}$

$T = \mathbf{6}$

7. $\begin{array}{r} 8 \\ +\ 5 \\ \hline 13 \end{array}$

$N = \mathbf{5}$

8. $\begin{array}{r} \$25 \\ +\ \$14 \\ \hline \mathbf{\$39} \end{array}$

9. $\begin{array}{r} \$85 \\ +\ \$14 \\ \hline \mathbf{\$99} \end{array}$

10. $\begin{array}{r} \$22 \\ +\ \ \$6 \\ \hline \mathbf{\$28} \end{array}$

11. $\begin{array}{r} \$40 \\ +\ \$38 \\ \hline \mathbf{\$78} \end{array}$

12. $\begin{array}{r} 13 \\ -\ 9 \\ \hline \mathbf{4} \end{array}$

13. $\begin{array}{r} 17 \\ -\ 5 \\ \hline \mathbf{12} \end{array}$

14. $\begin{array}{r} 17 \\ -\ 8 \\ \hline \mathbf{9} \end{array}$

15. $\begin{array}{r} 14 \\ -\ 6 \\ \hline \mathbf{8} \end{array}$

16.   Pattern: $\begin{array}{l} \text{Some} \\ +\ \text{Some more} \\ \hline \text{Total} \end{array}$

Problem: $\begin{array}{r} \$23 \\ +\ \$42 \\ \hline \mathbf{\$65} \end{array}$

17.   One hundred and eight ones is 108.
One hundred eight

18.   **8/5/94**

19.   The rule is **"Count up by threes."**
21, 24, 27

20.   The rule is **"Count up by sevens."**
49, 56, 63

21. $\begin{array}{r} 5 \\ 8 \\ 7 \\ 6 \\ 4 \\ +\ 3 \\ \hline \mathbf{33} \end{array}$

22. $\begin{array}{r} 9 \\ 7 \\ 6 \\ 4 \\ 8 \\ +\ 7 \\ \hline \mathbf{41} \end{array}$

23. $\begin{array}{r} 2 \\ 5 \\ 7 \\ 3 \\ 5 \\ +\ 4 \\ \hline \mathbf{26} \end{array}$

24.   **5 + 6 + 7 = 18,**
5 + 7 + 6 = 18,
6 + 5 + 7 = 18,
6 + 7 + 5 = 18,
7 + 5 + 6 = 18,
7 + 6 + 5 = 18

25. 7 + 8 = 15
8 + 7 = 15
15 − 7 = 8
15 − 8 = 7

26. 7 + ◆ = 15
7 + 8 = 15, so ◆ = 8
8 − 7 ≠ 15, so ◆ − 7 = 15 is not true
A. ◆ − 7 = 15

LESSON 9, WARM-UP

a. 56

b. 55

c. 67

d. 66

e. 44

f. 43

Patterns

10 days after Saturday: Count forward by 7 days and then by 3 days: **Tuesday.**

10 days before Saturday: Count backward by 7 days and then by 3 days: **Wednesday.**

70 days after Saturday: Count up by 7's to 70. There are no days "left over," so 70 days after Saturday is **Saturday.**

LESSON 9, LESSON PRACTICE

a.
$$\begin{array}{r} \overset{1}{\$36} \\ + \ \$29 \\ \hline \$65 \end{array}$$

b.
$$\begin{array}{r} \overset{1}{\$47} \\ + \ \ \$8 \\ \hline \$55 \end{array}$$

c.
$$\begin{array}{r} \overset{1}{\$57} \\ + \ \$13 \\ \hline \$70 \end{array}$$

d.
$$\begin{array}{r} \overset{1}{68} \\ + \ 24 \\ \hline 92 \end{array}$$

e.
$$\begin{array}{r} \overset{1}{\$59} \\ + \ \ \$8 \\ \hline \$67 \end{array}$$

f.
$$\begin{array}{r} \overset{1}{46} \\ + \ 25 \\ \hline 71 \end{array}$$

LESSON 9, MIXED PRACTICE

1. 613

2. 901

3. Nine hundred forty-one

4.
$$\begin{array}{r} 6 \\ + \ 5 \\ \hline 11 \end{array}$$
$F = 5$

5.
$$\begin{array}{r} 7 \\ + \ 6 \\ \hline 13 \end{array}$$
$G = 6$

6.
$$\begin{array}{r} 4 \\ + \ 11 \\ \hline 15 \end{array}$$
$H = 4$

7.
$$\begin{array}{r} 9 \\ + \ 7 \\ \hline 16 \end{array}$$
$N = 7$

8.
$$\begin{array}{r} \overset{1}{33} \\ + \ 8 \\ \hline 41 \end{array}$$

9.
$$\begin{array}{r} \overset{1}{\$47} \\ + \ \$18 \\ \hline \$65 \end{array}$$

10.
$$\begin{array}{r} \overset{1}{2}7 \\ + 69 \\ \hline 96 \end{array}$$

11.
$$\begin{array}{r} \overset{1}{\$}49 \\ + \$25 \\ \hline \$74 \end{array}$$

12.
$$\begin{array}{r} 17 \\ - 8 \\ \hline 9 \end{array}$$

13.
$$\begin{array}{r} 12 \\ - 6 \\ \hline 6 \end{array}$$

14.
$$\begin{array}{r} 9 \\ - 7 \\ \hline 2 \end{array}$$

15.
$$\begin{array}{r} 13 \\ - 6 \\ \hline 7 \end{array}$$

16. **Sum**

17. **Difference**

18. The twelfth month is December. Two months after December is **February.**

19. The rule is **"Count up by sixes."**
48, 54, 60

20. The rule is **"Count up by sevens."**
49, 56, 63

21. **8**

22.
$$\begin{array}{r} \overset{1}{2}8 \\ + 6 \\ \hline 34 \end{array}$$

23.
$$\begin{array}{r} \overset{1}{\$}47 \\ + \$28 \\ \hline \$75 \end{array}$$

24.
$$\begin{array}{r} \overset{1}{3}5 \\ + 27 \\ \hline 62 \end{array}$$

25. Pattern:
$$\begin{array}{r} \text{Some} \\ + \text{ Some more} \\ \hline \text{Total} \end{array}$$

Problem:
$$\begin{array}{r} \overset{1}{\$}28 \\ + \$17 \\ \hline \$45 \end{array}$$

$28 + $17 = $45

26. One hundred and three tens is 130.
D. 130

LESSON 10, WARM-UP

a. **37**

b. **55**

c. **52**

d. **44**

e. **65**

f. **64**

Problem Solving
Terrell has 7 coins in his right pocket and 3 coins in his left pocket.

LESSON 10, LESSON PRACTICE

a. **Odd**

b. **Even**

c. **Odd**

d. **Even**

e. **630, 632, 634, 636, 638**

LESSON 10, MIXED PRACTICE

1. 542

2. 619

3. $4 + 7 = 11$
$7 + 4 = 11$
$11 - 4 = 7$
$11 - 7 = 4$

4. Nine hundred three

5. Seven hundred forty-six

6. 501, 503, 505, 507, 509

7. $\begin{array}{r} 7 \\ + 7 \\ \hline 14 \end{array}$
$N = 7$

8. $\begin{array}{r} 7 \\ + 6 \\ \hline 13 \end{array}$
$P = 7$

9. $\begin{array}{r} 12 \\ + 2 \\ \hline 14 \end{array}$
$Q = 2$

10. $\begin{array}{r} 6 \\ + 5 \\ \hline 11 \end{array}$
$R = 6$

11. $\begin{array}{r} 15 \\ - 7 \\ \hline 8 \end{array}$

12. $\begin{array}{r} 14 \\ - 7 \\ \hline 7 \end{array}$

13. $\begin{array}{r} 17 \\ - 8 \\ \hline 9 \end{array}$

14. $\begin{array}{r} 11 \\ - 6 \\ \hline 5 \end{array}$

15. $\begin{array}{r} \overset{1}{\$}25 \\ + \$38 \\ \hline \$63 \end{array}$

16. $\begin{array}{r} \overset{1}{\$}19 \\ + \$34 \\ \hline \$53 \end{array}$

17. $\begin{array}{r} \overset{1}{4}2 \\ + \ 8 \\ \hline 50 \end{array}$

18. $\begin{array}{r} \overset{1}{1}7 \\ + 49 \\ \hline 66 \end{array}$

19. The rule is **"Count up by threes."**
27, 30, 33

20. The rule is "Count up by sixes."
6, 12, 18, 24, 30, 36, 42, **48**

21. Pattern: $\begin{array}{r} \text{Some} \\ + \text{ Some more} \\ \hline \text{Total} \end{array}$

    Problem: $\begin{array}{r} \$6 \\ \$12 \\ + \$20 \\ \hline \$38 \end{array}$

    $\$6 + \$12 + \$20 = \$38$

22. $2 + 3 + 5 + 7 + 8 + 4 + 5 =$ **34**

23. **Sample answer: 9/22/2004 or 9/22/04**

24. Two hundreds and three tens is 230.
Two hundred thirty

25. The smallest three-digit number is 100. The largest two-digit even number is the largest even number less than 100, which is **98.**

26. $\Delta + 4 = 12$
$8 + 4 = 12$, so $\Delta = 8$
$12 + 4 \neq 8$, so $12 + 4 = \Delta$ is not true
C. $12 + 4 = \Delta$

INVESTIGATION 1

1. 25

2. 16

3. 40

4. 85

5. (a) −15
    (b) negative fifteen

6. 0, −5, −10, −15

7. −3

8. −6

9. $-3 < 1$

10. $3 > 2$

11. $2 + 3 = 3 + 2$

12. $-4 > -5$

13. Negative one is less than zero.

14. $-2 > -3$

15. $1 > -1$

16. $4 = 2 + 2$

17. $-2 < 0$

18. $4 > 1 + 2$

LESSON 11, WARM-UP

a. 58

b. 57

c. 87

d. 86

e. 96

f. 95

Problem Solving

An even number of objects can be divided into two equal groups. The stacks were of equal height, which meant the number of boxes in each stack was equal.

LESSON 11, LESSON PRACTICE

a.
$$\begin{array}{r} 4 \text{ marigolds} \\ + \ N \text{ marigolds} \\ \hline 12 \text{ marigolds} \end{array}$$
$N = $ **8 marigolds**

b.
$$\begin{array}{r} N \text{ agates} \\ + \ 8 \text{ agates} \\ \hline 15 \text{ agates} \end{array}$$
$N = $ **7 agates**

LESSON 11, MIXED PRACTICE

1. Pattern:
$$\begin{array}{l} \text{Some} \\ + \ \text{Some more} \\ \hline \text{Total} \end{array}$$

    Problem:
$$\begin{array}{r} 4 \text{ horses} \\ + \ 13 \text{ horses} \\ \hline \textbf{17 horses} \end{array}$$

2. Pattern: Some + some more = total
    Problem: 6 pages + N pages = 13 pages
    $N = $ **7 pages**

3. 642

4. $-12 < 0$

5.

    $-2 \ \boxed{<} \ 2$

6. 571, 573, 575, 577, 579

7. (a) **15**

    (b) **−4**

8. An **even number** of objects can be separated into two equal groups.

9.
$$\begin{array}{r} 12 \\ + 6 \\ \hline 18 \end{array}$$
$B = 6$

10.
$$\begin{array}{r} 7 \\ + 8 \\ \hline 15 \end{array}$$
$N = 7$

11.
$$\begin{array}{r} 11 \\ + 1 \\ \hline 12 \end{array}$$
$A = 1$

12.
$$\begin{array}{r} 4 \\ + 10 \\ \hline 14 \end{array}$$
$M = 4$

13.
$$\begin{array}{r} 12 \\ - 3 \\ \hline 9 \end{array}$$

14.
$$\begin{array}{r} 14 \\ - 7 \\ \hline 7 \end{array}$$

15.
$$\begin{array}{r} 12 \\ - 8 \\ \hline 4 \end{array}$$

16.
$$\begin{array}{r} 13 \\ - 6 \\ \hline 7 \end{array}$$

17.
$$\begin{array}{r} \overset{1}{7}4 \\ + 18 \\ \hline 92 \end{array}$$

18.
$$\begin{array}{r} \overset{1}{9}3 \\ + 39 \\ \hline 132 \end{array}$$

19.
$$\begin{array}{r} \overset{1}{2}8 \\ + 45 \\ \hline 73 \end{array}$$

20.
$$\begin{array}{r} \overset{1}{2}8 \\ + 47 \\ \hline 75 \end{array}$$

21. The rule is "Count down by threes."
3, 0, −3

22. The rule is "Count up by sixes."
48, 54, 60

23.   **5 + 9 = 14**
    9 + 5 = 14
  14 − 5 = 9
  14 − 9 = 5

24. 4 + 3 + 5 + 8 + 7 + 6 + 2 = **35**

25. **7 + 8 + 9 = 24,**
    7 + 9 + 8 = 24,
    8 + 7 + 9 = 24,
    8 + 9 + 7 = 24,
    9 + 7 + 8 = 24,
    9 + 8 + 7 = 24

26.   3 + ▲ = 7
    3 + 4 = 7, so ▲ = 4
  ▲ + ■ = 4 + 5 = 9
  D. 9

LESSON 12, WARM-UP

a. **81**

b. **72**

c. **53**

d. **75**

e. **76**

f. 91

Patterns

1	2	3	4	5	6	7	8	9	10
11	12	13	14	15	16	17	18	19	20
21	22	23	24	25	26	27	28	29	30
31	32	33	34	35	36	37	38	39	40
41	42	43	44	45	46	47	48	49	50
51	52	53	54	55	56	57	58	59	60
61	62	63	64	65	66	67	68	69	70
71	72	73	74	75	76	77	78	79	80
81	82	83	84	85	86	87	88	89	90
91	92	93	94	95	96	97	98	99	100

LESSON 12, LESSON PRACTICE

a.
$$
\begin{array}{r} 6 \\ +\ 8 \\ \hline 14 \end{array}
\qquad
\text{check:}
\qquad
\begin{array}{r} 14 \\ -\ 8 \\ \hline 6 \end{array}
$$
$N = \mathbf{8}$

b.
$$
\begin{array}{r} 5 \\ +\ 2 \\ \hline 7 \end{array}
\qquad
\text{check:}
\qquad
\begin{array}{r} 7 \\ -\ 5 \\ \hline 2 \end{array}
$$
$N = \mathbf{7}$

c.
$$
\begin{array}{r} 2 \\ +\ 7 \\ \hline 9 \end{array}
\qquad
\text{check:}
\qquad
\begin{array}{r} 9 \\ -\ 7 \\ \hline 2 \end{array}
$$
$N = \mathbf{7}$

d.
$$
\begin{array}{r} 5 \\ +\ 7 \\ \hline 12 \end{array}
\qquad
\text{check:}
\qquad
\begin{array}{r} 12 \\ -\ 7 \\ \hline 5 \end{array}
$$
$N = \mathbf{12}$

LESSON 12, MIXED PRACTICE

1. Pattern: Some + some more = total
Problem: 9 acorns + N acorns = 17 acorns
$N = \mathbf{8\ acorns}$

2. Pattern:
$$
\begin{array}{l} \text{Some} \\ +\ \text{Some more} \\ \hline \text{Total} \end{array}
$$
Problem:
$$
\begin{array}{r} \overset{1}{3}5\ \text{butterflies} \\ +\ 27\ \text{butterflies} \\ \hline \mathbf{62\ butterflies} \end{array}
$$

3. 715

4. One hundred and four ones is 104.
One hundred four

5. 6/7/02

6. The largest three-digit number will have a 9 in the hundreds place.
946

7. 70

8.
$$
\begin{array}{r} 11 \\ +\ 4 \\ \hline 15 \end{array}
$$
$N = \mathbf{4}$

9.
$$
\begin{array}{r} 8 \\ +\ 7 \\ \hline 15 \end{array}
$$
$A = \mathbf{8}$

10.
$$
\begin{array}{r} 9 \\ +\ 6 \\ \hline 15 \end{array}
$$
$N = \mathbf{6}$

11.
$$
\begin{array}{r} 6 \\ +\ 9 \\ \hline 15 \end{array}
$$
$A = \mathbf{9}$

12.
$$
\begin{array}{r} 14 \\ -\ 6 \\ \hline 8 \end{array}
$$
$N = \mathbf{14}$

13.
$$
\begin{array}{r} 16 \\ -\ 8 \\ \hline \mathbf{8} \end{array}
$$

14.
$$
\begin{array}{r} 14 \\ -\ 7 \\ \hline \mathbf{7} \end{array}
$$

15.
$$
\begin{array}{r} 12 \\ -\ 5 \\ \hline 7 \end{array}
$$
$A = \mathbf{5}$

16.
$$\begin{array}{r} 12 \\ -\ 6 \\ \hline 6 \end{array}$$
$B = \mathbf{12}$

17.
$$\begin{array}{r} 13 \\ -\ 5 \\ \hline 8 \end{array}$$
$C = \mathbf{5}$

18.
$$\begin{array}{r} \overset{1}{\$48} \\ +\ \$16 \\ \hline \mathbf{\$64} \end{array}$$

19.
$$\begin{array}{r} \overset{1}{\$37} \\ +\ \$14 \\ \hline \mathbf{\$51} \end{array}$$

20. The rule is "Count up by sevens."
49, 56, 63

21. The rule is "Count up by threes."
27, 30, 33

22. 5¢, 10¢, 15¢, 20¢, 25¢, 30¢, 35¢, 40¢, **45¢**

23.

$-3 \;\text{\small\textcircled{$>$}}\; -5$

24. (a) **Negative eleven**

(b) **−11**

25. $7 + 3 + 8 + 5 + 4 + 3 + 2 = \mathbf{32}$

26. **B.** $N - 5$

LESSON 13, WARM-UP

a. **90**

b. **93**

c. **55**

d. **92**

e. **92**

f. **55**

Problem Solving
Counting by 2's: 22, 24, 26, 28
Counting by 3's: 21, 24, 27
24 dominoes

LESSON 13, LESSON PRACTICE

a.
$$\begin{array}{r} \overset{1\ 1}{\$579} \\ +\ \$186 \\ \hline \mathbf{\$765} \end{array}$$

b.
$$\begin{array}{r} \overset{1}{408} \\ +\ 243 \\ \hline \mathbf{651} \end{array}$$

c.
$$\begin{array}{r} \overset{1\ 1}{\$498} \\ +\ \ \$89 \\ \hline \mathbf{\$587} \end{array}$$

d.
$$\begin{array}{r} \overset{1}{\$458} \\ +\ \$336 \\ \hline \mathbf{\$794} \end{array}$$

e.
$$\begin{array}{r} \overset{1}{\underset{1}{56}} \\ +\ 569 \\ \hline \mathbf{625} \end{array}$$

LESSON 13, MIXED PRACTICE

1. Pattern:
$$\begin{array}{l} \text{Some} \\ +\ \text{Some more} \\ \hline \text{Total} \end{array}$$

Problem:
$$\begin{array}{r} \overset{1}{77}\ \text{children} \\ +\ 19\ \text{children} \\ \hline \mathbf{96\ children} \end{array}$$

2. Pattern: Some + some more = total
Problem: 5 girls + N boys = 12 children
$N = \mathbf{7\ boys}$

3. **Nine hundred thirteen**

4. 743

5. $75 > -80$

6. (a) 413 $\boxed{>}$ 314

(b)

-5 $\boxed{-4}$ -3 -2 -1 0 1 2 $\boxed{3}$ 4 5

-4 $\boxed{<}$ 3

7. $7 + 9 = 16$
$9 + 7 = 16$
$16 - 7 = 9$
$16 - 9 = 7$

8. (a) **84**

(b) **−5**

9.
$$\begin{array}{r} \overset{1}{}\$475 \\ + \ \$332 \\ \hline \mathbf{\$807} \end{array}$$

10.
$$\begin{array}{r} \overset{1}{}\$714 \\ + \ \$226 \\ \hline \mathbf{\$940} \end{array}$$

11.
$$\begin{array}{r} \overset{1\ 1}{743} \\ + \ 187 \\ \hline \mathbf{930} \end{array}$$

12.
$$\begin{array}{r} \overset{1\ 1}{576} \\ + \ 228 \\ \hline \mathbf{804} \end{array}$$

13.
$$\begin{array}{r} 13 \\ + \ \ 4 \\ \hline 17 \end{array}$$
$K = \mathbf{4}$

14.
$$\begin{array}{r} 10 \\ + \ \ 5 \\ \hline 15 \end{array}$$
$N = \mathbf{5}$

15.
$$\begin{array}{r} 15 \\ + \ \ 2 \\ \hline 17 \end{array}$$
$A = \mathbf{2}$

16.
$$\begin{array}{r} 6 \\ + \ 10 \\ \hline 16 \end{array}$$
$N = \mathbf{6}$

17.
$$\begin{array}{r} 8 \\ - \ 6 \\ \hline 2 \end{array}$$
$N = \mathbf{6}$

18.
$$\begin{array}{r} 17 \\ - \ \ 8 \\ \hline 9 \end{array}$$

19.
$$\begin{array}{r} 13 \\ - \ \ 7 \\ \hline \mathbf{6} \end{array}$$

20.
$$\begin{array}{r} 15 \\ - \ \ 8 \\ \hline 7 \end{array}$$
$N = \mathbf{15}$

21.
$$\begin{array}{r} 14 \\ - \ \ 8 \\ \hline 6 \end{array}$$
$N = \mathbf{8}$

22.
$$\begin{array}{r} 16 \\ - \ \ 7 \\ \hline 9 \end{array}$$
$A = \mathbf{7}$

23.
$$\begin{array}{r} 16 \\ - \ \ 9 \\ \hline 7 \end{array}$$
$N = \mathbf{16}$

24.
$$\begin{array}{r} \overset{1}{}\$49 \\ + \ \$76 \\ \hline \mathbf{\$125} \end{array}$$

25. (a) The rule is "Count up by sevens."
49, 56, 63

(b) The rule is "Count down by fives."
0, −5, −10

26. Five tens and eight ones is 58.
C. 58

22

LESSON 14, WARM-UP

a. 700

b. 900

c. 550

d. 92

e. 73

f. 77

Patterns

1	2	3	4	5	6	7	8	9	10
11	12	13	14	15	16	17	18	19	20
21	22	23	24	25	26	27	28	29	30
31	32	33	34	35	36	37	38	39	40
41	42	43	44	45	46	47	48	49	50
51	52	53	54	55	56	57	58	59	60
61	62	63	64	65	66	67	68	69	70
71	72	73	74	75	76	77	78	79	80
81	82	83	84	85	86	87	88	89	90
91	92	93	94	95	96	97	98	99	100

All shaded squares contain even numbers.

LESSON 14, LESSON PRACTICE

a.
$$\begin{array}{r} \$485 \\ -\ \$242 \\ \hline \mathbf{\$243} \end{array}$$

b.
$$\begin{array}{r} \$56 \\ -\ \$33 \\ \hline \mathbf{\$23} \end{array}$$

c.
$$\begin{array}{r} 97 \\ -\ 53 \\ \hline \mathbf{44} \end{array}$$

d.
$$\begin{array}{r} 54 \\ -\ 23 \\ \hline \mathbf{31} \end{array}$$

e.
$$\begin{array}{r} 24 \\ +\ 41 \\ \hline 65 \end{array}$$
$Q = \mathbf{41}$

f.
$$\begin{array}{r} 36 \\ +\ 31 \\ \hline 67 \end{array}$$
$M = \mathbf{36}$

g.
$$\begin{array}{r} 36 \\ +\ 63 \\ \hline 99 \end{array}$$
$W = \mathbf{63}$

h.
$$\begin{array}{r} 54 \\ +\ 45 \\ \hline 99 \end{array}$$
$Y = \mathbf{54}$

LESSON 14, MIXED PRACTICE

1. Pattern:
$$\begin{array}{l} \text{Some} \\ +\ \text{Some more} \\ \hline \text{Total} \end{array}$$

   Problem:
$$\begin{array}{l} \text{42 red surfboards} \\ +\ \text{17 red surfboards} \\ \hline \textbf{59 red surfboards} \end{array}$$

2. Pattern:
$$\begin{array}{l} \text{Some} \\ +\ \text{Some more} \\ \hline \text{Total} \end{array}$$

   Problem:
$$\begin{array}{l} \text{4 green grasshoppers} \\ +\ N \text{ green grasshoppers} \\ \hline \text{11 green grasshoppers} \end{array}$$

   $N = \textbf{7 green grasshoppers}$

3. A number less than 200 has the digit 1 in the hundreds place. A number that ends with 2 is even.
   132

4. $2 + 7 = 9$
   $7 + 2 = 9$
   $9 - 7 = 2$
   $9 - 2 = 7$

5.
$$\begin{array}{r} 824 \\ -\ 713 \\ \hline \mathbf{111} \end{array}$$

6. (a) 704 $\bigcirc\!\!\!>$ 407

   (b)
$$\overleftarrow{\underset{-5\ -4\ -3\ -2\ -1\ \ 0\ \ 1\ \ 2\ \ 3\ \ 4\ \ 5}{\rule{6cm}{0.4pt}}}\!\!\!\rightarrow$$

   $-3 \bigcirc\!\!\!> -5$

7. 1st month: January = 31 days

2nd month: February = 28 days

$$\begin{array}{r} 31 \text{ days} \\ + \ 28 \text{ days} \\ \hline \mathbf{59 \text{ days}} \end{array}$$

8. **45**

9.
$$\begin{array}{r} \overset{1\,1}{\$346} \\ + \ \$298 \\ \hline \mathbf{\$644} \end{array}$$

10.
$$\begin{array}{r} \overset{1\,1}{499} \\ + \ 275 \\ \hline \mathbf{774} \end{array}$$

11.
$$\begin{array}{r} \overset{1\,1}{\$421} \\ + \ \$389 \\ \hline \mathbf{\$810} \end{array}$$

12.
$$\begin{array}{r} 506 \\ + \ 210 \\ \hline \mathbf{716} \end{array}$$

13.
$$\begin{array}{r} \$438 \\ - \ \$206 \\ \hline \mathbf{\$232} \end{array}$$

14.
$$\begin{array}{r} 17 \\ - \ 8 \\ \hline 9 \end{array}$$

$A = \mathbf{8}$

15.
$$\begin{array}{r} 7 \\ + \ 7 \\ \hline 14 \end{array}$$

$B = \mathbf{7}$

16.
$$\begin{array}{r} 5 \\ - \ 3 \\ \hline 2 \end{array}$$

$C = \mathbf{3}$

17.
$$\begin{array}{r} 8 \\ + \ 7 \\ \hline 15 \end{array}$$

$D = \mathbf{7}$

18.
$$\begin{array}{r} 15 \\ - \ 6 \\ \hline 9 \end{array}$$

$K = \mathbf{6}$

19.
$$\begin{array}{r} 5 \\ + \ 8 \\ \hline 13 \end{array}$$

$N = \mathbf{8}$

20.
$$\begin{array}{r} 476 \\ - \ 252 \\ \hline \mathbf{224} \end{array}$$

21.
$$\begin{array}{r} 47 \\ - \ 16 \\ \hline \mathbf{31} \end{array}$$

22.
$$\begin{array}{r} 28 \\ - \ 13 \\ \hline \mathbf{15} \end{array}$$

23.
$$\begin{array}{r} 75 \\ + \ 12 \\ \hline 87 \end{array}$$

$T = \mathbf{12}$

24.
$$\begin{array}{r} 24 \\ + \ 43 \\ \hline 67 \end{array}$$

$E = \mathbf{43}$

25. (a) The rule is "Count down by nines."
54, 45, 36

(b) The rule is "Count down by fours."
0, −4, −8

26. $\square - 7 = 2$
$9 - 7 = 2$, so $\square = 9$
$7 - 9 \neq 2$, so $7 - \square = 2$ is not true
A. $7 - \square = 2$

LESSON 15, WARM-UP

a. **900**

b. **920**

c. **354**

d. 93

e. 64

f. 46

Patterns

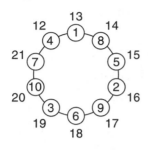

LESSON 15, LESSON PRACTICE

a. See lesson for model of how to illustrate the subtraction.

$$\begin{array}{r} \$53 \\ -\ \$29 \\ \hline \mathbf{\$24} \end{array}$$

b. See lesson for model of how to illustrate the subtraction.

$$\begin{array}{r} \$56 \\ -\ \$27 \\ \hline \mathbf{\$29} \end{array}$$

c. See lesson for model of how to illustrate the subtraction.

$$\begin{array}{r} \$42 \\ -\ \$24 \\ \hline \mathbf{\$18} \end{array}$$

d. See lesson for model of how to illustrate the subtraction.

$$\begin{array}{r} \$60 \\ -\ \$27 \\ \hline \mathbf{\$33} \end{array}$$

e.

$$\begin{array}{r} \overset{5}{\cancel{6}}{}^{1}3 \\ -\ 3\ 6 \\ \hline \mathbf{2\ 7} \end{array}$$

f.

$$\begin{array}{r} \overset{3}{\cancel{4}}{}^{1}0 \\ -\ 1\ 3 \\ \hline \mathbf{2\ 7} \end{array}$$

g.

$$\begin{array}{r} \overset{6}{\cancel{7}}{}^{1}2 \\ -\ 2\ 4 \\ \hline \mathbf{4\ 8} \end{array}$$

h.

$$\begin{array}{r} \overset{1}{\cancel{2}}{}^{1}4 \\ -\ 1\ 8 \\ \hline \mathbf{6} \end{array}$$

LESSON 15, MIXED PRACTICE

1. Pattern:

$$\begin{array}{r} \text{Some} \\ +\ \text{Some more} \\ \hline \text{Total} \end{array}$$

Problem:

$$\begin{array}{r} \overset{1}{6}18 \text{ acorns} \\ +\ 117 \text{ acorns} \\ \hline \mathbf{735 \text{ acorns}} \end{array}$$

2.

$$\begin{array}{r} 76 \text{ knights} \\ -\ 16 \text{ knights} \\ \hline \mathbf{60 \text{ knights}} \end{array}$$

3. Three hundreds are less than four hundreds, so 3 is in the hundreds place. An even number may end in 6.
376

4. **Six hundred five**

5. **10**

6. (a) 75 $\textcircled{>}$ 57

(b) $5 + 7 = 12$ and $4 + 8 = 12$
$$12 = 12$$
$$5 + 7 \textcircled{=} 4 + 8$$

7.

$$\begin{array}{r} 375 \\ -\ 245 \\ \hline \mathbf{130} \end{array}$$

8. **34**

9.

$$\begin{array}{r} \overset{1\,1}{\$426} \\ +\ \$298 \\ \hline \mathbf{\$724} \end{array}$$

10.

$$\begin{array}{r} \overset{1\,1}{\$278} \\ +\ \$456 \\ \hline \mathbf{\$734} \end{array}$$

11.
$$\begin{array}{r} \overset{1\ 1}{721} \\ +\ 189 \\ \hline 910 \end{array}$$

12.
$$\begin{array}{r} \overset{1\ 1}{409} \\ +\ 198 \\ \hline 607 \end{array}$$

13.
$$\begin{array}{r} 5 \\ +\ 7 \\ \hline 12 \end{array}$$
$D = 5$

14.
$$\begin{array}{r} 18 \\ -\ 9 \\ \hline 9 \end{array}$$
$A = 9$

15.
$$\begin{array}{r} 38 \\ +\ 21 \\ \hline 59 \end{array}$$
$B = 21$

16.
$$\begin{array}{r} 5 \\ -\ 4 \\ \hline 1 \end{array}$$
$C = 5$

17.
$$\begin{array}{r} \$456 \\ -\ \$120 \\ \hline \$336 \end{array}$$

18.
$$\begin{array}{r} \$\overset{4}{\cancel{5}}{}^{1}4 \\ -\ \$2\ 7 \\ \hline \$2\ 7 \end{array}$$

19.
$$\begin{array}{r} \overset{3}{\cancel{4}}{}^{1}6 \\ -\ 2\ 8 \\ \hline 1\ 8 \end{array}$$

20.
$$\begin{array}{r} \overset{2}{\cancel{3}}{}^{1}5 \\ -\ 1\ 6 \\ \hline 1\ 9 \end{array}$$

21. November $= 30$ days
December $= 31$ days

$$\begin{array}{r} 30 \text{ days} \\ +\ 31 \text{ days} \\ \hline 61 \textbf{ days} \end{array}$$

22. $5 + 6 = 11$
$6 + 5 = 11$
$11 - 6 = 5$
$11 - 5 = 6$

23. $3 + 6 + 7 + 5 + 4 + 8 = \mathbf{33}$

24. The rule is "Count down by nines."
45, 36, 27

25. The rule is "Count down by sevens."
−28, −35, −42

26. $6 + \Delta = 10$
$6 + 4 = 10$, so $\Delta = 4$
B. 4

LESSON 16, WARM-UP

a. **90**

b. **900**

c. **9**

d. **55**

e. **66**

f. **73**

Patterns

1	2	3	4	5	6	7	8	9	10
11	12	13	14	15	16	17	18	19	20
21	22	23	24	25	26	27	28	29	30
31	32	33	34	35	36	37	38	39	40
41	42	43	44	45	46	47	48	49	50
51	52	53	54	55	56	57	58	59	60
61	62	63	64	65	66	67	68	69	70
71	72	73	74	75	76	77	78	79	80
81	82	83	84	85	86	87	88	89	90
91	92	93	94	95	96	97	98	99	100

10, 20, 30, 40, 50, 60, 70, 80, 90, 100

LESSON 16, LESSON PRACTICE

a. $80 + 6$

b. $300 + 20 + 5$

c. $500 + 7$

d.
$$\begin{array}{r} 36 \\ -\ 15 \\ \hline 21 \end{array}$$
$P = \mathbf{15}$

e.
$$\begin{array}{r} 47 \\ -\ 23 \\ \hline 24 \end{array}$$
$Q = \mathbf{23}$

f.
$$\begin{array}{r} 38 \\ -\ 22 \\ \hline 16 \end{array}$$
$M = \mathbf{38}$

g.
$$\begin{array}{r} 75 \\ -\ 32 \\ \hline 43 \end{array}$$
$W = \mathbf{75}$

h.
$$\begin{array}{r} 43 \\ -\ 11 \\ \hline 32 \end{array}$$
$X = \mathbf{11}$

LESSON 16, MIXED PRACTICE

1.
$$\begin{array}{r} 23 \text{ horses} \\ +\ \textbf{66 horses} \\ \hline 89 \text{ horses} \end{array}$$

2. Pattern:
$$\begin{array}{r} \text{Some} \\ +\ \text{Some more} \\ \hline \text{Total} \end{array}$$

Problem:
$$\begin{array}{r} \overset{1}{3}75 \text{ bats} \\ +\ 107 \text{ bats} \\ \hline \textbf{482 bats} \end{array}$$

3. $22 + 33 = \mathbf{55}$
$33 + 22 = \mathbf{55}$
$55 - 22 = \mathbf{33}$
$55 - 33 = \mathbf{22}$

4. $700 + 80 + 2$

5. The smallest three-digit number is 100.
The number 100 is even.
100

6. (a) $918 \;\boxed{>}\; 819$

(b)
$-7 \;\boxed{<}\; -5$

7. There are 7 days in one week. Count by sevens six times: 7, 14, 21, 28, 36, 42.
42 days

8. **475**

9.
$$\begin{array}{r} \overset{1\,1}{\$576} \\ +\ \$128 \\ \hline \mathbf{\$704} \end{array}$$

10.
$$\begin{array}{r} \overset{1\,1}{\$243} \\ +\ \$578 \\ \hline \mathbf{\$821} \end{array}$$

11.
$$\begin{array}{r} \overset{1\,1}{186} \\ +\ 285 \\ \hline \mathbf{471} \end{array}$$

12.
$$\begin{array}{r} \overset{1\,1}{329} \\ +\ 186 \\ \hline \mathbf{515} \end{array}$$

13.
$$\begin{array}{r} 5 \\ +\ 12 \\ \hline 17 \end{array}$$
$D = \mathbf{5}$

14.
$$\begin{array}{r} 17 \\ -\ 8 \\ \hline 9 \end{array}$$
$A = \mathbf{8}$

15.
$$\begin{array}{r} 8 \\ +\ 6 \\ \hline 14 \end{array}$$
$B = \mathbf{6}$

16.
$$
\begin{array}{r}
9 \\
-\ 7 \\
\hline
2
\end{array}
$$
$C = 9$

17.
$$
\begin{array}{r}
\overset{1}{2}{}^{1}5 \\
-\ 1\ 9 \\
\hline
6
\end{array}
$$

18.
$$
\begin{array}{r}
\overset{3}{4}{}^{1}2 \\
-\ 2\ 8 \\
\hline
1\ 4
\end{array}
$$

19.
$$
\begin{array}{r}
\overset{3}{4}{}^{1}6 \\
-\ 1\ 8 \\
\hline
2\ 8
\end{array}
$$

20.
$$
\begin{array}{r}
\overset{3}{4}{}^{1}2 \\
-\ 1\ 6 \\
\hline
2\ 6
\end{array}
$$

21.
$$
\begin{array}{r}
68 \\
-\ 34 \\
\hline
34
\end{array}
$$
$D = 34$

22.
$$
\begin{array}{r}
49 \\
-\ 34 \\
\hline
15
\end{array}
$$
$B = 49$

23.
$$
\begin{array}{r}
62 \\
-\ 41 \\
\hline
21
\end{array}
$$
$H = 41$

24.
$$
\begin{array}{r}
78 \\
-\ 46 \\
\hline
32
\end{array}
$$
$L = 78$

25. (a) The rule is "Count up by fours."
28, 32, 36

(b) The rule is "Count down by fours."
4, 0, −4

26. $N - 3 = 6$
$9 - 3 = 6$, so $N = 9$
$6 - 3 \neq 9$, so $6 - 3 = N$ is not true
C. 6 − 3 = N

LESSON 17, WARM-UP

a. **900**

b. **540**

c. **65**

d. **84**

e. **95**

f. **73**

Patterns

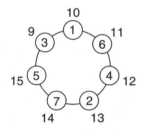

The pattern inside the circles is "1, skip, skip, 2, etc." Outside the circles the numbers count up by one from 9 to 15, starting at the upper left.

LESSON 17, LESSON PRACTICE

a.
$$
\begin{array}{r}
\overset{2}{4}7 \\
29 \\
46 \\
+\ 95 \\
\hline
217
\end{array}
$$

b.
$$
\begin{array}{r}
\overset{2}{2}8 \\
47 \\
+\ 65 \\
\hline
140
\end{array}
$$

c.
$$
\begin{array}{r}
\overset{1}{3}8 \\
22 \\
31 \\
+\ 46 \\
\hline
137
\end{array}
$$

d. $\begin{array}{r} {}^{1\,1}438 \\ 76 \\ +5 \\ \hline 519 \end{array}$

e. $\begin{array}{r} {}^{2}15 \\ 24 \\ 11 \\ 25 \\ +36 \\ \hline 111 \end{array}$

LESSON 17, MIXED PRACTICE

1. Pattern: $\begin{array}{l} \text{Some} \\ +\text{Some more} \\ \hline \text{Total} \end{array}$

Problem: $\begin{array}{l} \text{24 stitches} \\ +N \text{ stitches} \\ \hline \text{75 stitches} \end{array}$

$N = $ **51 stitches**

2. Pattern: $\begin{array}{l} \text{Some} \\ +\text{Some more} \\ \hline \text{Total} \end{array}$

Problem: $\begin{array}{l} \text{407 roses} \\ +\text{362 roses} \\ \hline \textbf{769 roses} \end{array}$

3. Two hundreds are less than three hundreds, so 2 is in the hundreds place. An even number may end in 8.
298

4. 800 + 10 + 3; eight hundred thirteen

5. The smallest two-digit number is 10.
The smallest two-digit odd number is **11**.

6. −30

7. $\begin{array}{r} {}^{1\,1}294 \\ 312 \\ +5 \\ \hline 611 \end{array}$

8. $\begin{array}{r} {}^{1\,1}\$189 \\ +\$298 \\ \hline \$487 \end{array}$

9. $\begin{array}{r} {}^{1\,1}\$378 \\ +\$496 \\ \hline \$874 \end{array}$

10. $\begin{array}{r} {}^{1}109 \\ +486 \\ \hline 595 \end{array}$

11. $\begin{array}{r} {}^{3}14 \\ 28 \\ 35 \\ {}_{1}16 \\ +227 \\ \hline 320 \end{array}$

12. $\begin{array}{r} 14 \\ -7 \\ \hline 7 \end{array}$

$A = $ **7**

13. $\begin{array}{r} 8 \\ +6 \\ \hline 14 \end{array}$

$B = $ **6**

14. $\begin{array}{r} 18 \\ -13 \\ \hline 5 \end{array}$

$C = $ **18**

15. $\begin{array}{r} 11 \\ -2 \\ \hline 9 \end{array}$

$D = $ **2**

16. $\begin{array}{r} 13 \\ -5 \\ \hline 8 \end{array}$

$E = $ **13**

17. $\begin{array}{r} {}^{2}\cancel{3}{}^{1}8 \\ -2\,9 \\ \hline 9 \end{array}$

18. $\begin{array}{r} {}^{4}\cancel{5}{}^{1}7 \\ -3\,8 \\ \hline 1\,9 \end{array}$

19.
$$\begin{array}{r} 34 \\ + 52 \\ \hline 86 \end{array}$$

B = 52

20.
$$\begin{array}{r} 48 \\ - 23 \\ \hline 25 \end{array}$$

C = 23

21.
$$\begin{array}{r} 58 \\ - 46 \\ \hline 12 \end{array}$$

D = 58

22.
$$\begin{array}{r} 39 \\ - 15 \\ \hline 24 \end{array}$$

Y = 39

23. The rule is "Count down by fours."
36, 32, 28

24. The rule is "Count up by threes."
21, 24, 27

25. **6 + 9 = 15**
9 + 6 = 15
15 − 6 = 9
15 − 9 = 6

26. 6 + 4 = 10
6 − 4 = 2
C. 6 and 4

LESSON 18, WARM-UP

a. **650**

b. **75**

c. **94**

d. **83**

e. **66**

f. **627**

Patterns

LESSON 18, LESSON PRACTICE

a. Count up by twos from 90: 90, 92, 94, 96, 98.
98°F

b. Count up by twos from 0: 0, 2, 4, 6, 8.
8°C

c. Count up by fives from 40: 40, 45
45 mph

LESSON 18, MIXED PRACTICE

1.
$$\begin{array}{r} 21 \text{ seconds} \\ + \textbf{37 seconds} \\ \hline 58 \text{ seconds} \end{array}$$

2. Pattern:
$$\begin{array}{r} \text{Some} \\ + \text{ Some more} \\ \hline \text{Total} \end{array}$$

Problem:
$$\begin{array}{r} {}^{1\ 1} \\ 297 \text{ boys} \\ + 315 \text{ girls} \\ \hline \textbf{612 children} \end{array}$$

297 + 315 = 612

3. **8 + 9 = 17**
9 + 8 = 17
17 − 9 = 8
17 − 8 = 9

4. **249**

5. Count up by fours to the eighth number:
4, 8, 12, 16, 20, 24, 28, **32.**

6. 475

7.
$$
\begin{array}{r}
\overset{1\ 1}{\$392} \\
+\ \$278 \\
\hline
\$670
\end{array}
$$

8.
$$
\begin{array}{r}
\overset{1}{\$439} \\
+\ \$339 \\
\hline
\$778
\end{array}
$$

9.
$$
\begin{array}{r}
\overset{1}{774} \\
+\ 174 \\
\hline
948
\end{array}
$$

10.
$$
\begin{array}{r}
\overset{1\ 1}{389} \\
+\ 398 \\
\hline
787
\end{array}
$$

11.
$$
\begin{array}{r}
\overset{2}{13} \\
25 \\
46 \\
25 \\
+\ 29 \\
\hline
138
\end{array}
$$

12.
$$
\begin{array}{r}
18 \\
-\ 6 \\
\hline
12
\end{array}
$$
$A = 6$

13.
$$
\begin{array}{r}
8 \\
+\ 8 \\
\hline
16
\end{array}
$$
$B = 8$

14.
$$
\begin{array}{r}
8 \\
-\ 5 \\
\hline
3
\end{array}
$$
$C = 8$

15.
$$
\begin{array}{r}
\overset{5}{6}{}^{1}2 \\
-\ 4\ 8 \\
\hline
1\ 4
\end{array}
$$

16.
$$
\begin{array}{r}
\overset{7}{8}{}^{1}2 \\
-\ 5\ 8 \\
\hline
2\ 4
\end{array}
$$

17.
$$
\begin{array}{r}
\overset{2}{28} \\
36 \\
57 \\
+\ 47 \\
\hline
168
\end{array}
$$

18.
$$
\begin{array}{r}
35 \\
-\ 21 \\
\hline
14
\end{array}
$$
$Y = 21$

19.
$$
\begin{array}{r}
45 \\
+\ 10 \\
\hline
55
\end{array}
$$
$P = 10$

20.
$$
\begin{array}{r}
75 \\
-\ 33 \\
\hline
42
\end{array}
$$
$L = 33$

21.
$$
\begin{array}{r}
78 \\
-\ 47 \\
\hline
31
\end{array}
$$
$C = 78$

22.
$$
\begin{array}{r}
22 \\
+\ 15 \\
\hline
37
\end{array}
$$
$E = 22$

23. $400 + 90 + 8$

24. (a) $423 \lessdot 432$

(b)
$$\xleftarrow{\hspace{1cm}} \begin{array}{ccccccccccc} \text{-5} & \text{-4} & \text{-3} & \text{-2} & \text{-1} & 0 & 1 & 2 & 3 & 4 & 5 \end{array} \xrightarrow{\hspace{1cm}}$$

$3 \gtrdot -3$

25. (a) Count up by twos from 50: 50, 52, 54, 56.
56°F

(b) Count up by twos from -10: -10, -8, -6.
−6°C

26. 903 and 309 are odd numbers. 903 is greater than 750.
C. 903

LESSON 19, WARM-UP

a. 127

b. 263

c. 348

d. 245

e. 164

f. 377

g. 8; 5; 4; 7; 9

Problem Solving

$$
\begin{array}{r}
3N = 15¢ \\
+ \ 1D = 10¢ \\
\hline
25¢
\end{array}
$$

3 nickels and 1 dime

LESSON 19, LESSON PRACTICE

a. Hour: 8
Minute: 5, 10, 15, 20, 25, 30
8:30 a.m.

b. Hour: 7
Minute: 5, 10, 11, 12
7:12 a.m.

c. Hour: 10
Minute: 5, 10, 15, 20, 25, 30, 35, 40
10:40 a.m.

d. **8:50 p.m.**

e. **24 hours**

f. **60 minutes**

g. **60 seconds**

LESSON 19, MIXED PRACTICE

1.
$$
\begin{array}{r}
51 \text{ pencils} \\
+ \ \textbf{25 pencils} \\
\hline
76 \text{ pencils}
\end{array}
$$

2.
$$
\begin{array}{r}
12 \text{ boys} \\
+ \ \textbf{15 girls} \\
\hline
27 \text{ children}
\end{array}
$$

3. $B + A = 9$
$9 - A = B$
$9 - B = A$

4. **900 + 5; nine hundred five**

5. **120 > 112**

6. Hour: 4
Minute: 5, 10, 15, 20, 25, 30
4:30 p.m.

7. **0°C**

8.
$$
\begin{array}{r}
\overset{1\ 1}{\$468} \\
+ \ \$293 \\
\hline
\mathbf{\$761}
\end{array}
$$

9.
$$
\begin{array}{r}
\overset{1\ 1}{468} \\
+ \ 185 \\
\hline
\mathbf{653}
\end{array}
$$

10.
$$
\begin{array}{r}
\overset{1\ 1}{\$187} \\
+ \ \$698 \\
\hline
\mathbf{\$885}
\end{array}
$$

11.
$$
\begin{array}{r}
14 \\
- \ 7 \\
\hline
7
\end{array}
$$
$A = \mathbf{7}$

12.
$$
\begin{array}{r}
8 \\
+ \ 8 \\
\hline
16
\end{array}
$$
$B = \mathbf{8}$

13.
$$\begin{array}{r} 15 \\ -\ 8 \\ \hline 7 \end{array}$$

$C = 15$

14.
$$\begin{array}{r} 14 \\ -\ 5 \\ \hline 9 \end{array}$$

$D = 5$

15.
$$\begin{array}{r} \overset{6}{7}{}^14 \\ -\ 5\ 8 \\ \hline 1\ 6 \end{array}$$

16.
$$\begin{array}{r} \$\overset{3}{4}{}^14 \\ -\ \$2\ 8 \\ \hline \$1\ 6 \end{array}$$

17.
$$\begin{array}{r} \overset{1}{2}{}^13 \\ -\ 1\ 8 \\ \hline 5 \end{array}$$

18.
$$\begin{array}{r} \$\overset{5}{6}{}^12 \\ -\ \$4\ 3 \\ \hline \$1\ 9 \end{array}$$

19.
$$\begin{array}{r} \overset{2}{2}5 \\ 28 \\ 46 \\ +\ 88 \\ \hline 187 \end{array}$$

20.
$$\begin{array}{r} 45 \\ -\ 24 \\ \hline 21 \end{array}$$

$P = 24$

21.
$$\begin{array}{r} 13 \\ +\ 24 \\ \hline 37 \end{array}$$

$B = 24$

22.
$$\begin{array}{r} 77 \\ -\ 45 \\ \hline 32 \end{array}$$

$F = 77$

23. 4 quarters = 1 dollar
Count up by fours: 4, 8, 12, 16.
16 quarters

24. **3 + 6 = 9 (or 6 + 3 = 9)**

25. (a) The rule is "Count up by eights."
32, 40, 48
(b) The rule is "Count down by twos."
2, 0, −2

26. $9 - \Delta = 4$
$9 - 5 = 4$, so $\Delta = 5$
$5 - 4 \neq 9$, so $\Delta - 4 = 9$ is not true
B. $\Delta - 4 = 9$

LESSON 20, WARM-UP

a. **456**

b. **354**

c. **83**

d. **244**

e. **267**

f. **477**

g. **1; 3; 5; 6; 2**

Patterns

1	2	3	4	5	6	7	8	9	10
11	12	13	14	15	16	17	18	19	20
21	22	23	24	25	26	27	28	29	30
31	32	33	34	35	36	37	38	39	40
41	42	43	44	45	46	47	48	49	50
51	52	53	54	55	56	57	58	59	60
61	62	63	64	65	66	67	68	69	70
71	72	73	74	75	76	77	78	79	80
81	82	83	84	85	86	87	88	89	90
91	92	93	94	95	96	97	98	99	100

Multiples of 5: 5, 10, 15, 20, 25, 30, 35, 40, 45, 50, 55, 60, 65, 70, 75, 80, 85, 90, 95, 100
30, 60, 90

LESSON 20, LESSON PRACTICE

a.

70 71 72 73 74 75 76 77 (78) 79 80

80

b.

40 41 42 (43) 44 45 46 47 48 49 50

40

c.

60 (61) 62 63 64 65 66 67 68 69 70

60

d.

40 41 42 43 44 (45) 46 47 48 49 50

50

e. 29 cents is less than 50. Round down to **$14.**

f. 95 cents is more than 50. Round up to **$9.**

g. 45 cents is less than 50. Round down to **$21.**

h. 89 cents is more than 50. Round up to **$30.**

_____

LESSON 20, MIXED PRACTICE

1. Pattern: Some
          + Some more
           Total

Problem: N eggs
         + 21 eggs
          72 eggs

N = **51 eggs**

2. Pattern: Some
          + Some more
           Total

Problem: $\overset{1\,1}{476}$ children
         + 397 children
          873 children

3. **665**

4. **500 + 9; five hundred nine**

5. **−20 < 10**

6. Count up by ones from 30°F: 30, 31, 32.
32°F; 0°C

7. Hour: 4
Minute: 5, 10, 15
4:15 p.m.

8. (a)

40 41 42 43 44 45 46 (47) 48 49 50

50

(b)

70 71 72 73 (74) 75 76 77 78 79 80

70

9.
$$\begin{array}{r} \overset{1\,1}{\$476} \\ + \ \$285 \\ \hline \mathbf{\$761} \end{array}$$

10.
$$\begin{array}{r} \overset{1\,1}{\$185} \\ + \ \$499 \\ \hline \mathbf{\$684} \end{array}$$

11.
$$\begin{array}{r} \overset{1\,1}{568} \\ + \ 397 \\ \hline \mathbf{965} \end{array}$$

12.
$$\begin{array}{r} \overset{1\,1}{478} \\ + \ 196 \\ \hline \mathbf{674} \end{array}$$

13.
$$\begin{array}{r} 17 \\ - \ 8 \\ \hline 9 \end{array}$$
A = **8**

14.
$$\begin{array}{r} 14 \\ - \ 0 \\ \hline 14 \end{array}$$
B = **0**

15.
$$\begin{array}{r} 13 \\ - \ 7 \\ \hline 6 \end{array}$$
C = **7**

16.
$$\begin{array}{r} \$\overset{2}{3}\overset{1}{}5 \\ - \ \$2\ 8 \\ \hline \mathbf{\$7} \end{array}$$

17.
$$\begin{array}{r} \overset{1}{2}\overset{1}{3} \\ -\ 1\ 5 \\ \hline 8 \end{array}$$

18.
$$\begin{array}{r} \overset{5}{\cancel{6}}\overset{1}{3} \\ -\ 3\ 6 \\ \hline 2\ 7 \end{array}$$

19.
$$\begin{array}{r} \overset{6}{\cancel{7}}\overset{1}{4} \\ -\ 5\ 9 \\ \hline 1\ 5 \end{array}$$

20.
$$\begin{array}{r} 23 \\ +\ 22 \\ \hline 45 \end{array}$$
$M\ =\ \mathbf{23}$

21.
$$\begin{array}{r} 47 \\ -\ 15 \\ \hline 32 \end{array}$$
$K\ =\ \mathbf{47}$

22.
$$\begin{array}{r} 47 \\ -\ 13 \\ \hline 34 \end{array}$$
$K\ =\ \mathbf{13}$

23.
$$\begin{array}{r} \overset{2}{2}8 \\ 36 \\ 44 \\ +\ 58 \\ \hline 166 \end{array}$$

24.
$$\begin{array}{r} \overset{2}{4}9 \\ 28 \\ 32 \\ +\ 55 \\ \hline 164 \end{array}$$

25. (a) 67 cents is more than 50. Round up to **$26.**
(b) 42 cents is less than 50. Round down to **$14.**

26. 3 tens, 7 ones $=$ 37
7 tens, 3 ones $=$ 73
$$\begin{array}{r} \overset{1}{3}7 \\ +\ 73 \\ \hline 110 \end{array}$$

C. $37\ +\ 73\ =\ 110$

INVESTIGATION 2

1. Jones

2. Answers vary.

3. 100 cm

4. inch

5. 30 cm

6. probably about 11 in.

7. probably about 28 cm

8. (a) **4 in.**
   (b) **10 cm**

9. (a) **6 in.**
   (b) **between 15 cm and 16 cm**

10. about 1000 big steps

11. mile

12. 1000 mm

13. (a) **2 in.**
    (b) **5 cm**
    (c) **50 mm**

14. (a) **3 cm**
    (b) **2 cm**

15. $3\text{ cm}\ +\ 2\text{ cm}\ +\ 3\text{ cm}\ +\ 2\text{ cm}\ =\ \mathbf{10\ cm}$

16. $80\text{ yd}\ +\ 40\text{ yd}\ +\ 80\text{ yd}\ +\ 40\text{ yd}\ =\ \mathbf{240\ yd}$

17. $2\text{ cm}\ +\ 2\text{ cm}\ +\ 2\text{ cm}\ +\ 2\text{ cm}\ =\ \mathbf{8\ cm}$

18. $10\text{ in.}\ +\ 10\text{ in.}\ +\ 10\text{ in.}\ +\ 10\text{ in.}\ =\ \mathbf{40\ in.}$

19. $3\text{ cm}\ +\ 4\text{ cm}\ +\ 5\text{ cm}\ =\ \mathbf{12\ cm}$

20. (a) **3 ft**

(b) **2 ft**

(c) 3 ft + 2 ft + 3 ft + 2 ft = **10 ft**

21. wire fence

22. wooden frame

23. Answers vary.

___

LESSON 21, WARM-UP

a. **76**

b. **77**

c. **76**

d. **175**

e. **289**

f. **185**

g. **3; 8; 1; 5; 4**

Patterns

1	2	3	4	5	6	7	8	9	10
11	12	13	14	15	16	17	18	19	20
21	22	23	24	25	26	27	28	29	30
31	32	33	34	35	36	37	38	39	40
41	42	43	44	45	46	47	48	49	50
51	52	53	54	55	56	57	58	59	60
61	62	63	64	65	66	67	68	69	70
71	72	73	74	75	76	77	78	79	80
81	82	83	84	85	86	87	88	89	90
91	92	93	94	95	96	97	98	99	100

Only numbers that end in 5 or 0 are multiples of 5. Numbers ending in 5 are not even, so the only shaded number that is even and a multiple of 5 is **70.**

LESSON 21, LESSON PRACTICE

a. **Sample answer:**

b. **Proportions should be similar to this figure:**

c.

d. The diameter of a circle equals two radii.
3 cm + 3 cm = **6 cm**

e. **Square**

___

LESSON 21, MIXED PRACTICE

1. Pattern: Some
            + Some more
            Total

Problem: 417 marbles
            + 222 marbles
            639 marbles

2. Pattern: Some
            + Some more
            Total

Problem: 40 jacks
            + *N* jacks
            72 jacks

$N = $ **32 jacks**

3. **645**

4. **700 + 50 + 3**

5. $y + x = 10$
10 − *x* = *y*
10 − *y* = *x*

6. Count up by fifties from 200: 200, 250, 300, **350.**

7. (a) **4 cm**

(b) **2 cm**

(c) 4 cm + 2 cm + 4 cm + 2 cm = **12 cm**

8.
$$\begin{array}{r} {}^{1\,1}493 \\ +\ 278 \\ \hline 771 \end{array}$$

9.
$$\begin{array}{r} {}^{1\,1}\$486 \\ +\ \$378 \\ \hline \$864 \end{array}$$

10.
$$\begin{array}{r} {}^{1}\$524 \\ +\ \$109 \\ \hline \$633 \end{array}$$

11.

2 cm + 2 cm + 2 cm = **6 cm**

12.

2 in. + 2 in. + 2 in. + 2 in. = **8 in.**

13.
$$\begin{array}{r} 17 \\ -\ 8 \\ \hline 9 \end{array}$$
$A = $ **8**

14.
$$\begin{array}{r} {}^{3}\cancel{4}{}^{1}5 \\ -\ 2\,9 \\ \hline 1\,6 \end{array}$$

15.
$$\begin{array}{r} 15 \\ -\ 9 \\ \hline 6 \end{array}$$
$B = $ **9**

16.
$$\begin{array}{r} {}^{5}\cancel{6}{}^{1}2 \\ -\ 4\,5 \\ \hline 1\,7 \end{array}$$

17.
$$\begin{array}{r} 24 \\ +\ 21 \\ \hline 45 \end{array}$$
$D = $ **21**

18.
$$\begin{array}{r} 14 \\ -\ 12 \\ \hline 2 \end{array}$$
$B = $ **12**

19.
$$\begin{array}{r} 89 \\ -\ 36 \\ \hline 53 \end{array}$$
$Y = $ **89**

20.
$$\begin{array}{r} 75 \\ -\ 30 \\ \hline 45 \end{array}$$
$P = $ **30**

21.
$$\begin{array}{r} {}^{2}46 \\ 35 \\ 27 \\ +\ 39 \\ \hline 147 \end{array}$$

22.
$$\begin{array}{r} {}^{2}14 \\ 28 \\ 77 \\ +\ 23 \\ \hline 142 \end{array}$$

23.
$$\begin{array}{r} {}^{1}14 \\ 23 \\ 38 \\ +\ 64 \\ \hline 139 \end{array}$$

24.
$$\begin{array}{r} {}^{2}15 \\ 24 \\ 36 \\ +\ 99 \\ \hline 174 \end{array}$$

25. (a) The rule is "Count up by sevens."
49, 56, 63

(b) The rule is "Count down by tens."
10, 0, −10

26. The diameter of a circle equals two radii.
4 cm + 4 cm = 8 cm
C. 8 cm

LESSON 22, WARM-UP

a. 84

b. 68

c. 95

d. 267

e. 95

f. 630

g. 2; 6; 7; 1; 5

Problem Solving

$2 + 2 + 2 + 2 + 2 + 2 + 2$
$= $ **14 times**

LESSON 22, LESSON PRACTICE

a. There are three equal parts and one is shaded.
$\dfrac{1}{3}$

b. There are nine equal parts and five are shaded.
$\dfrac{5}{9}$

c. There are eight equal parts and three are shaded.
$\dfrac{3}{8}$

d. There are four equal parts and one is shaded.
$\dfrac{1}{4}$

e. Four quarters equal a dollar, so one quarter is $\dfrac{1}{4}$ of a dollar.

f. Twenty nickels equal a dollar, so one nickel is $\dfrac{1}{20}$ of a dollar.

g. Ten dimes equal one dollar, so each dime is $\dfrac{1}{10}$ of a dollar. Three dimes are $\dfrac{3}{10}$ of a dollar.

h.
$$\begin{array}{r} {\scriptstyle 1\ 1} \\ \$2.75 \\ +\ \$2.75 \\ \hline \$5.50 \end{array}$$

i.
$$\begin{array}{r} {\scriptstyle 1} \\ \$3.65 \\ +\ \$4.28 \\ \hline \$7.93 \end{array}$$

LESSON 22, MIXED PRACTICE

1.
$$\begin{array}{r} 1 \\ 3 \\ 5 \\ +\ 7 \\ \hline 16 \end{array}$$

2. Pattern:
$$\begin{array}{r} \text{Some} \\ +\ \text{Some more} \\ \hline \text{Total} \end{array}$$

Problem:
$$\begin{array}{r} {\scriptstyle 1} \\ 49 \text{ inches} \\ +\ \ 2 \text{ inches} \\ \hline 51 \text{ inches} \end{array}$$

3. A three-digit number less than 200 has the digit 1 in the hundreds place. An odd number may end with 3.
123

4. The rule is "Count down by eights."
56, 48, 40

5. The rule is "Count down by sixes."
42, 36, 30

6.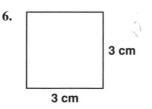

3 cm + 3 cm + 3 cm + 3 cm = **12 cm**

7. 1 yd = **3 ft**

8. **Tens**

9. $100 + 6$; one hundred six

10. $6 + 9 = 15$
$9 + 6 = 15$
$15 - 9 = 6$
$15 - 6 = 9$

11. $18 > -20$

12. (a)

30

(b) 95 cents is more than 50. Round up to **$6.**

13. About 2 m for a full-size bicycle; about $1\frac{1}{2}$ m for a small bicycle

14. Count up by twos from 80: 80, 82, 84, 86, 88, 90, 92, 94, **96.**

15.

The diameter of a circle equals two radii.
$1\,\text{cm} + 1\,\text{cm} = 2\,\text{cm}$
1 cm

16. There are six equal parts and one is shaded.

$\dfrac{1}{6}$

17. $1\,\text{m} = 100\,\text{cm}$
$2\,\text{m} = \textbf{200 cm}$

18.
$$\begin{array}{r} \overset{4}{\cancel{5}}{}^{1}1 \\ -\ 4\ 3 \\ \hline \textbf{8} \end{array}$$

19.
$$\begin{array}{r} \overset{6}{\cancel{7}}{}^{1}0 \\ -\ 4\ 4 \\ \hline \textbf{2 6} \end{array}$$

20.
$$\begin{array}{r} \overset{2}{\cancel{3}}{}^{1}7 \\ -\quad 9 \\ \hline \textbf{2 8} \end{array}$$

21.
$$\begin{array}{r} \overset{1\ 1}{\ \$8.79} \\ +\ \$0.64 \\ \hline \mathbf{\$9.43} \end{array}$$

22.
$$\begin{array}{r} \overset{1\ 1}{\ \$5.75} \\ +\ \$2.75 \\ \hline \mathbf{\$8.50} \end{array}$$

23.
$$\begin{array}{r} 4 \\ +\ 13 \\ \hline 17 \end{array}$$
$N = \textbf{4}$

24.
$$\begin{array}{r} 69 \\ -\ 42 \\ \hline 27 \end{array}$$
$X = \textbf{69}$

25.
$$\begin{array}{r} 37 \\ -\ 23 \\ \hline 14 \end{array}$$
$P = \textbf{23}$

26.
$$\begin{array}{r} 20 \\ +\ 40 \\ \hline 60 \end{array}$$
$N = 40$
$40 - 20 \neq 60$, so $N - 20 = 60$ is not true
C. $N - 20 = 60$

LESSON 23, WARM-UP

a. 576

b. 677

c. 395

d. 687

e. 296

f. 599

g. 8; 4; 3; 9; 5

Patterns

1	2	3	4	5	6	7	8	9	10
11	12	13	14	15	16	17	18	19	20
21	22	23	24	25	26	27	28	29	30
31	32	33	34	35	36	37	38	39	40
41	42	43	44	45	46	47	48	49	50
51	52	53	54	55	56	57	58	59	60
61	62	63	64	65	66	67	68	69	70
71	72	73	74	75	76	77	78	79	80
81	82	83	84	85	86	87	88	89	90
91	92	93	94	95	96	97	98	99	100

9
18
27
36
45
54
63
72
81
90

The tens digit goes up one each time, the ones digit goes down one, and the sum of the digits is nine.

LESSON 23, LESSON PRACTICE

a. Sample answer:

b.

c.

d. **Parallel**

e. **3 angles**

f. **B.**

LESSON 23, MIXED PRACTICE

1. Pattern:
Some
+ Some more
Total

Problem:
28 children
+ 42 children
70 children

2. Pattern:
Some
+ Some more
Total

Problem:
12 books
+ N books
28 books
N = **16 books**

3. A three-digit number greater than 300 may have the digit 3 in the hundreds place. An odd number may end in 1.
321

4. (a) The rule is "Count down by fours."
28, 24, 20

(b) The rule is "Count down by threes."
21, 18, 15

5. 15 + 16 = 31
16 + 15 = 31
31 − 16 = 15
31 − 15 = 16

6. 638 **<** 683

7. (a)

90 91 (92) 93 94 95 96 97 98 99 100
90

(b) 67 cents is more than 50. Round up to **$20.**

8. Count up by twos ten times: 2, 4, 6, 8, 10, 12, 14, 16, 18, 20.
20 cm

9. (a) **3 cm**

(b) **1 cm**

(c) 3 cm + 1 cm + 3 cm + 1 cm = **8 cm**

10. **B.**

11. There are three equal parts and one is shaded.
$\frac{1}{3}$

12. Hour: 2
Minute: 5, 10, 15
2:15 p.m.

13.

$$\begin{array}{r} \overset{7}{\cancel{8}}{}^{1}3 \\ -\ \$2\ 7 \\ \hline \$5\ 6 \end{array}$$

14.

$$\begin{array}{r} \overset{3}{\cancel{4}}{}^{1}2 \\ -\ 2\ 7 \\ \hline 1\ 5 \end{array}$$

15.

$$\begin{array}{r} \overset{6}{\cancel{7}}{}^{1}2 \\ -\ 3\ 6 \\ \hline 3\ 6 \end{array}$$

16.

$$\begin{array}{r} \overset{1}{\ }\overset{1}{\ }\ \\ \$4.28 \\ +\ \$1.96 \\ \hline \$6.24 \end{array}$$

17.

$$\begin{array}{r} \overset{1}{\ }\overset{1}{\ }\ \\ \$4.36 \\ +\ \$2.95 \\ \hline \$7.31 \end{array}$$

18.

$$\begin{array}{r} 57 \\ +\ 31 \\ \hline 88 \end{array}$$

$K = 31$

19.

$$\begin{array}{r} 67 \\ -\ 51 \\ \hline 16 \end{array}$$

$B = 51$

20.

$$\begin{array}{r} 44 \\ -\ 22 \\ \hline 22 \end{array}$$

$K = 44$

21.

$$\begin{array}{r} \overset{3}{\cancel{4}}{}^{1}2 \\ -\ \ \ 7 \\ \hline 3\ 5 \end{array}$$

22.

$$\begin{array}{r} \overset{4}{\cancel{5}}{}^{1}5 \\ -\ 4\ 8 \\ \hline 7 \end{array}$$

23.

$$\begin{array}{r} 31 \\ -\ 20 \\ \hline 11 \end{array}$$

24.

$$\begin{array}{r} \overset{2}{25} \\ 25 \\ 25 \\ +\ 25 \\ \hline 100 \end{array}$$

25. (a) **20 nickels**

(b) $\dfrac{1}{20}$

(c) $\dfrac{7}{20}$

26. The fact family that includes the equation
$26 + M = 63$ also includes the equations
in answer choices A, C, and D: $M + 26 = 63$,
$63 - M = 26$, and $63 - 26 = M$.
B. $M - 63 = 26$

27. C. ◄———•

LESSON 24, WARM-UP

a. **686**

b. **695**

c. **365**

d. **556**

e. **362**

f. **491**

g. **5**

Problem Solving
   2 and 7; 3 and 6; 4 and 5

LESSON 24, LESSON PRACTICE

a. $23 + M = 42$

$$\begin{array}{r} \overset{3}{\cancel{4}}{}^{1}2 \\ -\ 2\ 3 \\ \hline M = 1\ 9 \end{array}$$

b. $Q + 17 = 45$

$$\begin{array}{r} \overset{3}{\cancel{4}}{}^{1}5 \\ -\ 1\ 7 \\ \hline Q = \mathbf{2\ 8} \end{array}$$

c. $53 - W = 28$

$$\begin{array}{r} \overset{4}{\cancel{5}}{}^{1}3 \\ -\ 2\ 8 \\ \hline W = \mathbf{2\ 5} \end{array}$$

d. $N - 26 = 68$

$$\begin{array}{r} \overset{1}{6}8 \\ +\ 2\ 6 \\ \hline N = \mathbf{94} \end{array}$$

e. $36 + Y = 63$

$$\begin{array}{r} \overset{5}{\cancel{6}}{}^{1}3 \\ -\ 3\ 6 \\ \hline Y = \mathbf{2\ 7} \end{array}$$

f. $62 - A = 26$

$$\begin{array}{r} \overset{5}{\cancel{6}}{}^{1}2 \\ -\ 2\ 6 \\ \hline A = \mathbf{3\ 6} \end{array}$$

LESSON 24, MIXED PRACTICE

1. 1 ft. = 12 in.

$$\begin{array}{r} 12 \text{ in.} \\ +\ 12 \text{ in.} \\ \hline \mathbf{24 \text{ in.}} \end{array}$$

2. Pattern:

$$\begin{array}{r} \text{Some} \\ +\ \text{Some more} \\ \hline \text{Total} \end{array}$$

Problem:

$$\begin{array}{r} 47 \text{ apples} \\ +\ N \text{ apples} \\ \hline 82 \text{ apples} \end{array}$$

$N = \mathbf{35 \text{ apples}}$

3. Odd numbers cannot form two equal rows.
B. 45

4. The rule is "Count up by nines."
27, 36, 45, **54**

5. Count up by sevens to the sixth number:
7, 14, 21, 28, 35, **42**

6. $15 - 9 = 6$
$13 - 8 = 5$
$6 > 5$
$15 - 9 \; \text{⟩} \; 13 - 8$

7. (a)

80

(b) 39 cents is less than 50. Round down to **$29.**

8. About 2 m

9. Hour: 7
Minute: 5, 10
7:10 a.m.

10. Oak

11. (a) **10 dimes**

(b) $\dfrac{1}{10}$

(c) $\dfrac{9}{10}$

12.

5 cm

$5 \text{ cm} + 2 \text{ cm} + 5 \text{ cm} + 2 \text{ cm} = \mathbf{14 \text{ cm}}$

13. (a) **Right angle**

(b) **Obtuse angle**

(c) **Acute angle**

14.

$$\begin{array}{r} \overset{2}{\$\cancel{3}}{}^{1}1 \\ -\ \$1\ 4 \\ \hline \$1\ 7 \end{array}$$

15.

$$\begin{array}{r} \overset{1\ 1}{\$468} \\ +\ \$247 \\ \hline \$715 \end{array}$$

16.

$$\begin{array}{r} 57 \\ -\ 37 \\ \hline 20 \end{array}$$

17.
$$\begin{array}{r} {\scriptstyle 1\ 1} \\ \$4.97 \\ +\ \$2.58 \\ \hline \$7.55 \end{array}$$

18. $36 - C = 19$
$36 - 19 = C$

$$\begin{array}{r} \overset{2}{\cancel{3}}{}^{1}6 \\ -\ 1\ 9 \\ \hline C = 1\ 7 \end{array}$$

19. $B + 65 = 82$
$82 - 65 = B$

$$\begin{array}{r} \overset{7}{\cancel{8}}{}^{1}2 \\ -\ 6\ 5 \\ \hline B = 1\ 7 \end{array}$$

20. $87 + D = 93$
$93 - 87 = D$

$$\begin{array}{r} \overset{8}{\cancel{9}}{}^{1}3 \\ -\ 8\ 7 \\ \hline D = 6 \end{array}$$

21. $N - 32 = 19$
$19 + 32 = N$

$$\begin{array}{r} {\scriptstyle 1} \\ 19 \\ +\ 32 \\ \hline N = 51 \end{array}$$

22.
$$\begin{array}{r} 48 \\ -\ 28 \\ \hline 20 \end{array}$$

23.
$$\begin{array}{r} \overset{3}{\cancel{4}}{}^{1}1 \\ -\ 3\ 2 \\ \hline 9 \end{array}$$

24.
$$\begin{array}{r} \overset{6}{\cancel{7}}{}^{1}6 \\ -\ 5\ 8 \\ \hline 1\ 8 \end{array}$$

25.
$$\begin{array}{r} {\scriptstyle 1\ 2} \\ 416 \\ 35 \\ 27 \\ 43 \\ +\ \ 5 \\ \hline 526 \end{array}$$

26. -3 is between -10 and 0 on a number line.
B. Point x

27. **Sample answer: A segment has two endpoints and is part of a line. A line has no endpoints.**

LESSON 25, WARM-UP

a. **590**

b. **682**

c. **785**

d. **275**

e. **187**

f. **45**

g. **4; 3**

Problem Solving
7 hours of a.m. sunlight $+$ 7 hours of p.m. sunlight $-$ **14 hours**

LESSON 25, LESSON PRACTICE

a. Pattern:
$$\begin{array}{l} \text{Some} \\ -\ \text{Some went away} \\ \hline \text{What is left} \end{array}$$

Problem:
$$\begin{array}{r} \textbf{42 marbles} \\ -\ \boldsymbol{N}\ \textbf{marbles} \\ \hline \textbf{26 marbles} \end{array}$$

$$\begin{array}{r} \overset{3}{\cancel{4}}{}^{1}\textbf{2 marbles} \\ -\ 2\ \textbf{6 marbles} \\ \hline N = \textbf{1 6 marbles} \end{array}$$

b. Pattern:
$$\begin{array}{l} \text{Some} \\ -\ \text{Some went away} \\ \hline \text{What is left} \end{array}$$

Problem:
$$\begin{array}{r} \boldsymbol{N}\ \textbf{marbles} \\ -\ \textbf{42 marbles} \\ \hline \textbf{26 marbles} \end{array}$$

$$\begin{array}{r} \textbf{26 marbles} \\ +\ \textbf{42 marbles} \\ \hline N = \textbf{68 marbles} \end{array}$$

c. Pattern: Some
                − Some went away
                What is left

      Problem: **75 cents**
                   − **27 cents**
                   N cents

$$\begin{array}{r} \overset{6}{\cancel{7}}{}^{1}5 \text{ cents} \\ -\ 2\ 7 \text{ cents} \\ \hline N =\ \mathbf{4\ 8} \text{ cents} \end{array}$$

LESSON 25, MIXED PRACTICE

1. Pattern: Some
                − Some went away
                What is left

      Problem: **75 rocks**
                   − **N rocks**
                   27 rocks

$$\begin{array}{r} \overset{6}{\cancel{7}}{}^{1}5 \text{ rocks} \\ -\ 2\ 7 \text{ rocks} \\ \hline \mathbf{4\ 8} \text{ rocks} \end{array}$$

2. Pattern: Some
                − Some went away
                What is left

      Problem: **63 birds**
                   − **14 birds**
                   N birds

$$\begin{array}{r} \overset{5}{\cancel{6}}{}^{1}3 \text{ birds} \\ -\ 1\ 4 \text{ birds} \\ \hline \mathbf{4\ 9} \text{ birds} \end{array}$$

3. Pattern: Some
                − Some went away
                What is left

      Problem: **N cats**
                   − **75 cats**
                   47 cats

$$\begin{array}{r} \overset{1}{\ }47 \text{ cats} \\ +\ 75 \text{ cats} \\ \hline \mathbf{122} \text{ cats} \end{array}$$

4. **6 months**

5. (a) The rule is "Count up by fives."
      15, 20, 25, **30**

    (b) The rule is "Count down by fives."
      −5, −10, −15, **−20**

6. **762 < 826**

7. (a)
      70 71 72 73 74 75 76 77 ⑦⑧ 79 80

    80

    (b) 80 cents is more than 50. Round up to **$8.**

8. The diameter of a circle equals two radii.
    10 in. + 10 in. = 20 in.
    10 in.

9. Hour: 1
    Minute: 5, 10, 15, 20
    1:20 p.m.

10. **Broadway**

11. There are twelve equal parts and five are shaded.
$$\frac{5}{12}$$

12.

    4 cm + 4 cm + 4 cm + 4 cm = **16 cm**

13. **286**

14.
$$\begin{array}{r} \$\overset{4}{\cancel{5}}{}^{1}2 \\ -\ \$1\ 4 \\ \hline \mathbf{\$3\ 8} \end{array}$$

15.
$$\begin{array}{r} \overset{1\ 1}{476} \\ +\ 177 \\ \hline \mathbf{653} \end{array}$$

16.
$$\begin{array}{r} \overset{5}{\cancel{6}}{}^{1}2 \\ -\ 3\ 8 \\ \hline \mathbf{2\ 4} \end{array}$$

17. $\overset{1\ 1}{}$ $\$4.97$
$+\ \$2.03$
$\overline{\ \ \$7.00\ \ }$

18. 36
$-\ \ G$
$\overline{\ \ \ 18\ \ }$

$\overset{2}{\cancel{3}}{}^{1}6$
$-\ 1\ 8$
$\overline{G\ =\ \mathbf{1\ 8}}$

19. 55
$+\ \ B$
$\overline{\ \ \ 87\ \ }$

87
$-\ 55$
$\overline{B\ =\ \mathbf{32}}$

20. D
$-\ 23$
$\overline{\ \ \ 58\ \ }$

$\overset{1}{5}8$
$+\ 23$
$\overline{D\ =\ \mathbf{81}}$

21. Y
$+\ 14$
$\overline{\ \ \ 32\ \ }$

$\overset{2}{\cancel{3}}{}^{1}2$
$-\ 1\ 4$
$\overline{Y\ =\ \mathbf{1\ 8}}$

22. $\overset{3}{\cancel{4}}{}^{1}2$
$-\ 3\ 7$
$\overline{\ \ \ \ \ \mathbf{5}\ }$

23. 52
$-\ 22$
$\overline{\ \ \ \mathbf{30}\ \ }$

24. $\overset{6}{\cancel{7}}{}^{1}3$
$-\ 5\ 9$
$\overline{\ \ \ \mathbf{1\ 4}\ }$

25. 900
90
$+\ \ \ \ 9$
$\overline{\ \ \mathbf{999}\ \ }$

26. C. 1000 km

27. Sample answer: A ray has one endpoint.
A segment has two endpoints.

LESSON 26, WARM-UP

a. 80

b. 62

c. 70

d. 82

e. 380

f. 482

g. 7; 5

Problem Solving

$1HD\ =\ 50\cent$
$1D\ =\ 10\cent$
$+\ \ 1N\ =\ \ \ 5\cent$
$\overline{\ \ \ \ \ \ \ \ \ \ \ 65\cent\ }$

1 half-dollar, 1 dime, and 1 nickel

LESSON 26, LESSON PRACTICE

a. One possibility:

b. One possibility:

c. One possibility:

d. One possibility:

e. No. The circle is divided into two parts, but
the parts are unequal. The shaded part is
smaller than the unshaded part.

SOLUTIONS

LESSON 26, MIXED PRACTICE

1. Pattern:
Some
− Some went away
What is left

Problem:
42 pebbles
− N pebbles
27 pebbles

$\overset{3}{\cancel{4}}{}^{1}2$ pebbles
− 2 7 pebbles
1 5 pebbles

2. Pattern:
Some
− Some went away
What is left

Problem:
N pebbles
− 17 pebbles
46 pebbles

$\overset{1}{4}6$ pebbles
+ 17 pebbles
63 pebbles

3.
112 stars
+ N stars
317 stars

317 stars
− 112 stars
205 stars

4. A three-digit number less than 500 may have the digit 4 in the hundreds place. An even number may end in 6.
456

5. **One possibility:**

6.
6 cm
8 cm
+ 10 cm
24 cm

7. **−20 < −12**

8. (a)
10 11 12 13 14 15 16 17 18 ⑲ 20
20

(b) 90 cents is more than 50. Round up to **$11.**

9. 1 m = **100 cm**

10. Hour: 11
Minute: 5, 10, 15, 20, 25, 30, 35
11:35 a.m.

11. **Broadway**

12. There are ten equal parts and three are shaded.
$\dfrac{3}{10}$

13. Count up by twenties from 300: 300, 320, 340, 360.
360 pounds

14.
Y
+ 63
81

$\overset{7}{\cancel{8}}{}^{1}1$
− 6 3
Y = **1 8**

15.
$\overset{1\,1}{\$486}$
+ $277
$763

16.
$\$\overset{5}{\cancel{6}}{}^{1}8$
− $3 9
$2 9

17.
$\overset{1\ 1}{\$5.97}$
+ $2.38
$8.35

18.
N
+ 42
71

$\overset{6}{\cancel{7}}{}^{1}1$
− 4 2
N = **2 9**

19.
87
− N
65

87
− 65
N = **22**

46   *Saxon Math 5/4—Homeschool*

20.
$$\begin{array}{r} 27 \\ + \ C \\ \hline 48 \end{array}$$

$$\begin{array}{r} 48 \\ - \ 27 \\ \hline C = \mathbf{21} \end{array}$$

21.
$$\begin{array}{r} E \\ - \ 14 \\ \hline 28 \end{array}$$

$$\begin{array}{r} \overset{1}{2}8 \\ + \ 14 \\ \hline E = \mathbf{42} \end{array}$$

22.
$$\begin{array}{r} \overset{3}{\cancel{4}}{}^{1}2 \\ - \ 2 \ 9 \\ \hline \mathbf{1 \ 3} \end{array}$$

23.
$$\begin{array}{r} 77 \\ - \ 37 \\ \hline \mathbf{40} \end{array}$$

24.
$$\begin{array}{r} \overset{3}{\cancel{4}}{}^{1}1 \\ - \ 1 \ 9 \\ \hline \mathbf{2 \ 2} \end{array}$$

25.
$$\begin{array}{r} 4 \\ {}_{2}7 \\ 15 \\ 21 \\ 5 \\ 4 \\ + \ 3 \\ \hline \mathbf{59} \end{array}$$

26. D.

27. Obtuse

LESSON 27, WARM-UP

a. 61

b. 372

c. 81

d. 382

e. 84

f. 392

g. 6; 5

Problem Solving

18, **27**, **36**, **45**, **54**, **63**, **72**, 81

LESSON 27, LESSON PRACTICE

a. 4×3

b. 3×9

c. 6×7

d. 8×5

e. Step 1: Counting forward 25 minutes from 10:35 a.m. makes it 11:00 a.m.
Step 2: Counting forward 2 hours from 11:00 a.m. makes it **1:00 p.m.**

f. Step 1: Counting back 30 minutes from 10:35 a.m. makes it 10:05 a.m.
Step 2: Counting back 6 hours from 10:05 a.m. makes it **4:05 a.m.**

LESSON 27, MIXED PRACTICE

1. Pattern:
$$\begin{array}{l} \text{Some} \\ - \ \text{Some went away} \\ \hline \text{What is left} \end{array}$$

Problem:
$$\begin{array}{r} \textbf{78 kittens} \\ - \ \textbf{\textit{N} kittens} \\ \hline \textbf{42 kittens} \end{array}$$

$$\begin{array}{r} 78 \text{ kittens} \\ - \ 42 \text{ kittens} \\ \hline \textbf{36 kittens} \end{array}$$

2. (a) 1 ft = **12 in.**

(b)
$$\begin{array}{r} 12 \text{ in.} \\ 12 \text{ in.} \\ 12 \text{ in.} \\ + \ 12 \text{ in.} \\ \hline \textbf{48 in.} \end{array}$$

SOLUTIONS

3. **32, 34, 36, 38**

4. The rule is "Count up by threes."
21, 24, 27

5. The rule is "Count up by twelves."
48, 60, 72

6. **200 + 60 + 5**

7. **Negative nineteen**

8. (a)

$$\leftarrow\!\!-\!\!\overset{|}{60}\,\overset{|}{61}\,\overset{|}{62}\,\overset{\textcircled{\tiny 63}}{}\,\overset{|}{64}\,\overset{|}{65}\,\overset{|}{66}\,\overset{|}{67}\,\overset{|}{68}\,\overset{|}{69}\,\overset{|}{70}\!\!-\!\!\rightarrow$$

60

(b) 30 cents is less than 50. Round down to **$6.**

9. (a) 392 $\boxed{>}$ 329

(b)

$$\leftarrow\!\!-\!\!\overset{|}{-20}\;\overset{\textcircled{\tiny -15}}{}\;\overset{|}{-10}\;\overset{|}{0}\!\!-\!\!\rightarrow$$

-15 $\boxed{>}$ -20

10. **550**

11. **One possibility:**

2 cm (right side)
2 cm (bottom)

12. There are six equal parts and five are shaded.
$$\frac{5}{6}$$

13. Counting forward 3 hours from 1:20 p.m.
makes it **4:20 p.m.**

14.
$$\begin{array}{r} \overset{5}{\$\cancel{6}}{}^{1}7 \\ -\ \$2\ 9 \\ \hline \$3\ 8 \end{array}$$

15.
$$\begin{array}{r} \overset{1\ 1}{483} \\ +\ 378 \\ \hline 861 \end{array}$$

16.
$$\begin{array}{r} \overset{6}{\cancel{7}}{}^{1}1 \\ -\ 3\ 9 \\ \hline 3\ 2 \end{array}$$

17.
$$\begin{array}{r} \overset{1\ 1}{\$5.88} \\ +\ \$2.39 \\ \hline \$8.27 \end{array}$$

18.
$$\begin{array}{r} D \\ +\ 19 \\ \hline 36 \end{array}$$

$$\begin{array}{r} \overset{2}{\cancel{3}}{}^{1}6 \\ -\ 1\ 9 \\ \hline D\ =\ \mathbf{1\ 7} \end{array}$$

19.
$$\begin{array}{r} 66 \\ +\ F \\ \hline 87 \end{array}$$

$$\begin{array}{r} 87 \\ -\ 66 \\ \hline F\ =\ \mathbf{21} \end{array}$$

20.
$$\begin{array}{r} 87 \\ -\ R \\ \hline 67 \end{array}$$

$$\begin{array}{r} 87 \\ -\ 67 \\ \hline R\ =\ \mathbf{20} \end{array}$$

21.
$$\begin{array}{r} B \\ -\ 14 \\ \hline 27 \end{array}$$

$$\begin{array}{r} \overset{1}{27} \\ +\ 14 \\ \hline B\ =\ \mathbf{41} \end{array}$$

22.
$$\begin{array}{r} 400 \\ -\ 300 \\ \hline 100 \end{array}$$

23.
$$\begin{array}{r} 663 \\ -\ 363 \\ \hline 300 \end{array}$$

24. **4 × 9**

25. (a) **100 pennies**

(b) $\dfrac{1}{100}$

(c) $\dfrac{11}{100}$

26. $3 + 4 + 3 = 10$
C. **10**

27.

_____

LESSON 28, WARM-UP

a. **90**

b. **73**

c. **174**

d. **580**

e. **270**

f. **77**

g. **4; 2**

Problem Solving

3:00   **9:00**

_____

LESSON 28, LESSON PRACTICE

a. The row for 9 and the column for 3 meet at **27.**

b. The row for 3 and the column for 9 meet at **27.**

c. The row for 6 and the column for 4 meet at **24.**

d. The row for 4 and the column for 6 meet at **24.**

e. The row for 7 and the column for 8 meet at **56.**

f. The row for 8 and the column for 7 meet at **56.**

g. The row for 5 and the column for 8 meet at **40.**

h. The row for 8 and the column for 5 meet at **40.**

i. The row for 10 and the column for 10 meet at **100.**

j. The row for 10 and the column for 8 meet at **80.**

k. The row for 11 and the column for 9 meet at **99.**

l. The row for 12 and the column for 12 meet at **144.**

m. **Commutative property of multiplication**

n. **0**

o. **25**

_____

LESSON 28, MIXED PRACTICE

1. Pattern:
$$\begin{array}{r} \text{Some} \\ + \ \text{Some more} \\ \hline \text{Total} \end{array}$$

Problem:
$$\begin{array}{r} 72 \text{ pieces of gingerbread} \\ + \ 42 \text{ pieces of gingerbread} \\ \hline \textbf{114 pieces of gingerbread} \end{array}$$

2. $\begin{array}{r} \$18 \\ + \ N \\ \hline \$35 \end{array}$

$$\begin{array}{r} \$\overset{2}{\cancel{3}}{}^{1}5 \\ - \ \$1\ 8 \\ \hline N = \$1\ 7 \end{array}$$

3.

☐	3 cm
4 cm	

$4\,\text{cm} + 3\,\text{cm} + 4\,\text{cm} + 3\,\text{cm} = \textbf{14 cm}$

4. The rule is "Count up by sixes."
12, **18, 24**, 30, 36, **42**

5. The rule is "Count down by fours."
36, **32, 28**, 24, 20, **16**

6. 7×6
The row for 7 and the column for 6 meet at **42.**

7. (a)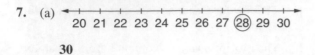

30

(b) 29 cents is less than 50. Round down to **$12.**

8.

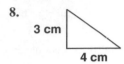

3 cm

4 cm

9. Counting forward 10 minutes from 10:15 a.m. makes it **10:25 a.m.**

10. There are twelve equal-sized circles and five are shaded.

$$\frac{5}{12}$$

11. **400 + 10 + 7; four hundred seventeen**

12. Count up by twos from −10: −10, −8.
−8°C

13.
$$\begin{array}{r} \overset{6}{\cancel{7}}{}^{1}6 \\ -\ 2\ 9 \\ \hline \mathbf{4\ 7} \end{array}$$

14.
$$\begin{array}{r} {}^{1\,1}\$286 \\ +\ \$388 \\ \hline \mathbf{\$674} \end{array}$$

15.
$$\begin{array}{r} \$\overset{6}{\cancel{7}}{}^{1}3 \\ -\ \$3\ 9 \\ \hline \mathbf{\$3\ 4} \end{array}$$

16.
$$\begin{array}{r} {}^{1\ 1}\$5.87 \\ +\ \$2.43 \\ \hline \mathbf{\$8.30} \end{array}$$

17.
$$\begin{array}{r} 46 \\ -\ C \\ \hline 19 \end{array}$$

$$\begin{array}{r} \overset{3}{\cancel{4}}{}^{1}6 \\ -\ 1\ 9 \\ \hline C =\ \mathbf{2\ 7} \end{array}$$

18.
$$\begin{array}{r} N \\ +\ 48 \\ \hline 87 \end{array}$$

$$\begin{array}{r} \overset{7}{\cancel{8}}{}^{1}7 \\ -\ 4\ 8 \\ \hline N =\ \mathbf{3\ 9} \end{array}$$

19.
$$\begin{array}{r} 29 \\ +\ Y \\ \hline 57 \end{array}$$

$$\begin{array}{r} \overset{4}{\cancel{5}}{}^{1}7 \\ -\ 2\ 9 \\ \hline Y =\ \mathbf{2\ 8} \end{array}$$

20.
$$\begin{array}{r} D \\ -\ 14 \\ \hline 37 \end{array}$$

$$\begin{array}{r} {}^{1}37 \\ +\ 14 \\ \hline D =\ \mathbf{51} \end{array}$$

21.
$$\begin{array}{r} 78 \\ -\ 43 \\ \hline \mathbf{35} \end{array}$$

22.
$$\begin{array}{r} 77 \\ -\ 17 \\ \hline \mathbf{60} \end{array}$$

23.
$$\begin{array}{r} \overset{4}{\cancel{5}}{}^{1}3 \\ -\ 1\ 9 \\ \hline \mathbf{3\ 4} \end{array}$$

24. (a) The row for 8 and the column for 11 meet at **88.**

(b) The row for 7 and the column for 10 meet at **70.**

(c) The row for 5 and the column for 12 meet at **60.**

25. 1 yard $\lessdot$ 1 meter

26. 4 hundreds and 3 ones is 403.
B. 403

27. The product of 9 and 3 appears where the row for 9 and the column for 3 meet (9 × 3) and where the row for 3 and the column for 9 meet (3 × 9).
2 times; commutative property of multiplication

LESSON 29, WARM-UP

a. 42

b. 34

c. 43

d. 53

e. 75

f. 53

Problem Solving

Lowest value: 7P = **7¢**
Highest value: 7Q = **$1.75**

LESSON 29, LESSON PRACTICE

For answers, please refer to the Facts Practice Tests section of this manual.

LESSON 29, MIXED PRACTICE

1. Pattern: Some
   − Some went away
   What is left

   Problem: **92 blackbirds**
   − **N blackbirds**
   24 blackbirds

   $\overset{8}{\cancel{9}}{}^{1}2$ blackbirds
   − 2 4 blackbirds
   6 8 blackbirds

2.   42 seashells
   + N seashells
   83 seashells

   83 seashells
   − 42 seashells
   41 seashells

3. (a) 1 yd = 3 ft
   2 yd = **6 ft**

   (b) The diameter of a circle equals two radii.
   6 ft + 6 ft = **12 ft**

4. The rule is "Count up by eights."
   8, **16, 24,** 32, 40, **48**

5. The rule is "Count up by sevens."
   14, **21, 28,** 35, 42

6. An odd number may end in 5. 645 is greater than 640.
   465

7. **209 > 190**

8. Counting forward 6 hours from 3:25 p.m. makes it **9:25 p.m.**

9. **One possibility:**

   1 cm
   3 cm

10. (a) 2 × 8 = **16**

    (b) 5 × 7 = **35**

    (c) 2 × 7 = **14**

    (d) 5 × 8 = **40**

11. **Obtuse angle**

12. **−15**

13. (a) **32°F**

    (b) **0°C**

14. $\overset{7}{\cancel{8}}{}^{1}3$
    − $1 9
    $6 4

15. $\overset{1\,1}{\$286}$
    + $387
    $673

16. $\overset{6}{\cancel{7}}{}^{1}2$
    − 3 8
    3 4

17. $\overset{1\ 1}{\$5.87}$
    + $2.79
    $8.66

18.
$$\begin{array}{r} 19 \\ +\ Q \\ \hline 46 \end{array}$$

$$\begin{array}{r} \overset{3}{\cancel{4}}{}^{1}6 \\ -\ 1\ 9 \\ \hline Q = \mathbf{2\ 7} \end{array}$$

19.
$$\begin{array}{r} 88 \\ -\ N \\ \hline 37 \end{array}$$

$$\begin{array}{r} 88 \\ -\ 37 \\ \hline N = \mathbf{51} \end{array}$$

20.
$$\begin{array}{r} 88 \\ -\ M \\ \hline 47 \end{array}$$

$$\begin{array}{r} 88 \\ -\ 47 \\ \hline M = \mathbf{41} \end{array}$$

21.
$$\begin{array}{r} G \\ +\ 14 \\ \hline 47 \end{array}$$

$$\begin{array}{r} 47 \\ -\ 14 \\ \hline G = \mathbf{33} \end{array}$$

22.
$$\begin{array}{r} 870 \\ -\ 470 \\ \hline \mathbf{400} \end{array}$$

23.
$$\begin{array}{r} 525 \\ -\ 521 \\ \hline \mathbf{4} \end{array}$$

24. 3×8
$3 \times 8 = \mathbf{24}$

25.
$$\begin{array}{r} 1 \\ 9 \\ 2 \\ 8 \\ 3 \\ 7 \\ 4 \\ 6 \\ {}_{4}5 \\ +\ 10 \\ \hline \mathbf{55} \end{array}$$

26. D. 8×4

27. (a) **Zero property of multiplication**

(b) **Commutative property of multiplication**

(c) **Identity property of multiplication**

LESSON 30, WARM-UP

a. **44**

b. **54**

c. **56**

d. **186**

e. **590**

f. **283**

Patterns

$$\overset{+3}{\frown}\ \overset{+5}{\frown}\ \overset{+7}{\frown}\ \overset{+9}{\frown}\ \overset{+11}{\frown}\ \overset{+13}{\frown}\ \overset{+15}{\frown}\ \overset{+17}{\frown}\ \overset{+19}{\frown}$$

$1,\ 4,\ 9,\ 16,\ \mathbf{\underline{25}},\ \mathbf{\underline{36}},\ \mathbf{\underline{49}},\ \mathbf{\underline{64}},\ \mathbf{\underline{81}},\ \mathbf{\underline{100}}$

LESSON 30, LESSON PRACTICE

a.
$$\begin{array}{r} \$\overset{2}{\cancel{3}}\overset{1}{\cancel{6}}5 \\ -\ \$2\ 8\ 7 \\ \hline \mathbf{\$7\ 8} \end{array}$$

b.
$$\begin{array}{r} \$4.\ \overset{2}{\cancel{3}}{}^{1}0 \\ -\ \$1.\ 1\ 8 \\ \hline \mathbf{\$3.\ 1\ 2} \end{array}$$

c.
$$\begin{array}{r} 5\ \overset{5}{\cancel{6}}{}^{1}3 \\ -\ 3\ 5\ 6 \\ \hline \mathbf{2\ 0\ 7} \end{array}$$

d.
$$\begin{array}{r} \overset{1}{\cancel{2}}\ \overset{1}{\cancel{4}}{}^{1}0 \\ -\ \ 6\ 5 \\ \hline \mathbf{1\ 7\ 5} \end{array}$$

e. $\overset{3}{\cancel{4}}{}^{1}5\,9$
 $-\;1\,7\,6$
 $\overline{\quad 2\,8\,3}$

f. $\overset{0}{\cancel{1}}\;\overset{1}{\cancel{5}}{}^{1}7$
 $-\quad\;9\,8$
 $\overline{\qquad 5\,9}$

LESSON 30, MIXED PRACTICE

1.   Pattern: Some
           $-$ Some went away
              What is left

   Problem: **N horses**
            $-$ **47 horses**
             22 horses

              22 horses
            $+$ 47 horses
             69 horses

2.   56 children
          $+$ N children
          73 children

          $\overset{6}{\cancel{7}}{}^{1}3$ children
        $-\,5\,6$ children
         1 7 children

3.   45¢ is 9 nickels. 9 is not an even number.
    A. 45¢

4.   Counting forward 15 minutes from 9:20 a.m.
    makes it **9:35 a.m.**

5.   Count up by sixes six times: 6, 12, 18, 24, 30, **36.**

6.   **480**

7.   **One possibility:**

8.   **800 + 40 + 3; eight hundred forty-three**

9.   (a) 6 × 8 = **48**

    (b) 4 × 2 = **8**

    (c) 4 × 5 = **20**

    (d) 6 × 10 = **60**

10.   10 + 20 = 30
     20 + 10 = 30
     30 − 20 = 10
     30 − 10 = 20

11.   (a) **6 cm**

     (b) **2 cm**

     (c) 6 cm + 2 cm + 6 cm + 2 cm = **16 cm**

12.   **Right angle**

13. $\overset{6}{\cancel{7}}{}^{1}4\,6$
     $-\;2\,9\,5$
     $\overline{\quad 4\,5\,1}$

14. $\overset{1\;\;1}{\$3.86}$
     $+\;\$2.78$
     $\overline{\;\;\$6.64}$

15. $\overset{5}{\cancel{6}}{}^{1}1$
     $-\;4\,8$
     $\overline{\quad 1\,3}$

16. $\$4.86$
     $-\;\$2.75$
     $\overline{\;\;\$2.11}$

17.   51
          $+\;M$
          70

          $\overset{6}{\cancel{7}}{}^{1}0$
        $-\;5\,1$
     $M = \mathbf{1\,9}$

18.   86
          $-\;A$
          43

             86
          $-\;43$
     $A = \mathbf{43}$

19.   25
          $+\;Y$
          36

             36
          $-\;25$
     $Y = \mathbf{11}$

20.
$$
\begin{array}{r}
Q \\
-\ 24 \\
\hline
37
\end{array}
$$

$$
\begin{array}{r}
\overset{1}{3}7 \\
+\ 24 \\
\hline
Q\ =\ \mathbf{61}
\end{array}
$$

21. (a)

80 81 82 83 84 85 86 87 88 (89) 90

90

(b) 90 cents is more than 50. Round up to **$9.**

22.
$$
\begin{array}{r}
\overset{2}{2}5¢ \\
25¢ \\
25¢ \\
+\ 25¢ \\
\hline
\mathbf{100¢}\ \ \text{or}\ \ \mathbf{\$1.00}
\end{array}
$$

23. **50 cents**

24. 7 × 7

7 × 7 = **49**

25.
$$
\begin{array}{r}
4 \\
3 \\
8 \\
4 \\
2 \\
5 \\
+\ 7 \\
\hline
\mathbf{33}
\end{array}
$$

26. 3 + 4 ≠ 5

D. 3, 4, 5

27. (a) **100**

(b) **121**

(c) **144**

INVESTIGATION 3

1. **3 rows**

2. **4 columns**

3. **12 X's**

4. **3 × 4 = 12**

5.
X X X X X X
X X X X X

6. **6 columns**

7. **2 × 6 = 12**

8. **2 and 5**

9. **no**

10. **no**

11. **yes**

12.
X X X X X X
X X X X X X
X X X X X X

3 × 6 = 18

13.

6 × 4 = 24

14.

24 squares

15.

24 squares; 3 × 8 = 24

16.

12 cm; 2 × 12 = 24

17. Observe student action. Student should trace the perimeter of the tabletop.

18. Observe student action. Student should sweep his/her hand over the area of the tabletop.

19. 3 cm × 2 cm = **6 sq. cm**

20.

12 sq. cm

21. 2 in. × 2 in. = **4 sq. in.**

22. See student work. Rectangle should measure 3 in. by 3 in. The area is 3 in. × 3 in., or **9 sq. in.**

23.

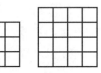

1 × 1 = 1   2 × 2 = 4   3 × 3 = 9   4 × 4 = 16

24.

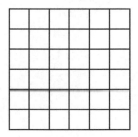

6 × 6 = **36**

25.

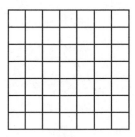

7 × 7 = **49**

26.
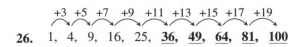

+3 +5 +7 +9 +11 +13 +15 +17 +19

1, 4, 9, 16, 25, **36**, **49**, **64**, **81**, **100**

27. They are aligned diagonally from upper left to lower right.

28. (a) 9 × 9 = **81**

(b) The square root of 9 is **3**.

29. (a) $\sqrt{4}$ = **2**

(b) $\sqrt{16}$ = **4**

(c) $\sqrt{64}$ = **8**

30. The square root of 49 is 7. So the side length of a square whose area is 49 sq. cm is **7 cm.**

LESSON 31, WARM-UP

a. 65

b. 84

c. 53

d. 890

e. 282

f. 592

Patterns

25   36

LESSON 31, LESSON PRACTICE

a. Pattern: Larger
           − Smaller
           Difference

Problem: **43**
         − 27
         N

$$\overset{3}{\cancel{4}}{}^{1}3$$
− 2 7
1 6

b. Pattern: Larger
 − Smaller
 Difference

Problem: **42 peanuts**
 − **22 peanuts**
 N peanuts

 42 peanuts
 − 22 peanuts
 20 peanuts

c. Pattern: Larger
 − Smaller
 Difference

Problem: **95 shells**
 − **53 shells**
 N shells

 95 shells
 − 53 shells
 42 shells

LESSON 31, MIXED PRACTICE

1. Pattern: Some
 − Some went away
 What is left

Problem: **43 parrots**
 − **N parrots**
 27 parrots

 $\overset{3}{\cancel{4}}{}^{1}3$ parrots
 − 2 7 parrots
 1 6 parrots

2. Pattern: Larger
 − Smaller
 Difference

Problem: **150**
 − **23**
 N

 1 $\overset{4}{\cancel{5}}{}^{1}0$
 − 2 3
 1 2 7

 N = **127**

3. Pattern: Larger
 − Smaller
 Difference

Problem: **75 apples**
 − **23 apples**
 N apples

 75 apples
 − 23 apples
 52 apples

4. Counting forward 3 hours from 8:05 p.m. makes it **11:05 p.m.**

5. **400 + 10 + 2**
 Four hundred twelve

6. There are twelve equal parts and five are shaded.
 $\dfrac{5}{12}$

7. (a) $4\,\text{cm} + 2\,\text{cm} + 4\,\text{cm} + 2\,\text{cm} =$ **12 cm**

 (b) $4\,\text{cm} \times 2\,\text{cm} =$ **8 sq. cm**

8. (a) $2 \times 5 =$ **10**

 (b) $5 \times 7 =$ **35**

 (c) $2 \times 7 =$ **14**

 (d) On the multiplication table, the row for 4 and the column for 11 meet at **44.**

9. **20 + 30 = 50**
 30 + 20 = 50
 50 − 20 = 30
 50 − 30 = 20

10. **−5°C**

11. **110**

12. (a) $5 \times 8 =$ **40**

 (b) $2 \times 8 =$ **16**

 (c) $5 \times 9 =$ **45**

13. (a) **4 quarters**

 (b) $\dfrac{1}{4}$

 (c) $\dfrac{3}{4}$

14. **309 < 390**

15. Pattern: Larger
− Smaller
Difference

Problem: $3\overset{8}{\cancel{9}}{}^{1}0$
$-\ 3\ 0\ 9$
$\overline{\ \ \ 8\ 1}$

16. $\$\overset{3}{\cancel{4}}.\overset{1}{\cancel{2}}{}^{1}2$
$-\ \$2.\ 9\ 5$
$\overline{\ \$1.\ 2\ 7}$

17. $\overset{8}{\cancel{9}}{}^{1}0\ 9$
$-\ \ \ \ 2\ 7$
$\overline{\ 8\ 8\ 2}$

18. $\$\overset{3}{\cancel{4}}\ \overset{1}{\cancel{2}}{}^{1}2$
$-\ \$1\ 4\ 4$
$\overline{\ \$2\ 7\ 8}$

19. $\overset{6}{\cancel{7}}{}^{1}0\ 3$
$-\ \ 4\ 7\ 1$
$\overline{\ \ 2\ 3\ 2}$

20. $\overset{1}{\ }\ \overset{1}{\ }$
$\$4.86$
$|\ \$2.95$
$\overline{\ \$7.81}$

21. $3\ \overset{6}{\cancel{7}}{}^{1}0$
$-\ 2\ 0\ 9$
$\overline{\ 1\ 6\ 1}$

22. $\overset{\prime\prime}{2}2$
$+\ \ N$
$\overline{\ 3\ 7}$

$\ 3\ 7$
$-\ 2\ 2$
$\overline{\ 1\ 5}$

$N\ =\ \textbf{15}$

23. $\ 7\ 6$
$-\ \ C$
$\overline{\ 2\ 8}$

$\overset{6}{\cancel{7}}{}^{1}6$
$-\ 2\ 8$
$\overline{\ 4\ 8}$

$C\ =\ \textbf{48}$

24. $3\ \times\ 3\ =\ \textbf{9}$

25. (a) $3\ \times\ 3\ =\ 9$
$\sqrt{9}\ =\ \textbf{3}$

(b) $5\ \times\ 5\ =\ 25$
$\sqrt{25}\ =\ \textbf{5}$

26. $9\ \times\ 9\ =\ 81$, so $9\ =\ \sqrt{81}$, not $\sqrt{18}$

C. $\sqrt{18}$

27. (a) $1\ \times\ 1\ =\ \textbf{1}$

(b) $5\ \times\ 5\ =\ \textbf{25}$

(c) On the multiplication table, the row for 8 and the column for 8 meet at **64.**

(d) On the multiplication table, the row for 9 and the column for 9 meet at **81.**

LESSON 32, WARM-UP

a. 55

b. 73

c. 46

d. 965

e. 186

f. 890

Problem Solving

1Q = 25¢		2Q = 50¢
2D = 20¢		1D = 10¢
2N = 10¢		2N = 10¢
+ 2P = 2¢		+ 2P = 2¢
	57¢ or **$0.57**	72¢ or **$0.72**
2Q = 50¢		2Q = 50¢
2D = 20¢		2D = 20¢
1N = 5¢		2N = 10¢
+ 2P = 2¢		+ 1P = 1¢
	77¢ or **$0.77**	81¢ or **$0.81**

LESSON 32, LESSON PRACTICE

a. $9\ \times\ 6\ =\ \textbf{54}$

b. $5\ \times\ 9\ =\ \textbf{45}$

c. $9 \times 8 = \textbf{72}$

d. $3 \times 9 = \textbf{27}$

e. $9 \times 4 = \textbf{36}$

f. $7 \times 9 = \textbf{63}$

g. $9 \times 2 = \textbf{18}$

h. $9 \times 9 = \textbf{81}$

LESSON 32, MIXED PRACTICE

1. Pattern:
$$\begin{array}{r} \text{Some} \\ - \text{ Some went away} \\ \hline \text{What is left} \end{array}$$

Problem:
$$\begin{array}{r} \textbf{215 pages} \\ - \quad \textbf{86 pages} \\ \hline \textbf{\textit{N} pages} \end{array}$$

$$\begin{array}{r} \overset{1}{\cancel{2}}\,\overset{\;1}{\cancel{1}}\text{5 pages} \\ - \quad 8\;6 \text{ pages} \\ \hline \textbf{1 2 9 pages} \end{array}$$

2. An even number may end with 8. A three-digit number greater than 800 may have a 9 in the hundreds place.
978

3. $\textbf{485} < \textbf{690}$

4. $5 \times 5 = \textbf{25}$
$6 \times 6 = \textbf{36}$
$7 \times 7 = \textbf{49}$

5. Hour: 6
Minute: 5, 6
6:06 a.m.

6. $\textbf{7} \times \textbf{6};$ The row for 7 and the column for 6 meet at **42.**

7. $\textbf{700} + \textbf{20} + \textbf{9}$
Seven hundred twenty-nine

8. (a)

70

(b) 60 cents is more than 50. Round up to **$7.**

9. (a) **2 cm**

(b) $2 \text{ cm} + 2 \text{ cm} + 2 \text{ cm} + 2 \text{ cm} = \textbf{8 cm}$

10. **L and T**

11. $62 - W = 38$

$$\begin{array}{r} \overset{5}{\cancel{6}}\,{}^{1}2 \\ - \quad 3\;8 \\ \hline 2\;4 \end{array}$$

$W = \textbf{24}$

12. There are eight equal parts and seven are shaded.
$$\dfrac{\textbf{7}}{\textbf{8}}$$

13.
× × × × ×
× × × × ×
× × × × ×
× × × × ×
× × × × ×

14. Two dimes and three nickels: 10¢, 20¢, 25¢, 30¢, 35¢. 35 is an **odd number.**

15. **125**

16. (a) $9 \times 6 = \textbf{54}$

(b) $9 \times 8 = \textbf{72}$

(c) $9 \times 4 = \textbf{36}$

(d) $9 \times 10 = \textbf{90}$

17. (a) $6 \times 6 = \textbf{36}$

(b) $4 \times 4 = \textbf{16}$

(c) $7 \times 7 = \textbf{49}$

(d) $10 \times 10 = \textbf{100}$

18. $\textbf{5} \times \textbf{5} = \textbf{25}$

19. $9 \times 9 = 81$
$\sqrt{81} = \textbf{9}$

20.
$$\begin{array}{r} \$3.\,\overset{5}{\cancel{6}}{}^{1}0 \\ -\ \$1.\,3\,7 \\ \hline \$2.\,2\,3 \end{array}$$

21.
$$\begin{array}{r} \overset{3}{\cancel{4}}{}^{1}1\,3 \\ -\ 3\,8\,0 \\ \hline 3\,3 \end{array}$$

22.
$$\begin{array}{r} 8\,\overset{6}{\cancel{7}}{}^{1}5 \\ -\ 2\,1\,8 \\ \hline 6\,5\,7 \end{array}$$

23.
$$\begin{array}{r} \overset{1}{}47 \\ +\ 36 \\ \hline 83 \end{array}$$

$$\begin{array}{r} \overset{1}{}57 \\ +\ 26 \\ \hline 83 \end{array}$$

83 $\boxed{=}$ 83

24. $5 \times 5 = 25$

Five squared $= 25$

The square root of $25 = 5$

$$\begin{array}{r} 25 \\ -\ 5 \\ \hline 20 \end{array}$$

25. $5\ cm \times 2\ cm = $ **10 sq. cm**

26. $4 \times 4 = 16$
$16 - 1 = 15$
D. 15

27. $4 \times 6 = 6 \times 4$
Commutative property of multiplication

_____

LESSON 33, WARM-UP

a. 20

b. 50

c. 500

d. 74

e. 186

f. 490

Patterns

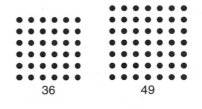

36   49

_____

LESSON 33, LESSON PRACTICE

a. Read aloud:
"one hundred twenty-five thousand"

b. Read aloud:
"four hundred thirty-five million"

c. Read aloud:
"twelve thousand, five hundred"

d. Read aloud:
"twenty-five million, three hundred seventy-five thousand"

e. Read aloud:
"four thousand, eight hundred seventy-five"

f. Read aloud:
"nine million, two hundred fifty thousand, six hundred twenty-five"

g. Two thousand, seven hundred fifty

h. Fourteen thousand, five hundred eighteen

i. Sixteen million

j. Three million, five hundred thousand

k. $5000 + 200 + 80$

l. $2000 + 40$

m. 8

n. 8

o. 2,760,000 $>$ 2,670,000

p. **1903, 1927, 1957, 1969**

LESSON 33, MIXED PRACTICE

1. Pattern:

    Larger
  − Smaller
    Difference

Problem:

   272 rabbits
  − **211 rabbits**
    N rabbits

   272 rabbits
  − 211 rabbits
   61 rabbits

2. **3000 + 400 + 20 + 5**
Three thousand, four hundred twenty-five

3. **One possibility:**

4. $7 \times 7 = 49$
The square root of $49 = 7$
  $4 \times 4 = 16$
$16 - 7 = $ **9**

5.

6 cm × 2 cm

(a) 6 cm + 2 cm + 6 cm + 2 cm = **16 cm**

(b) 6 cm × 2 cm = **12 sq. cm**

6. **1,250,000**
One million, two hundred fifty thousand

7. The rule is "Count up by tens."
270, 280, 290, 300

8. **2**

9. $9 \times 4 = 36$
$6 \times 6 = 36$
The square root of $36 = 6$
    36 $>$ 6
9×4 $>$ $\sqrt{36}$

10. Step 1: Counting forward 25 minutes from
5:35 p.m. makes it 6:00 p.m.
Step 2: Counting forward 2 hours from 6:00 p.m.
makes it **8:00 p.m.**

11. **2260**

12. (a) $5 \times 8 = $ **40**

(b) $4 \times 4 = $ **16**

(c) $8 \times 8 = $ **64**

(d) $12 \times 12 = $ **144**

13. (a) $9 \times 3 = $ **27**

(b) $9 \times 4 = $ **36**

(c) $9 \times 5 = $ **45**

(d) $9 \times 0 = $ **0**

14.   **40 + 60 = 100**
  60 + 40 = 100
100 − 40 = 60
100 − 60 = 40

15. $20 + 20 + 20 + 20 + 20 = $ **5 × 20**

16.
$$\begin{array}{r} \$7.\overset{6}{\cancel{3}}\overset{1}{\cancel{3}}7 \\ -\ \$2.6\,8 \\ \hline \$4.6\,9 \end{array}$$

17.
$$\begin{array}{r} \overset{8}{\cancel{9}}\,\overset{1}{\cancel{2}}\,1 \\ -\ \ \ 5\,8 \\ \hline 8\,6\,3 \end{array}$$

18.
$$\begin{array}{r} {}^{1\ 1}464 \\ +\ 247 \\ \hline 711 \end{array}$$

19.
$$\begin{array}{r} 329 \\ +\ \ \ Z \\ \hline 547 \end{array}$$

$$\begin{array}{r} 5\,\overset{3}{\cancel{4}}7 \\ -\ 3\,2\,9 \\ \hline 2\,1\,8 \end{array}$$

$Z = $ **218**

20.
$$\begin{array}{r} {}^{1\ 1}\$4.88 \\ +\ \$2.69 \\ \hline \$7.57 \end{array}$$

21.
$$
\begin{array}{r}
555 \\
- \quad C \\
\hline
222
\end{array}
$$

$$
\begin{array}{r}
555 \\
- \ 222 \\
\hline
333
\end{array}
$$

$C = \textbf{333}$

22. The fifth month is **May.**

23.

The diameter of a circle equals two radii.

1 in. + 1 in. = **2 in.**

24.
$$
\begin{array}{r}
4 \\
{}_3 8 \\
12 \\
16 \\
14 \\
28 \\
+ \ 37 \\
\hline
119
\end{array}
$$

25.
$$
\begin{array}{r}
5 \\
8 \\
{}_3 7 \\
14 \\
6 \\
21 \\
+ \ 15 \\
\hline
76
\end{array}
$$

26. 3,025,000 $\lt$ 3,250,000

27. $7 \times 7 = \textbf{49}$
$8 \times 8 = \textbf{64}$
$9 \times 9 = \textbf{81}$
$10 \times 10 = \textbf{100}$

LESSON 34, WARM-UP

a. **300**

b. **35**

c. **350**

d. **162**

e. **265**

f. **682**

Problem Solving
$$
\begin{array}{r}
2Q = 50¢ \\
1D = 10¢ \\
2N = 10¢ \\
+ \ 2P = \ \ 2¢ \\
\hline
72¢ \text{ or } \$0.72
\end{array}
$$

LESSON 34, LESSON PRACTICE

a. **121,340**

b. **12,507**

c. **5075**

d. **25,000,000**

e. **12,500,000**

f. **280,000,000**

LESSON 34, MIXED PRACTICE

1. Pattern:
$$
\begin{array}{r}
\text{Larger} \\
- \ \text{Smaller} \\
\hline
\text{Difference}
\end{array}
$$

Problem:
$$
\begin{array}{r}
\textbf{465} \\
- \ \ \textbf{24} \\
\hline
N
\end{array}
$$

$$
\begin{array}{r}
465 \\
- \ \ 24 \\
\hline
441
\end{array}
$$

$N = \textbf{441}$

2. Pattern:
$$
\begin{array}{r}
\text{Larger} \\
- \ \text{Smaller} \\
\hline
\text{Difference}
\end{array}
$$

Problem:
$$
\begin{array}{r}
\textbf{420 marbles} \\
- \ \textbf{123 marbles} \\
\hline
N \textbf{ marbles}
\end{array}
$$

$$
\begin{array}{r}
{}_3 \, {}^1\!\!4 \ {}^1 2 \, 0 \text{ marbles} \\
- \ 1 \ 2 \ 3 \text{ marbles} \\
\hline
\textbf{2 9 7 marbles}
\end{array}
$$

3.
4 cm

4 cm

(a) 4 cm + 4 cm + 4 cm + 4 cm = **16 cm**

(b) 4 cm × 4 cm = **16 sq. cm**

4. **20,000 + 5000 + 400 + 60 + 3**

5.
4 cm

The diameter of a circle equals two radii.
2 cm + 2 cm = 4 cm
2 cm

6. Step 1: A half hour is 30 minutes. Counting forward 30 minutes from 4:10 p.m. makes it 4:40 p.m.

Step 2: Counting forward 4 hours from 4:40 p.m. makes it **8:40 p.m.**

7. There are eleven equal-sized circles and four are shaded.

$$\frac{4}{11}$$

8. **5 × 8; 40**

9. 76 rounds up to 80.
59 rounds up to 60.

$$\begin{array}{r} 80 \\ + 60 \\ \hline 140 \end{array}$$

10. (a) 3 $\gtrdot$ −4

(b) 2,000,000 $\gtrdot$ 200,000

11. **2140**

12. (a) 5 × 7 = **35**

(b) 6 × 6 = **36**

(c) 9 × 9 = **81**

(d) 10 × 10 = **100**

13. (a) 3 × 9 = **27**

(b) 9 × 7 = **63**

(c) 8 × 9 = **72**

(d) 9 × 1 = **9**

14. **Three million, five hundred thousand**

15. **750,000**

16.
$$\begin{array}{r} \overset{4}{\cancel{5}}\,\overset{\overset{12}{}}{\cancel{3}}{}^{1}5 \\ - \quad 2\,6\,8 \\ \hline \mathbf{2\,6\,7} \end{array}$$

17.
$$\begin{array}{r} \overset{8}{\cancel{9}}{}^{1}0\,8 \\ - \quad \quad 4\,3 \\ \hline \mathbf{8\,6\,5} \end{array}$$

18.
$$\begin{array}{r} \$4\,\overset{6}{\cancel{7}}{}^{1}1 \\ - \quad \$3\,4\,6 \\ \hline \mathbf{\$1\,2\,5} \end{array}$$

19. C + 329 = 715

$$\begin{array}{r} \overset{6}{\cancel{7}}\,\overset{10}{\cancel{1}}{}^{1}5 \\ - \quad 3\,2\,9 \\ \hline 3\,8\,6 \end{array}$$

C = **386**

20. C − 127 = 398

$$\begin{array}{r} \overset{1\,1}{398} \\ + \quad 127 \\ \hline 525 \end{array}$$

C = **525**

21. 12 in. = 1 ft

The diameter of a circle equals two radii.

1 ft + 1 ft = **2 ft**

22. 5 × 5 = 25
5 + 5 = 10

$$\begin{array}{r} 25 \\ - \quad 10 \\ \hline 15 \end{array}$$

23. **Sample answer: 1 + 3 = 4**
3 + 1 = 4
4 − 1 = 3
4 − 3 = 1

24. 3 × 3 = 9, so $\sqrt{9}$ = 3
4 × 4 = 16, so $\sqrt{16}$ = 4
3 + 4 = 7

25. One possibility:

26. D. 6

27. California   33,871,648
Texas   20,851,820
New York   18,976,457
Florida   15,982,378

LESSON 35, WARM-UP

a. 550

b. 36

c. 45

d. 95

e. 162

f. 885

Patterns

$5 \times 5 = 25$
$6 \times 6 = 36$
$7 \times 7 = 49$
$8 \times 8 = 64$
$9 \times 9 = 81$
1, 4, 9, 16, **25, 36, 49, 64, 81**, 100

LESSON 35, LESSON PRACTICE

a. $2\frac{1}{2}$

b. $3\frac{1}{3}$

c. Twelve and three fourths

d. Two and seven tenths

e. Six and nine hundredths

f. 17¢

g. 5¢

h. $0.08

i. $0.30

j. $0.25 + $0.25 + $0.10 + $0.10 + $0.05
= **$0.75 = 75¢**

k. Twelve dollars and twenty-five cents

l. Twenty dollars and five cents

LESSON 35, MIXED PRACTICE

1. Pattern: Larger
              − Smaller
              ‾‾‾‾‾‾‾‾‾
              Difference

Problem: **270 peasants**
            − 155 peasants
            ‾‾‾‾‾‾‾‾‾‾‾‾‾‾
            N peasants

            2 7̶¹0 peasants
            − 1 5 5 peasants
            ‾‾‾‾‾‾‾‾‾‾‾‾‾‾‾
            1 1 5 peasants

2. 300 yards + 100 yards + 300 yards
       + 100 yards = **800 yards**

3. Pattern: Some
              Some more
            + Some more
            ‾‾‾‾‾‾‾‾‾‾
              Total

Problem: 97 oranges
            57 oranges
          + 48 oranges
          ‾‾‾‾‾‾‾‾‾‾‾
            202 oranges

4. $2\frac{2}{3}$

5. $4.65; 465¢

6. Count up by twos from 30: 30, 32, 34, 38.
38°F

7. C.

8. $9 \times 9 = 81$, so $\sqrt{81} = 9$
    $7 \times 7 = 49$, so $\sqrt{49} = 7$
    $49 - 9 = \mathbf{40}$

9. Step 1: Counting forward 20 minutes from
    9:25 p.m. makes it 9:45 p.m.
    Step 2: Counting forward 2 hours from
    9:45 p.m. makes it **11:45 p.m.**

10. **Two and three tenths**

11. **420**

12. **One dollar and forty-three cents**

13. (a) $6 \times 9 = \mathbf{54}$

    (b) $4 \times 9 = \mathbf{36}$

    (c) $3 \times 9 = \mathbf{27}$

    (d) $10 \times 9 = \mathbf{90}$

14. (a) $6 \times 6 = \mathbf{36}$

    (b) $7 \times 7 = \mathbf{49}$

    (c) $8 \times 8 = \mathbf{64}$

    (d) $1 \times 1 = \mathbf{1}$

15. $5 \times 5 = 25$, so $\sqrt{25} = 5$
    $4 \times 4 = 16$, so $\sqrt{16} = 4$
    $5 - 4 = \mathbf{1}$

16. **One possibility:**

    3 cm

    3 cm

17.
$$
\begin{array}{r}
\$\overset{5}{\cancel{6}}.\overset{1}{0}5 \\
- \ \$2.53 \\
\hline
\mathbf{\$3.52}
\end{array}
$$

18.
$$
\begin{array}{r}
489 \\
+ \ \ Z \\
\hline
766
\end{array}
$$

$$
\begin{array}{r}
\overset{6}{\cancel{7}}\overset{15}{\cancel{6}}6 \\
- \ 489 \\
\hline
277
\end{array}
$$

    $Z = \mathbf{277}$

19.
$$
\begin{array}{r}
\$5.32 \\
+ \ \$3.44 \\
\hline
\mathbf{\$8.76}
\end{array}
$$

20.
$$
\begin{array}{r}
C \\
+ \ 294 \\
\hline
870
\end{array}
$$

$$
\begin{array}{r}
\overset{7}{\cancel{8}}\overset{16}{\cancel{7}}0 \\
- \ 294 \\
\hline
576
\end{array}
$$

    $C = \mathbf{576}$

21.
$$
\begin{array}{r}
\overset{3}{\cancel{4}}\overset{11}{\cancel{2}}3 \\
- \ 245 \\
\hline
\mathbf{178}
\end{array}
$$

22.
$$
\begin{array}{r}
670 \\
- \ \ Z \\
\hline
352
\end{array}
$$

$$
\begin{array}{r}
6\overset{6}{\cancel{7}}0 \\
- \ 352 \\
\hline
318
\end{array}
$$

    $Z = \mathbf{318}$

23. **250,000,000**

24. The rule is "Count up by hundreds."
    3800, 3900, 4000

25. (a) 77 rounds up to **80**

    (b) 82 cents is more than 50. Round up to **$7.**

26. $7 + 3 = 10$, so $\square = 3$
    $7 - 3 = 4$
    B. 4

27. (a) 30,000 $\gtrdot$ 13,000

    (b) 74¢ $\eqdot$ $0.74

LESSON 36, WARM-UP

a. **640**

b. **820**

c. **625**

d. 282

e. 85

f. 970

Problem Solving
36 = 6 × 6
6 squares

LESSON 36, LESSON PRACTICE

a. $0.25 + $0.25 + $0.25 = **$0.75**

One quarter is $\frac{1}{4}$ of a dollar, so three quarters are $\frac{3}{4}$ of a dollar.

b. One nickel is $\frac{1}{20}$ of a dollar, so three nickels are $\frac{3}{20}$ of a dollar.

$0.05 + $0.05 + $0.05 = **$0.15**

c. One penny is $\frac{1}{100}$ of a dollar, so fifty pennies are $\frac{50}{100}$ of a dollar $\left(\text{or } \frac{1}{2}\right)$.

$0.50

d. A dime is $\frac{1}{10}$ of a dollar and a quarter is $\frac{1}{4}$ of a dollar. A dime is less than a quarter.

$\frac{1}{10}$ of a dollar $\bigotimes$ $\frac{1}{4}$ of a dollar

e. $\frac{1}{2}$ of a dollar $\bigoplus$ $0.20

$0.50

LESSON 36, MIXED PRACTICE

1. Pattern: Larger
              − Smaller
                 Difference

Problem: 70 in.
           − 49 in.
                 N in.

$\overset{6}{\cancel{7}}{}^{1}0$ in.
           − 4 9 in.
              2 1 in.

2. Pattern: Some
              − Some went away
                 What is left

Problem: $36.49
           −   N
                 $11.80

$3\,\overset{5}{\cancel{6}}.{}^{1}4\,9$
           − $1 1 . 8 0
              $2 4. 6 9

3. Pattern: Some
              − Some went away
                 What is left

Problem: 25 questions
           − N questions
              11 questions

25 questions
           − 11 questions
              14 questions

4. $3\frac{1}{6}$

5. **N**

6. **Two million, seven hundred thousand**

7. **82,500**

8. Step 1: Counting forward 20 minutes from 10:40 a.m. makes it 11:00 a.m.
Step 2: Counting forward 5 hours from 11:00 a.m. makes it **4:00 p.m.**

9. 4 + 4 + 4 + 4 + 4 + 4
        + 4 + 4 = **8 × 4**

10. (a) 176 rounds up to **180**

(b) 60 cents is more than 50. Round up to **$18.**

11. **575**

12. (a) 2 × 8 = **16**

(b) 5 × 6 = **30**

(c) 4 × 5 = **20**

13. (a) $3 \times 3 = $ **9**

(b) $5 \times 5 = $ **25**

(c) $9 \times 9 = $ **81**

14. (a) $9 \times 7 = $ **63**

(b) $9 \times 4 = $ **36**

(c) $9 \times 8 = $ **72**

15. $6 \times 6 = 36$, so $\sqrt{36} = 6$
$7 \times 7 = 49$, so $\sqrt{49} = 7$
$6 + 7 = $ **13**

16.
$$\begin{array}{r} \$7.\overset{6}{\cancel{7}}\overset{12}{\cancel{3}}2 \\ - \ \$3.4\,5 \\ \hline \$3.8\,7 \end{array}$$

17.
$$\begin{array}{r} \overset{1}{\ }\overset{1}{\ }\$4.89 \\ + \ \$2.57 \\ \hline \$7.46 \end{array}$$

18.
$$\begin{array}{r} 4\overset{5}{\cancel{6}}{}^{1}4 \\ - \ 2\,3\,8 \\ \hline 2\,2\,6 \end{array}$$

19.
$$\begin{array}{r} \overset{1}{\ }\overset{1}{\ }548 \\ + \ 999 \\ \hline 1547 \end{array}$$

20.
$$\begin{array}{r} 487 \\ + \quad Z \\ \hline 721 \end{array}$$

$$\begin{array}{r} \overset{6}{\cancel{7}}\,\overset{1}{\cancel{2}}{}^{1}1 \\ - \ 4\,8\,7 \\ \hline 2\,3\,4 \end{array}$$

$Z = $ **234**

21.
$$\begin{array}{r} 250 \\ - \quad C \\ \hline 122 \end{array}$$

$$\begin{array}{r} 2\,\overset{4}{\cancel{5}}{}^{1}0 \\ - \ 1\,2\,2 \\ \hline 1\,2\,8 \end{array}$$

$C = $ **128**

22. $C - 338 = 238$

$$\begin{array}{r} \overset{1}{\ }238 \\ + \ 338 \\ \hline 576 \end{array}$$

$C = $ **576**

23. $87 - B = 54$

$$\begin{array}{r} 87 \\ - \ 54 \\ \hline 33 \end{array}$$

$B = $ **33**

24. **6**

25. (a) One dime is $\frac{1}{10}$ of a dollar, so seven dimes are $\frac{7}{10}$ of a dollar.

(b) **$0.70**

26.

4 in.

5 in.

5 in. $\times$ 4 in. $= 20$ sq. in.

C. 20 sq. in.

27. (a) $-12 \ \boxed{>} \ -21$

(b) $\$0.25 \ \boxed{=} \ \$0.25$

LESSON 37, WARM-UP

a. **580**

b. **640**

c. **428**

d. **64**

e. **176**

f. **670**

Problem Solving

A is $\frac{1}{2}$, B is $\frac{1}{4}$, C is $\frac{1}{8}$, and D is $\frac{1}{8}$.

LESSON 37, LESSON PRACTICE

a. There are four segments between 0 and 1, so each segment equals $\frac{1}{4}$. The arrow points to $\frac{3}{4}$.

b. There are four segments between 2 and 3, so each segment equals $\frac{1}{4}$. The arrow points to $2\frac{1}{4}$.

c. There are three segments between 0 and 1, so each segment equals $\frac{1}{3}$. The arrow points to $\frac{2}{3}$.

d. There are three segments between 2 and 3, so each segment equals $\frac{1}{3}$. The arrow points to $2\frac{1}{3}$.

e. There are seven segments between 26 and 27, so each segment equals $\frac{1}{7}$. The arrow points to $26\frac{3}{7}$.

f. There are four segments between 1 and 2, so each segment equals $\frac{1}{4}$. The arrow points to $1\frac{1}{4}$.

LESSON 37, MIXED PRACTICE

1. Pattern:
$$\begin{array}{r} \text{Larger} \\ - \ \text{Smaller} \\ \hline \text{Difference} \end{array}$$

Problem:
$$\begin{array}{r} 715 \text{ petunia blossoms} \\ - \ 427 \text{ petunia blossoms} \\ \hline N \text{ petunia blossoms} \end{array}$$

$$\begin{array}{r} {}^{6}{}^{1}0 \\ \cancel{7}\ \cancel{1}5 \text{ petunia blossoms} \\ - \ 4\ 2\ 7 \text{ petunia blossoms} \\ \hline \mathbf{2\ 8\ 8 \text{ petunia blossoms}} \end{array}$$

2.
$$\begin{array}{r} 275 \text{ fans} \\ + \ N \text{ fans} \\ \hline 297 \text{ fans} \end{array}$$

$$\begin{array}{r} 297 \text{ fans} \\ - \ 275 \text{ fans} \\ \hline \mathbf{22 \text{ fans}} \end{array}$$

3. 4⑦5,342

4.

9 ft

9 ft

$9 \times 6 = \mathbf{81 \text{ tiles}}$

5. There are eight segments between 11 and 12, so each segment equals $\frac{1}{8}$. The arrow points to $11\frac{3}{8}$.

6.

3 cm

5 cm

$5 \text{ cm} + 3 \text{ cm} + 5 \text{ cm} + 3 \text{ cm} = \mathbf{16 \text{ cm}}$

7. $5\frac{1}{2}$

8. **Twelve and three tenths**

9. **$7000 + 20 + 6$**
Seven thousand, twenty-six

10. Step 1: Counting forward 35 minutes from 4:25 a.m. makes it 5:00 a.m.
Step 2: Counting forward 2 hours from 5:00 a.m. makes it **7:00 a.m.**

11. (a) One quarter is $\frac{1}{4}$ of a dollar, so three quarters are $\frac{3}{4}$ of a dollar.

(b) $\$0.25 + \$0.25 + \$0.25 = \mathbf{\$0.75}$

12. $\mathbf{4 \times 6 = 24}$

13. (a) $9 \times 6 = \mathbf{54}$

(b) $9 \times 5 = \mathbf{45}$

(c) $9 \times 0 = \mathbf{0}$

14. (a) $10 \times 10 = \mathbf{100}$

(b) $7 \times 7 = \mathbf{49}$

(c) $8 \times 8 = \mathbf{64}$

15. (a) $5 \times 7 = \mathbf{35}$

(b) $6 \times 5 = \mathbf{30}$

(c) $2 \times 8 = \mathbf{16}$

16. $9 \times 9 = 81$, so $\sqrt{81} = 9$
$7 \times 7 = 49$, so $\sqrt{49} = 7$
$9 + 7 = \mathbf{16}$

17.
$$\begin{array}{r} {}^{5} \\ \$6.\ \cancel{6}{}^{1}3 \\ - \ \$3.\ 5\ 5 \\ \hline \mathbf{\$3.\ 0\ 8} \end{array}$$

18.
$$\begin{array}{r} \overset{1\ \ 1}{\$4.99} \\ +\ \$2.88 \\ \hline \$7.87 \end{array}$$

19. $A - 247 = 321$

$$\begin{array}{r} 321 \\ +\ 247 \\ \hline 568 \end{array}$$

$A = \mathbf{568}$

20. $Z + 296 = 531$

$$\begin{array}{r} \overset{4}{\cancel{5}}\overset{12}{\cancel{3}}{}^{1}1 \\ -\ 2\ 9\ 6 \\ \hline 2\ 3\ 5 \end{array}$$

$Z = \mathbf{235}$

21. $523 - Z = 145$

$$\begin{array}{r} \overset{4}{\cancel{5}}\overset{1}{2}{}^{1}3 \\ -\ 1\ 4\ 5 \\ \hline 3\ 7\ 8 \end{array}$$

$Z = \mathbf{378}$

22.
$$\begin{array}{r} \overset{3}{28} \\ 46 \\ 48 \\ 64 \\ \overset{2}{}32 \\ +\ 344 \\ \hline 562 \end{array}$$

23. The rule is "Count up by tens."
490, 500, 510

24. 1 ft = 12 in.
The diameter of a circle equals two radii.
6 in. + 6 in. = 12 in.
6 in.

25. (a) $\frac{1}{4}$ of a dollar is one quarter
One quarter $\,\bigotimes\,$ $\frac{1}{2}$ of a dollar

(b) 101,010 $\,\bigotimes\,$ 110,000

26. 1 yd = 3 ft = 36 in.
1 yd ≠ 1 m
C. 1 m

27. | | |
|---|---|
| **Wyoming** | **493,782** |
| **Vermont** | **608,827** |
| **Alaska** | **626,932** |
| **North Dakota** | **642,200** |

LESSON 38, WARM-UP

a. $4.45

b. $7.75

c. $2.85

d. 245

e. 173

f. 275

Problem Solving

$$\begin{array}{r} 1\text{HD} = 50\text{¢} \\ 1\text{Q} = 25\text{¢} \\ 2\text{D} = 20\text{¢} \\ +\ 1\text{N} = 5\text{¢} \\ \hline 100\text{¢ or }\mathbf{\$1.00} \end{array}$$

LESSON 38, LESSON PRACTICE

For answers, please refer to the Facts Practice Tests section of this manual.

LESSON 38, MIXED PRACTICE

1. Pattern:
$$\begin{array}{r} \text{Larger} \\ -\ \text{Smaller} \\ \hline \text{Difference} \end{array}$$

Problem:
$$\begin{array}{r} 405 \text{ toys} \\ -\ 220 \text{ toys} \\ \hline N \text{ toys} \end{array}$$

$$\begin{array}{r} \overset{3}{\cancel{4}}{}^{1}0\ 5 \text{ toys} \\ -\ 2\ 2\ 0 \text{ toys} \\ \hline \mathbf{1\ 8\ 5 \text{ toys}} \end{array}$$

2. **575,542 people**

3. 2000 + 500 + 3
Two thousand, five hundred three

4.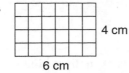

4 cm

6 cm

    (a) 6 cm + 4 cm + 6 cm + 4 cm = **20 cm**

    (b) 6 cm × 4 cm = **24 sq. cm**

5. There are four segments between 8 and 9, so each segment equals $\frac{1}{4}$. The arrow points to $\mathbf{8\frac{3}{4}}$.

6. Main Street

7. $\mathbf{2\frac{1}{8}}$

8. (a) 624 rounds down to **620**

    (b) 24 cents is less than 50. Round down to **$6.**

9. Step 1: Counting forward 15 minutes from 7:25 a.m. makes it 7:40 a.m.
Step 2: Counting forward 5 hours from 7:40 a.m. makes it **12:40 p.m.**

10. (a) $\dfrac{1}{2}$

    (b) **$0.50**

11. Two and eleven hundredths

12. 6 × 6 = 36

13. (a) 3 × 4 = **12**

    (b) 3 × 6 = **18**

    (c) 3 × 8 = **24**

14. (a) 4 × 6 = **24**

    (b) 4 × 7 = **28**

    (c) 4 × 8 = **32**

15. (a) 6 × 7 = **42**

    (b) 6 × 8 = **48**

    (c) 7 × 8 = **56**

16. $\frac{1}{10}$ of a dollar is one dime

One dime $\lessgtr$ $\frac{1}{2}$ of a dollar

17.
$$
\begin{array}{r}
\$\overset{6}{7}.\,\overset{1}{2}{}^{1}3 \\
-\ \ \$2.\,5\ 4 \\
\hline
\$4.\,6\ 9
\end{array}
$$

18.
$$
\begin{array}{r}
\overset{1}{\$5}.\overset{1}{4}2 \\
+\ \ \$2.69 \\
\hline
\$8.11
\end{array}
$$

19.
$$
\begin{array}{r}
\overset{8}{\cancel{9}}\ \overset{^13}{\cancel{4}}{}^13 \\
-\ \ 2\ 7\ 6 \\
\hline
6\ 6\ 7
\end{array}
$$

20. $Z - 581 = 222$

$$
\begin{array}{r}
\overset{1}{2}22 \\
+\ \ 581 \\
\hline
803
\end{array}
$$

$Z = \mathbf{803}$

21. $C + 843 = 960$

$$
\begin{array}{r}
9\ \overset{5}{\cancel{6}}{}^10 \\
-\ \ 8\ 4\ 3 \\
\hline
1\ 1\ 7
\end{array}
$$

$C = \mathbf{117}$

22. 100 cm = 1 m
The diameter of a circle equals two radii.
1 m + 1 m = **2 m**

23. $7 \times 7 = 49$, so $\sqrt{49} = 7$

$$
\begin{array}{r}
\overset{2}{2}8 \\
36 \\
78 \\
+\ \ \ 7 \\
\hline
149
\end{array}
$$

24.
$$
\begin{array}{r}
\overset{3}{1}4 \\
18 \\
6 \\
4 \\
18 \\
+\ \ 15 \\
\hline
75
\end{array}
$$

25.

$$\begin{array}{r} \overset{3}{29} \\ 5 \\ 13 \\ 27 \\ 63 \\ + \ 76 \\ \hline 213 \end{array}$$

26. D. 3

27. (a) **Right**

(b) **Obtuse**

(c) **Acute**

LESSON 39, WARM-UP

a. $4.44

b. $6.73

c. $1.80

d. 418

e. 264

f. 387

Problem Solving

One half of $\frac{1}{2}$ is $\frac{1}{4}$. So three slices is $\frac{1}{4} + \frac{1}{4} + \frac{1}{4}$, which is $\frac{3}{4}$ of the orange.

LESSON 39, LESSON PRACTICE

a. **One possibility:**

b. $\frac{3}{4}$ **in.**

c. $1\frac{1}{2}$ **in.**

d. $2\frac{1}{4}$ **in.**

e. $3\frac{3}{4}$ **in.**

f. Most notebook paper is 11 in. by $8\frac{1}{2}$ in.

LESSON 39, MIXED PRACTICE

1. Pattern:

$$\begin{array}{r} \text{Larger} \\ - \ \text{Smaller} \\ \hline \text{Difference} \end{array}$$

Problem:

$$\begin{array}{r} 35 \text{ years} \\ - \ 12 \text{ years} \\ \hline N \text{ years} \end{array}$$

$$\begin{array}{r} 35 \text{ years} \\ - \ 12 \text{ years} \\ \hline \mathbf{23 \text{ years}} \end{array}$$

2. **468,502 boxes**

3. **3000 + 900 + 5**
Three thousand, nine hundred five

4. Pattern:

$$\begin{array}{r} \text{Some} \\ + \ \text{Some more} \\ \hline \text{Total} \end{array}$$

Problem:

$$\begin{array}{r} 243 \text{ cans} \\ + \ 364 \text{ cans} \\ \hline 607 \text{ cans} \end{array}$$

A number that ends with 7 is an **odd number.**

5. **One hundred and one hundredth**

6. $-19 > -90$

7. $\$1.00 + \$1.00 + \$0.25 + \$0.10 + \$0.10 + \$0.05 + \$0.05 + \$0.05 = \mathbf{\$2.60}$

8. Counting forward 10 minutes from 8:17 a.m. makes it **8:27 a.m.**

9. (a) One dime is $\frac{1}{10}$ of a dollar, so nine dimes are $\frac{9}{10}$ of a dollar.

(b) **$0.90**

10. 1 km = **1000 m**

11. $2\frac{2}{3}$

12. $1\frac{1}{4}$ inches

13. (a) $4 \times 3 = \mathbf{12}$

(b) $8 \times 3 = \mathbf{24}$

(c) $8 \times 4 = \mathbf{32}$

14. (a) $6 \times 3 = \mathbf{18}$

(b) $6 \times 4 = \mathbf{24}$

(c) $7 \times 6 = \mathbf{42}$

15. (a) $7 \times 3 = \mathbf{21}$

(b) $7 \times 4 = \mathbf{28}$

(c) $8 \times 6 = \mathbf{48}$

16. $8 \times 8 = 64$, so $\sqrt{64} = 8$
$6 \times 6 = 36$, so $\sqrt{36} = 6$
$8 - 6 = \mathbf{2}$

17.
$$\begin{array}{r} \overset{1\ 1}{}\$4.86 \\ +\ \$2.47 \\ \hline \mathbf{\$7.33} \end{array}$$

18.
$$\begin{array}{r} \$4.\overset{7}{\cancel{8}}{}^{1}6 \\ -\ \$2.\ 4\ 7 \\ \hline \mathbf{\$2.\ 3\ 9} \end{array}$$

19.
$$\begin{array}{r} \overset{1\ 1}{293} \\ +\ 678 \\ \hline \mathbf{971} \end{array}$$

20.
$$\begin{array}{r} 8\ \overset{8}{\cancel{9}}{}^{1}3 \\ -\ 6\ 7\ 8 \\ \hline \mathbf{2\ 1\ 5} \end{array}$$

21.
$$\begin{array}{r} 463 \\ -\ \ Y \\ \hline 411 \end{array}$$

$$\begin{array}{r} 463 \\ -\ 411 \\ \hline 52 \end{array}$$

$Y = \mathbf{52}$

22.
$$\begin{array}{r} 463 \\ +\ \ Q \\ \hline 527 \end{array}$$

$$\begin{array}{r} \overset{4}{\cancel{5}}{}^{1}2\ 7 \\ -\ 4\ 6\ 3 \\ \hline 6\ 4 \end{array}$$

$Q = \mathbf{64}$

23. $\mathbf{8 \times 8 = 64}$

24. The rule is "Count up by tens."
510, 520, 530

25. One possibility:

26. $9 + 9 = 18$
$9 \times 9 \neq 18$

D. nine squared

27. **$1,280,000**
$476,000
$385,900
$299,000
$189,000

LESSON 40, WARM-UP

a. **$6.84**

b. **$9.61**

c. **$5.93**

d. **374**

e. **193**

f. **784**

Problem Solving

The two hands of the clock will be together between 1:05 p.m. and 1:06 p.m. So the hands will be together **65 or 66 minutes** after noon.

LESSON 40, ACTIVITY

a. **2 cups**

b. **2 pints**

c. **2 quarts**

d. **2 half gallons**

e. **4 quarters**

f. **4 quarts**

g. **U.S. LIQUID MEASURE**

8 fl oz = 1 c
2 c = 1 pt
2 pt = 1 qt
4 qt = 1 gal

h. 1 quart $<$ 1 liter

i. $\frac{1}{2}$ gallon $<$ 2 liters

j. One liter equals 1000 milliliters, so two liters equals 2000 milliliters, or **2000 mL.**

_____

LESSON 40, LESSON PRACTICE

There is no Lesson Practice for Lesson 40.

_____

LESSON 40, MIXED PRACTICE

1. Pattern: Larger
          $-$ Smaller
          Difference

    Problem: $\overset{0\ \ ^1 0}{\cancel{1}\ \cancel{1}^1 2}$ fish
           $-$ 2 5 fish
           8 7 fish

2. Pattern: Some
          $-$ Some went away
          What is left

    Problem: 36 inches
           $-$ 12 inches
           24 inches

3. Pattern: Some
          Some more
          $+$ Some more
          Total

    Problem: $\overset{1}{47}$ postcards
           62 postcards
           $+$ 75 postcards
           184 postcards

4. **7,000,000 + 500,000**
    Seven million, five hundred thousand

5. **4**

6. $1.00 + $1.00 + $1.00 + $0.25 + $0.25
    $+ $0.10 + $0.05 + $0.05 = **$3.70**
    Three dollars and seventy cents

7. 1 gallon = **4 quarts**

8. $4\frac{1}{2}$ **squares**

9. $1\frac{1}{2}$ **in.**

10. 1 gallon = 3.78 liters
     3.78 liters $>$ 3 liters
     1 gallon $\bigcirc\!\!\!>$ 3 liters

11. Step 1: Counting forward 50 minutes from
     10:10 p.m. makes it 11:00 p.m.
     Step 2: Counting forward 1 hour from
     11:00 p.m. makes it **12:00 a.m.**

12. **C. Obtuse**

13. (a) $-29 \bigcirc\!\!\!> -32$

     (b) One quarter is $\frac{1}{4}$ of a dollar, so three quarters
         are $\frac{3}{4}$ of a dollar. Three quarters equals $0.75.

         $0.75 \bigcirc\!\!\!= \frac{3}{4}$ of a dollar

14.

     The diameter of a circle equals two radii.
     1 cm + 1 cm = 2 cm
     1 cm

15. (a) $6 \times 6 = $ **36**

    (b) $7 \times 7 = $ **49**

    (c) $8 \times 8 = $ **64**

16. (a) $7 \times 9 = $ **63**

    (b) $6 \times 9 = $ **54**

    (c) $9 \times 9 = $ **81**

17. (a) $7 \times 8 = $ **56**

    (b) $6 \times 7 = $ **42**

    (c) $8 \times 4 = $ **32**

18.
$$\begin{array}{r} \overset{1\ 1}{\$4.98} \\ +\ \$7.65 \\ \hline \$12.63 \end{array}$$

19. $M - \$6.70 = \$3.30$

$$\begin{array}{r} \overset{1}{\$3.30} \\ +\ \$6.70 \\ \hline M = \$10.00 \end{array}$$

20. $416 - Z = 179$

$$\begin{array}{r} \overset{3\ \ {}^{1}0}{4\ \cancel{1}6} \\ -\ 1\ 7\ 9 \\ \hline 2\ 3\ 7 \end{array}$$
$$Z = \mathbf{237}$$

21. $536 + Z = 721$

$$\begin{array}{r} \overset{6\ {}^{1}1}{7\ \cancel{2}1} \\ -\ 5\ 3\ 6 \\ \hline 1\ 8\ 5 \end{array}$$
$$Z = \mathbf{185}$$

22. $1 \times 1 = 1$, so $\sqrt{1} = 1$
$2 \times 2 = 4$, so $\sqrt{4} = 2$
$3 \times 3 = 9$, so $\sqrt{9} = 3$
$1 + 2 + 3 = \mathbf{6}$

23. $\times\times\times\times\times\times$
$\times\times\times\times\times\times$
$\times\times\times\times\times\times$

24. **Ten and one tenth**

25. (a) One quarter is $\frac{1}{4}$ of a dollar, so two quarters are $\frac{2}{4}$ $\left(\text{or } \frac{1}{2}\right)$ of a dollar.

    (b) $\$0.25 + \$0.25 = $ **\$0.50**

26. An 8 in. by 3 in. rectangle contains 24 square inches.
D. 8 in. by 3 in.

27. $8\frac{1}{2}$ (or **8.5**)

_____

INVESTIGATION 4

1.

2. **\$2.50**

3. **2 dollars, 5 dimes, 0 pennies**

4. **two dollars and fifty cents**

5. **2.5, which is read "two point five."**

6. **\$2.50**

7. **The calculator does not "know" this is a money problem. However, 2.5 dollars equals \$2.50.**

8. **2 dollar bills and 5 dimes**

9.

10. **\$1.25**

11. **1 dollar, 2 dimes, 5 pennies**

12. **one dollar and twenty-five cents**

13. **C** ; **to clear previous entries**

14. **2** **·** **5**

15. **We can, but it is not necessary.**

16. [÷] [2] [=]

17. 1.25

18. $1.25

19. 1 dollar bill, 2 dimes, and 5 pennies

20. ⓓⓓⓓⓓⓓ ⓟⓟ   ⓓⓓⓓⓓⓓ ⓟⓟⓟ

21. One will receive $0.62, and the other will receive $0.63.

22. To divide the money equally, a penny needs to be divided, but there is no coin that has less value than a penny.

23. 5 mills

24. 6 dimes, 2 pennies, 5 mills

25. Accurate predictions would be .625 or 0.625.

26. [C] [1] [.] [2] [5] [÷] [2] [=]

27. 0.625

28. sixty-two and a half cents

29. Half of 1.25 cannot be expressed with only two places, just as $1.25 cannot be divided in half using only dimes and pennies.

30. 6 dimes, 2 pennies, 5 mills

31. $\frac{75}{100}$ = seventy-five hundredths

32. 0.75 = seventy-five hundredths

33. $\frac{50}{100}$ = fifty hundredths

34. 0.50 = fifty hundredths

35. $\frac{7}{100}$ = seven hundredths

36. 0.07 = seven hundredths

37. 0.05 = five hundredths

38. 0.03 = three hundredths

39. 0.30 = thirty hundredths

40. 0.21 = twenty-one hundredths

41. $\frac{3}{10}$ = three tenths

42. 0.3 = three tenths

43. $\frac{7}{10}$ = seven tenths

44. 0.7 = seven tenths

45. 0.9 = nine tenths

46. 0.09 = nine hundredths

47. $\frac{1}{1000}$ = one thousandth

48. 0.001 = one thousandth

49. $\frac{21}{1000}$ = twenty-one thousandths

50. 0.021 = twenty-one thousandths

51. $\frac{321}{1000}$ = three hundred twenty-one thousandths

52. 0.321 = three hundred twenty-one thousandths

53. 0.020 = twenty thousandths

54. 0.002 = two thousandths

55. 0.02 = two hundredths

56. 0.2 = two tenths

57. 10.75 = ten and seventy-five hundredths

58. 12.5 = twelve and five tenths

59. 6.42 = six and forty-two hundredths

60. 10.1 = ten and one tenth

61. 1.125 = one and one hundred twenty-five thousandths

62. 2.05 = two and five hundredths

63. 1.3

64. 2.25

65. 3.12

66. 4.5

67. 5.04

68. 0.15

69. 0.5

70. 0.05

71. 0.11; $\frac{11}{100}$

72. 0.01; $\frac{1}{100}$

73. 0.50; $\frac{50}{100}$ or $\frac{1}{2}$

74. 0.25; $\frac{25}{100}$ or $\frac{1}{4}$

75. 0.3; $\frac{3}{10}$

76. 0.9; $\frac{9}{10}$

77. 0.7; $\frac{7}{10}$

78. 0.5; $\frac{5}{10}$ or $\frac{1}{2}$

79. D. 0.05

80. 0.1 = 0.10

81. 0.2 > 0.02

82. 0.3 < 0.31

LESSON 41, WARM-UP

a. $7.64

b. $5.81

c. $9.22

d. 400

e. 785

f. 286

Problem Solving

 ±6

 ±8

LESSON 41, LESSON PRACTICE

a.
$$\begin{array}{r} \overset{29}{\$3.0^{1}0} \\ -\ \$1.3\ 2 \\ \hline \$1.6\ 8 \end{array}$$

b.
$$\begin{array}{r} \overset{39}{\$40^{1}5} \\ -\ \$15\ 6 \\ \hline \$24\ 9 \end{array}$$

c.
$$\begin{array}{r} \overset{19}{20^{1}1} \\ -\ 10\ 2 \\ \hline 9\ 9 \end{array}$$

d.
$$\begin{array}{r} \overset{39}{\$4.0^{1}0} \\ -\ \$0.8\ 6 \\ \hline \$3.1\ 4 \end{array}$$

e.
$$\begin{array}{r} \overset{29}{\$30^{1}4} \\ -\ \$12\ 8 \\ \hline \$17\ 6 \end{array}$$

f.
$$\begin{array}{r} \overset{69}{70^{1}3} \\ -\ 19\ 8 \\ \hline 50\ 5 \end{array}$$

g. $8W = 32$
$\quad 8 \times 4 = 32$
$\quad\quad\quad W = \mathbf{4}$

h. $p \times 3 = 12$
$\quad 4 \times 3 = 12$
$\quad\quad\quad p = \mathbf{4}$

i. $5m = 30$
$\quad 5 \times 6 = 30$
$\quad\quad\quad m = \mathbf{6}$

j. $Q \times 4 = 16$
$\quad 4 \times 4 = 16$
$\quad\quad\quad Q = \mathbf{4}$

LESSON 41, MIXED PRACTICE

1. (a) The square is divided into 100 equal parts and 31 parts are shaded.
$$\frac{31}{100}$$

(b) **0.31**

(c) **Thirty-one hundredths**

2. $\$0.10 + \$0.25 + \$0.01 = \mathbf{\$0.36}$

3. $\quad 1 \text{ gallon} = 4 \text{ quarts}$
$\quad 4 \text{ qt} - 1 \text{ qt} = \mathbf{3 \text{ qt}}$

4. The rule is "Count up by hundreds."
4500, 4600, 4700

5. $0.5 = \dfrac{1}{2}$

6. Counting forward 7 hours from 10:17 p.m. makes it **5:17 a.m.**

7. $\quad 5w = 45$
$\quad 5 \times 9 = 45$
$\quad\quad\quad w = \mathbf{9}$

8. **One and eighty-nine hundredths liters**

9. $1\dfrac{3}{4}$

10. **Z**

11. $\$1.25 + \$1.25 + \$1.25 + \$1.25$
$\quad = \mathbf{4 \times \$1.25}$

12. $1\dfrac{3}{4}$ **inches**

13. $1 \text{ m} = \mathbf{100 \text{ cm}}$

14. (a) One dime is $\frac{1}{10}$ of a dollar, so five dimes are $\frac{5}{10}$ $\left(\text{or } \frac{1}{2}\right)$ of a dollar.

(b) $\$0.10 + \$0.10 + \$0.10 + \$0.10$
$\quad\quad + \$0.10 = \mathbf{\$0.50}$

15. (a) $0.5 \;\boxed{=}\; 0.50$

(b) $\dfrac{1}{2} \;\boxed{>}\; \dfrac{1}{4}$

16. (a) $3 \times 8 = \mathbf{24}$

(b) $3 \times 7 = \mathbf{21}$

(c) $3 \times 6 = \mathbf{18}$

17. (a) $4 \times 8 = \mathbf{32}$

(b) $4 \times 7 = \mathbf{28}$

(c) $4 \times 6 = \mathbf{24}$

18. $\begin{array}{r} 8 \\ \times\, 8 \\ \hline 64 \end{array}$
$\quad M = \mathbf{8}$

19. $\begin{array}{r} 9 \\ \times\, 6 \\ \hline 54 \end{array}$
$\quad N = \mathbf{6}$

20. $\begin{array}{r} Z \\ +\, 179 \\ \hline 496 \end{array}$

$\begin{array}{r} \overset{8}{4}\,\overset{}{\cancel{9}}{}^{1}6 \\ -\, 1\,7\,9 \\ \hline 3\,1\,7 \end{array}$

$\quad Z = \mathbf{317}$

21. $\begin{array}{r} \overset{29}{\$\cancel{3}.\cancel{0}}{}^{1}0 \\ -\, \$1.8\,4 \\ \hline \mathbf{\$1.1\,6} \end{array}$

22. $$\overset{49}{\$\cancel{50}{}^{1}0}$$
$$-\ \ \$16\ 7$$
$$\overline{\$33\ 3}$$

23. $$W$$
$$-\ 297$$
$$\overline{486}$$

$$\overset{1\ 1}{486}$$
$$+\ 297$$
$$\overline{783}$$
$$W\ =\ \mathbf{783}$$

24. The rule is "Count down by sevens."
7, 0, −7, −14

25. **1,050,000**

26. 6 in. × 6 in. = 36 sq. in.
A. 6 in.

27. One and four tenths

LESSON 42, WARM-UP

a.   321

b. $3.25

c. $8.26

d. $6.84

e.   75

f.   393

Problem Solving
1:05, 2:11, 3:16, 4:22, 5:27, 7:38, 8:44, 9:49, 10:55

LESSON 42, LESSON PRACTICE

a. 50 × 7 = **350**

b. 600 × 3 = **1800**

c. 7 × 40 = **280**

d. 4 × 800 = **3200**

e. 813 is closer to 800 than 900, so it rounds to **800.**

f. 685 is closer to 700 than 600, so it rounds to **700.**

g. 427 is closer to 400 than 500, so it rounds to **400.**

h. 2573 is closer to 2600 than 2500, so it rounds to **2600.**

i. 297 rounds to 300 and 412 rounds to 400
300 + 400 = 700

j. 623 rounds to 600 and 287 rounds to 300
600 − 300 = 300

LESSON 42, MIXED PRACTICE

1. (a) 5 cm + 5 cm + 5 cm + 5 cm = **20 cm**

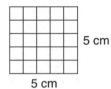

5 cm
5 cm

(b) 5 cm × 5 cm = **25 sq. cm**

2. Pattern: Some
         − Some went away
         What is left

  Problem: 67 grapes
         − N grapes
         38 grapes

         $$\overset{5}{\cancel{6}}{}^{1}7 \text{ grapes}$$
         − 3 8 grapes
         2 9 grapes

3. Pattern: Some
         + Some more
         Total

  Problem: 42 km
         + N km
         126 km

         $$\overset{0}{\cancel{1}}{}^{1}2\ 6 \text{ km}$$
         − 4 2 km
         8 4 km

4. Half an hour is 30 minutes. Counting forward 30 minutes from 4:12 p.m. makes it **4:42 p.m.**

5. $8 \times 8 = 64$
$9 \times 9 = 81$
$10 \times 10 = 100$
64, 81, 100

6. (a) 673 is closer to 700 than 600, so it rounds to **700.**

   (b) 673 is closer to 670 than 680, so it rounds to **670.**

7. $3\frac{3}{4}$ **squares**

8. (a) $1\frac{1}{4}$ **in.**

   (b) **3 cm**

9. $2.50 + $2.50 + $2.50 = **3 × $2.50**

10. **+** contains **perpendicular** line segments

11. **1425**

12. An odd number may end with 7. A number greater than 500 may have an 8 in the hundreds place. **847**

13. $6 \times 80 =$ **480**

14. $7 \times 700 =$ **4900**

15. $9 \times 80 =$ **720**

16. $7 \times 600 =$ **4200**

17.
$$\begin{array}{r} Z \\ +\ 338 \\ \hline 507 \end{array}$$
$$\begin{array}{r} {}^{4}\cancel{5}{}^{9}0{}^{1}7 \\ -\ 33\ 8 \\ \hline 16\ 9 \end{array}$$
$Z =$ **169**

18.
$$\begin{array}{r} {}^{3}\$\cancel{4}.{}^{9}\cancel{0}{}^{1}6 \\ -\ \$2.2\ 8 \\ \hline \$1.7\ 8 \end{array}$$

19.
$$\begin{array}{r} 7 \\ \times\ 6 \\ \hline 42 \end{array}$$
$W =$ **7**

20. $N - 422 = 305$
$$\begin{array}{r} 422 \\ +\ 305 \\ \hline 727 \end{array}$$
$N =$ **727**

21.
$$\begin{array}{r} {}^{1}\ 55 \\ {}_{1}55 \\ 555 \\ +\ 378 \\ \hline 988 \end{array}$$

22. (a) **Five thousand, two hundred eighty**

   (b) **8**

23. (a) One nickel is $\frac{1}{20}$ of a dollar, so ten nickels are $\frac{10}{20}$ (or $\frac{1}{2}$) of a dollar.

   (b) **$0.50**

24. (a) $0.5 \;\boxed{=}\; \frac{1}{2}$

   (b) $\frac{1}{4} \;\boxed{>}\; \frac{1}{10}$

25. $3 \times 3 = 9$
$4 \times 4 = 16$
$9 + 16 =$ **25**

26. One half of the rectangle is shaded.
$5.0 \neq$ one half.
C. 5.0

27. **Twenty-five hundredths**

LESSON 43, WARM-UP

a. **75**

b. **55**

c. **$9.21**

d. **582**

e. 273

Patterns

Sequence: $\frac{1}{2}$, 1, $1\frac{1}{2}$, 2, $2\frac{1}{2}$, 3, $3\frac{1}{2}$, 4, $4\frac{1}{2}$, 5, $5\frac{1}{2}$, 6,

$6\frac{1}{2}$, 7, $7\frac{1}{2}$, 8, $8\frac{1}{2}$, 9, $9\frac{1}{2}$, 10

Halfway between 2 and 5: $3\frac{1}{2}$

LESSON 43, LESSON PRACTICE

a.
$$\begin{array}{r} \$6.32 \\ + \ \$5.00 \\ \hline \mathbf{\$11.32} \end{array}$$

b.
$$\begin{array}{r} \$\overset{2}{\cancel{3}}.^{1}25 \\ - \ \$1.\ 75 \\ \hline \mathbf{\$1.\ 50} \end{array}$$

c.
$$\begin{array}{r} \overset{1}{}46¢ \\ + \ 64¢ \\ \hline \mathbf{110¢} \ \text{or} \ \mathbf{\$1.10} \end{array}$$

d.
$$\begin{array}{r} \overset{8}{\cancel{9}}^{1}8¢ \\ - \ 8\ 9¢ \\ \hline \mathbf{9¢} \end{array}$$

e.
$$\begin{array}{r} \overset{1\ 1}{\$1.46} \\ + \ \$0.87 \\ \hline \mathbf{\$2.33} \end{array}$$

f.
$$\begin{array}{r} 76¢ \\ - \ 5¢ \\ \hline \mathbf{71¢} \ \text{or} \ \mathbf{\$0.71} \end{array}$$

g.
$$\begin{array}{r} 3.47 \\ + \ 3.41 \\ \hline \mathbf{6.88} \end{array}$$

h.
$$\begin{array}{r} 0.75 \\ - \ 0.50 \\ \hline \mathbf{0.25} \end{array}$$

i.
$$\begin{array}{r} \overset{1}{}0.50 \\ + \ 1.75 \\ \hline \mathbf{2.25} \end{array}$$

j.
$$\begin{array}{r} \overset{3}{\cancel{4}}.^{1}25 \\ - \ 3.\ 75 \\ \hline \mathbf{0.\ 50} \ \text{or} \ \mathbf{0.5} \end{array}$$

k.
$$\begin{array}{r} \overset{1}{}5.6 \\ + \ 5.6 \\ \hline \mathbf{11.2} \end{array}$$

l.
$$\begin{array}{r} 2.75 \\ - \ 1.70 \\ \hline \mathbf{1.05} \end{array}$$

LESSON 43, MIXED PRACTICE

1. Pattern:
$$\begin{array}{l} \text{Some} \\ - \ \text{Some went away} \\ \hline \text{What is left} \end{array}$$

   Problem:
$$\begin{array}{r} \overset{09}{\cancel{10}}^{1}0 \ \text{pennies} \\ - \ \ 3\ 5 \ \text{pennies} \\ \hline \mathbf{6\ 5 \ pennies} \end{array}$$

2.

   3 cm (right side)
   3 cm (bottom)

   (a) 3 cm + 3 cm + 3 cm + 3 cm = **12 cm**

   (b) 3 cm × 3 cm = **9 sq. cm**

3. Pattern:
$$\begin{array}{l} \text{Some} \\ - \ \text{Some went away} \\ \hline \text{What is left} \end{array}$$

   Problem:
$$\begin{array}{r} 3.78 \ \text{liters} \\ - \ 1.50 \ \text{liters} \\ \hline 2.28 \ \text{liters} \end{array}$$

   About 2.28 liters

4. Pattern:
$$\begin{array}{l} \text{Some} \\ + \ \text{Some more} \\ \hline \text{Total} \end{array}$$

   Problem:
$$\begin{array}{r} 110 \ \text{miles} \\ + \ N \ \text{miles} \\ \hline 400 \ \text{miles} \end{array}$$

$$\begin{array}{r} \overset{3}{\cancel{4}}^{1}0\ 0 \ \text{miles} \\ - \ 1\ 1\ 0 \ \text{miles} \\ \hline \mathbf{2\ 9\ 0 \ miles} \end{array}$$

5. (a) 572 is closer to 600 than 500, so it rounds to **600.**

    (b) 572 is closer to 570 than 580, so it rounds to **570.**

6. (a) The square is divided into 100 equal parts and 33 parts are shaded.

    $$\frac{33}{100}$$

    (b) **0.33**

    (c) **Thirty-three hundredths**

7. **Parallel**

8. **One possibility:**

9. Counting back 2 hours from 8:05 a.m. makes it **6:05 a.m.**

10. **154**

11. $$\begin{array}{r} 2.45 \\ + \ 4.50 \\ \hline \mathbf{6.95} \end{array}$$

12. $$\begin{array}{r} {}^{2}\cancel{3}.\ \overset{1}{\cancel{2}}{}^{1}5 \\ - \ \$2.\ 4\ 7 \\ \hline \mathbf{\$0.\ 7\ 8} \end{array}$$

13. $$\begin{array}{r} \overset{1}{\$2.15} \\ \$3.00 \\ + \ \$0.07 \\ \hline \mathbf{\$5.22} \end{array}$$

14. $$\begin{array}{r} 3.75 \\ - \ 2.50 \\ \hline \mathbf{1.25} \end{array}$$

15. $$\begin{array}{r} 507 \\ - \ \ N \\ \hline 456 \end{array}$$

    $$\begin{array}{r} {}^{4}\cancel{5}{}^{1}0\ 7 \\ - \ 4\ 5\ 6 \\ \hline 5\ 1 \end{array}$$

    $N = \mathbf{51}$

16. $$\begin{array}{r} N \\ - \ 207 \\ \hline 423 \end{array}$$

    $$\begin{array}{r} \overset{1}{4}23 \\ + \ 207 \\ \hline 630 \end{array}$$

    $N = \mathbf{630}$

17. $$\begin{array}{r} \$\overset{49}{\cancel{5}}.\overset{1}{\cancel{0}}0 \\ - \ \$3.7\ 9 \\ \hline \mathbf{\$1.2\ 1} \end{array}$$

18. $6 \times 80 = \mathbf{480}$

19. $4 \times 300 = \mathbf{1200}$

20. $7 \times 90 = \mathbf{630}$

21. $8 \times 4 = 32$

    $N = \mathbf{4}$

22. $10 \times 10 = 100$

    $\sqrt{100} = \mathbf{10}$

23. ────────────

    About 5 cm

24. **One million, eighty thousand**

25. **C. Meters**

26. 2 liters $\bigcirc\!\!\!>$ $\frac{1}{2}$ gallon

    B. The carton will be full before the bottle is empty.

27. **$315,000**
    $248,000
    $232,000
    $219,900
    $179,500

────────────

LESSON 44, WARM-UP

a. **65**

b. **25**

c. **245**

d. $9.91

e. 285

f. 890

Vocabulary

$$\begin{array}{r} \boxed{\text{factor}} \\ \times\ \boxed{\text{factor}} \\ \hline \boxed{\text{product}} \end{array}$$

$$\boxed{\text{factor}} \ \times \ \boxed{\text{factor}} \ = \ \boxed{\text{product}}$$

LESSON 44, LESSON PRACTICE

a.
$$\begin{array}{r}3\,\textcircled{1}\\ \times\ \textcircled{2}\\ \hline 2\end{array} \longrightarrow \begin{array}{r}\textcircled{3}1\\ \times\ \textcircled{2}\\ \hline 62\end{array} \longrightarrow \begin{array}{r}31\\ \times\ 2\\ \hline \mathbf{62}\end{array}$$

b.
$$\begin{array}{r}3\,\textcircled{1}\\ \times\ \textcircled{4}\\ \hline 4\end{array} \longrightarrow \begin{array}{r}\textcircled{3}1\\ \times\ \textcircled{4}\\ \hline 124\end{array} \longrightarrow \begin{array}{r}31\\ \times\ 4\\ \hline \mathbf{124}\end{array}$$

c.
$$\begin{array}{r}4\,\textcircled{2}\\ \times\ \textcircled{4}\\ \hline 8\end{array} \longrightarrow \begin{array}{r}\textcircled{4}2\\ \times\ \textcircled{4}\\ \hline 168\end{array} \longrightarrow \begin{array}{r}42\\ \times\ 4\\ \hline \mathbf{168}\end{array}$$

d.
$$\begin{array}{r}3\,\textcircled{0}\\ \times\ \textcircled{2}\\ \hline 0\end{array} \longrightarrow \begin{array}{r}\textcircled{3}0\\ \times\ \textcircled{2}\\ \hline 60\end{array} \longrightarrow \begin{array}{r}30\\ \times\ 2\\ \hline \mathbf{60}\end{array}$$

e.
$$\begin{array}{r}3\,\textcircled{0}\\ \times\ \textcircled{4}\\ \hline 0\end{array} \longrightarrow \begin{array}{r}\textcircled{3}0\\ \times\ \textcircled{4}\\ \hline 120\end{array} \longrightarrow \begin{array}{r}30\\ \times\ 4\\ \hline \mathbf{120}\end{array}$$

f.
$$\begin{array}{r}2\,\textcircled{4}\\ \times\ \textcircled{0}\\ \hline 0\end{array} \longrightarrow \begin{array}{r}\textcircled{2}4\\ \times\ \textcircled{0}\\ \hline 0\end{array} \longrightarrow \begin{array}{r}24\\ \times\ 0\\ \hline \mathbf{0}\end{array}$$

LESSON 44, MIXED PRACTICE

1. **Three and seventy-eight hundredths liters**

2. **42,376 > 42,011**

3.
$$\begin{array}{r}\overset{49}{\$\cancel{5}.\cancel{0}^1 0}\\ -\ \$3.2\,5\\ \hline \mathbf{\$1.7\,5}\end{array}$$

4. $9 \times 9 = 81$
   $\sqrt{9} = 3$

   $$\begin{array}{r}\overset{7}{\cancel{8}}{}^1 1\\ -\quad 3\\ \hline \mathbf{7\,8}\end{array}$$

5. $8 \times 6 = 48$
   $M = \mathbf{6}$

6. 1 cup = 8 fl oz
   1 pint = 2 cups
   2 cups = **16 fl oz**

7. $3\dfrac{1}{3}$ circles

8. $\dfrac{3}{4}$ in.

9. (a) $-5 \ \textcircled{<}\ -2$

   (b) $4 \times 60 \ \textcircled{=}\ 3 \times 80$
   $\qquad 240 \qquad\qquad 240$

10.
$$\begin{array}{r}\overset{39}{\$\cancel{4}.\cancel{0}^1 3}\\ -\ \$1.6\,8\\ \hline \mathbf{\$2.3\,5}\end{array}$$

11.
$$\begin{array}{r}\overset{1}{\$4.33}\\ +\ \$5.28\\ \hline \mathbf{\$9.61}\end{array}$$

12.
$$\begin{array}{r}\overset{4}{\$}\overset{1}{\cancel{5}}.\,\overset{}{\cancel{2}}{}^1 2\\ -\ \$2.\,4\,6\\ \hline \mathbf{\$2.\,7\,6}\end{array}$$

13.
$$\begin{array}{r}\overset{69}{\$\cancel{7}.\cancel{0}^1 8}\\ -\ \$0.5\,9\\ \hline \mathbf{\$6.4\,9}\end{array}$$

14. $21 \times 6 = \mathbf{126}$

15. $40 \times 7 = \mathbf{280}$

16. $73 \times 2 = 146$

17. $51 \times 6 = 306$

18.
$$\begin{array}{r} \$2.00 \\ \$0.47 \\ + \ \$0.21 \\ \hline \$2.68 \end{array}$$

19.
$$\begin{array}{r} 8.7 \\ - \ 1.2 \\ \hline 7.5 \end{array}$$

20. $62 - N = 14$

$$\begin{array}{r} \overset{5}{\cancel{6}}{}^{1}2 \\ - \ 1\ 4 \\ \hline 4\ 8 \end{array}$$

$N = \textbf{48}$

21. $N - 472 = 276$

$$\begin{array}{r} \overset{1}{2}76 \\ + \ 472 \\ \hline 748 \end{array}$$

$N = \textbf{748}$

22. $2.1 + 2.1 + 2.1 + 2.1 + 2.1 + 2.1$
$= \mathbf{6 \times 2.1}$

23. (a) **7**

(b) **One thousand, seven hundred sixty**

(c) 1760 is closer to 1800 than 1700, so it rounds to **1800.**

24. 738 rounds to 700 and 183 rounds to 200
700 + 200 = 900

25.
$$\begin{array}{r} 1.50 \\ + \ 3.25 \\ \hline 4.75 \end{array}$$

26. $2 \text{ cm} \times 3 \text{ cm} = 6 \text{ sq. cm}$
A. 3 cm

27. (a) 75 cents is more than 50, so $5.75 is closer to **$6.**

(b) **6**

---

a. **$0.45**

b. **$1.35**

c. **$3.35**

d. **$7.27**

e. **94**

f. **382**

Patterns

$\frac{1}{4}, \frac{1}{2}, \frac{3}{4}, 1, 1\frac{1}{4}, 1\frac{1}{2}, 1\frac{3}{4}, 2, 2\frac{1}{4}, 2\frac{1}{2}, 2\frac{3}{4},$
$3, 3\frac{1}{4}, 3\frac{1}{2}, 3\frac{3}{4}, 4, 4\frac{1}{4}, 4\frac{1}{2}, 4\frac{3}{4}, 5$

LESSON 45, LESSON PRACTICE

a. $8 - (4 + 2)$
$8 - 6$
2

b. $(8 - 4) + 2$
$4 + 2$
6

c. $9 - (6 - 3)$
$9 - 3$
6

d. $(9 - 6) - 3$
$3 - 3$
0

e. $10 + (2 \times 3)$
$10 + 6$
16

f. $3 \times (10 + 20)$
3×30
90

g. $2 + (3 + 4) \; \ominus \; (2 + 3) + 4$
$\quad 2 + 7 \qquad\qquad 5 + 4$
$\qquad 9 \qquad\qquad\qquad 9$

h. $3 \times (4 \times 5) \; \ominus \; (3 \times 4) \times 5$
$\quad 3 \times 20 \qquad\qquad 12 \times 5$
$\qquad 60 \qquad\qquad\qquad 60$

i. Associative property

j. 10 cm − 4 cm = **6 cm**

k. $\overline{LJ}$ (or $\overline{JL}$)

LESSON 45, MIXED PRACTICE

1. 0.5 + 0.6 = 1.1
0.6 + 0.5 = 1.1
1.1 − 0.5 = 0.6
1.1 − 0.6 = 0.5

2. $2\overline{)60 \text{ minutes}}$ — **30 minutes**

3. Pattern: Larger
− Smaller
Difference

Problem: 155 miles
− 15 miles
140 miles

4. $\$5.0^{49}\!{}^{1}0$
− $3.8\ 5$
$1.1 5

5. Twelve and five tenths

6. $^-16 < {}^-6$

7. Counting back 20 minutes from 6:10 a.m. makes it **5:50 a.m.**

8. 4000 + 60
Four thousand, sixty

9. $2\frac{1}{4}$ circles

10. (a) 2 quarters ⊜ half-dollar
50¢   50¢

(b)
2,100,000 ⊙ one million, two hundred thousand
1,200,000

11. 6 × 7 = 42
W = **7**

12. (a) **4 in.**

(b) **10 cm**

13. 12 − (6 − 3) ⊙ (12 − 6) − 3
12 − 3   6 − 3
9   3

14. **No, the associative property does not apply to subtraction. Changing the order of the subtractions resulted in different answers.**

15. $4.^{3}0^{1}7$
− 2.2 6
1.8 1

16. $\$5.^{49}0^{1}2$
− $2.4 7
$2.5 5

17. $\$5.^{4}8^{1}{}^{7}3$
− $2.9 7
$2.8 6

18. $3.^{1}92
+ $5.14
$9.06

19. 42 × 3 − **126**

20. 83 × 2 = **166**

21. 40 × 4 = **160**

22. 41 × 6 = **246**

23. $2.^{1}75
$0.50
+ $3.00
$6.25

24. 3.^{1}50
+ 1.75
5.25

25. ⬜ 1 in.
　2 in.

    (a) 2 in. + 1 in. + 2 in. + 1 in. = **6 in.**

    (b) 2 in. × 1 in. = **2 sq. in.**

26. A. $\overline{RS}$

27. Pattern:　　Larger
　　　　　　　− Smaller
　　　　　　　　Difference

    Problem:　$\overset{2}{\cancel{3}}{}^{1}4$ seconds
　　　　　　− 2 8 seconds
　　　　　　　6 seconds

LESSON 46, WARM-UP

a. **250**

b. **$3.50**

c. **$3.50**

d. **$9.85**

e. **581**

f. **275**

Problem Solving

    56 = 7 × 8

LESSON 46, LESSON PRACTICE

a. $2\overline{)12}$ quotient **6**

b. $3\overline{)21}$ quotient **7**

c. $4\overline{)20}$ quotient **5**

d. $5\overline{)30}$ quotient **6**

e. $6\overline{)42}$ quotient **7**

f. $7\overline{)28}$ quotient **4**

g. $8\overline{)48}$ quotient **6**

h. $9\overline{)36}$ quotient **4**

LESSON 46, MIXED PRACTICE

1. Pattern:　　　　Some
　　　　　　　+ Some more
　　　　　　　　Total

    Problem:　495 oil drums
　　　　　　+ 　N oil drums
　　　　　　　762 oil drums

    　$\overset{6}{\cancel{7}}\,\overset{15}{\cancel{6}}{}^{1}2$ oil drums
　　　− 4 9 5 oil drums
　　　　2 6 7 oil drums

2. 　3.78
　− 2.12
　　1.66

3. Pattern:　　　Some
　　　　　　　Some more
　　　　　　+ Some more
　　　　　　　Total

    Problem:　$\overset{1}{8}2$ bales of hay
　　　　　　92 bales of hay
　　　　　+ 78 bales of hay
　　　　　252 bales of hay

4. (a) 786 is closer to 800 than 700, so it rounds to **800.**

    (a) 786 is closer to 790 than 780, so it rounds to **790.**

5. **One possibility:**

　▨ ▨ ▥

6. (a) $1 + 3 + 5 + 7 + 9 = \textbf{25}$

(b) $5 \times 5 = 25$

$\sqrt{25} = \textbf{5}$

7. Counting back 12 hours from 10:00 a.m. makes it **10:00 p.m.**

8. Acute angle

9. (a) $1\frac{1}{2}$ **in.**

(b) $\overline{DC}$ (or $\overline{CD}$)

10. One dozen $= 12$
Two dozen $= 24$
One big step is about 1 meter
24 big steps are **about 24 meters**

11. There are six segments between 10 and 11, so each segment equals $\frac{1}{6}$. The arrow points to $\mathbf{10\frac{5}{6}}$.

12. $64 + (9 \times 40)$
$64 + 360$
424

13.
$\overset{1}{\$6.25}$
$\$0.39$
$+ \ \$3.00$
$\overline{\textbf{\$9.64}}$

14.
$\$\overset{39}{4.0}{}^{1}2$
$- \ \$2.4\ 7$
$\overline{\textbf{\$1.5 5}}$

15.
$\$\overset{49}{5.0}{}^{1}0$
$- \ \$2.4\ 8$
$\overline{\textbf{\$2.5 2}}$

16.
N
$+ \ 2.5$
$\overline{3.7}$

3.7
$- \ 2.5$
$\overline{1.2}$

$N = \textbf{1.2}$

17.
4.3
$- \ \ C$
$\overline{3.2}$

4.3
$- \ 3.2$
$\overline{1.1}$

$C = \textbf{1.1}$

18. $42 \times 3 = \textbf{126}$

19. $81 \times 5 = \textbf{405}$

20. $6\overline{)30}$ → $\overset{5}{}$

21. $7\overline{)21}$ → $\overset{3}{}$

22. $8\overline{)56}$ → $\overset{7}{}$

23. $9\overline{)81}$ → $\overset{9}{}$

24. $7\overline{)28}$ → $\overset{4}{}$

25. $3\overline{)15}$ → $\overset{5}{}$

26. ▭ 1 in. / 3 in.

(a) 3 in. + 1 in. + 3 in. + 1 in. = **8 in.**

(b) 3 in. × 1 in. = **3 sq. in.**

27. The radius of the tetherball circle was about 6 ft.
$3\ ft = 1\ yd$
$6\ ft = 2\ yd$
D. 2 yd

LESSON 47, WARM-UP

a. 605

b. 708

c. **625**

d. **$4.50**

e. **$6.55**

f. **176**

Problem Solving

Sequence: $\frac{1}{4}, \frac{1}{2}, \frac{3}{4}, 1, 1\frac{1}{4}, 1\frac{1}{2}, 1\frac{3}{4}, 2, 2\frac{1}{4}, 2\frac{1}{2}, 2\frac{3}{4},$

$3, 3\frac{1}{4}, 3\frac{1}{2}, 3\frac{3}{4}, 4$

Halfway between $2\frac{1}{2}$ and 3: $2\frac{3}{4}$

Halfway between 3 and 4: $3\frac{1}{2}$

LESSON 47, LESSON PRACTICE

a. $49 \div 7 = $ **7**

b. $45 \div 9 = $ **5**

c. $40 \div 8 = $ **5**

d. $\frac{36}{6} = $ **6**

e. $\frac{32}{8} = $ **4**

f. $\frac{27}{3} = $ **9**

g. $9\overline{)27}, \ 27 \div 9, \ \frac{27}{9}$

h. $7\overline{)28}, \ 28 \div 7, \ \frac{28}{7}$

i. $3 \times 4 = 12$
$4 \times 3 = 12$
$12 \div 4 = 3$
$12 \div 3 = 4$

LESSON 47, MIXED PRACTICE

1. Pattern:
$$\begin{array}{r} \text{Larger} \\ - \ \text{Smaller} \\ \hline \text{Difference} \end{array}$$

   Problem:
$$\begin{array}{r} \overset{4}{\$\cancel{5}}.\overset{1}{0}\ 7 \\ - \ \$2.\ 4\ 3 \\ \hline \$2.\ 6\ 4 \end{array}$$

   Brand B costs **$2.64** more than Brand A.

2. $4 \times 5 = 20$
$5 \times 4 = 20$
$20 \div 4 = 5$
$20 \div 5 = 4$

3.
$$\begin{array}{r} \overset{1}{2}.3 \\ + \ 8.9 \\ \hline \mathbf{11.2} \end{array}$$

4. A four-digit number greater than 8420 may have an 8 in the thousands place. An even number may end with 6. A four-digit number greater than 8420 may have a 5 in the hundreds place. **8516**

5. (a) $1\frac{1}{2}$ $\bigotimes$ 1.75

   1.50

   (b) **One and seventy-five hundredths**

6. An 8-ft by 4-ft rectangle has an area of 32 square feet. One tile equals 1 square foot, so Chad will need **32 tiles.**

7. **−4**

8. (a) One dime is $\frac{1}{10}$ of a dollar, so five dimes are $\frac{5}{10}$ $\left(\text{or }\frac{1}{2}\right)$ of a dollar.

   (b) **$0.50**

9. $11\ \text{cm} - 2\ \text{cm} = $ **9 cm**

10. $\overline{BC}$ (or $\overline{CB}$)

11. 3296 is closer to 3300 than 3200, so it rounds to **3300.**

12. **Fifteen million**

13. $95 - (7 \times \sqrt{64})$
   $95 - (7 \times 8)$
   $95 - 56$
   39

14. $2.53
   $0.45
   + $3.00
   ‾‾‾‾‾‾‾
   $5.98

15. N
   $- \ 5.1$
   ‾‾‾‾‾
   2.3

   2.3
   $+ \ 5.1$
   ‾‾‾‾‾
   7.4

   $N = $ **7.4**

16. $40 \times 3 = $ **120**

17. $51 \times 5 = $ **255**

18. $28 \div 7 = $ **4**

19. $81 \div 9 = $ **9**

20. $35 \div 7 = $ **5**

21. $16 \div 4 = $ **4**

22. $\dfrac{28}{4} = $ **7**

23. $\dfrac{42}{7} = $ **6**

24. $\dfrac{48}{8} = $ **6**

25. $\dfrac{45}{5} = $ **9**

26. A. $24\overline{)4}$

27. (a) 90 cents is more than 50, so $12.90 is closer
   to **$13.**

   (b) **13**

LESSON 48, WARM-UP

a. 538

b. **$6.09**

c. 758

d. 590

e. **$7.05**

f. **$9.22**

Problem Solving
   **7:05, 8:11, 9:16, 10:22, 11:27, 1:38, 2:44, 3:49,
   4:55**

———————————

LESSON 48, LESSON PRACTICE

a. $\overset{2}{1}6$
   $\times \ \ \ 4$
   ‾‾‾‾‾
   64

b. $\overset{1}{2}4$
   $\times \ \ \ 3$
   ‾‾‾‾‾
   72

c. $\overset{3}{\$4}5$
   $\times \ \ \ 6$
   ‾‾‾‾‾
   $270

d. $\overset{2}{5}3$
   $\times \ \ \ 7$
   ‾‾‾‾‾
   371

e. $\overset{4}{3}5$
   $\times \ \ \ 8$
   ‾‾‾‾‾
   280

f. $\overset{3}{6}4$
   $\times \ \ \ 9$
   ‾‾‾‾‾
   576

g. **Guide and monitor student work.**

LESSON 48, MIXED PRACTICE

1. $3 \times 5 = 15$
$5 \times 3 = 15$
$15 \div 5 = 3$
$15 \div 3 = 5$

2. Pattern: Larger
$-$ Smaller
Difference

Problem: $4\,\overset{6}{\cancel{7}}{}^{1}2$ birds
$-\ 1\ 4\ 7$ birds
3 2 5 birds

3. Pattern: Some
$+$ Some more
Total

Problem: 42 miles
$+\ 75$ miles
117 miles

4. A number between 8000 and 8350 will have
an 8 in the thousands place. An odd number
between 8000 and 8350 can have 1 or 3 in the
hundreds place and 1 or 3 in the ones place.
Since $8361 > 8350$, the number is **8163.**

5. **300,000 + 6000 + 20**
Three hundred six thousand, twenty

6.

7. 1 mi = **5280 ft**

8. 5 in. + 3 in. + 3 in. + 6 in.
+ 5 in. = **22 in.**

9. 1 m = 100 cm

$\overset{09}{\cancel{10}}{}^{1}0$ cm
$-\quad 5\ 4$ cm
4 6 cm

10. **6 cm**

11. $100 + (4 \times 50)$
$100 + 200$
300

12. $\overset{1}{\$3.25}$
$\$0.37$
$+\ \$3.00$
$6.62

13. $\sqrt{4} \times \sqrt{9}$
2×3
6

14. $\overset{1}{33}$
$\times\ \ 6$
198

15. $\overset{2}{24}$
$\times\ \ 5$
120

16. 90
$\times\ \ 6$
540

17. $\overset{1}{\$42}$
$\times\ \ \ 7$
$294

18. $\$\overset{49}{\cancel{5}}.\overset{1}{\cancel{0}}6$
$-\ \$2.2\ 8$
$2.7 8

19. $\overset{1}{1.45}$
$+\ 2.70$
4.15

20. $\overset{2}{\cancel{3}}.{}^{1}2\ 5$
$-\ 1.\ 5\ 0$
1. 7 5

21. $\overset{3}{14}$
28
45
36
92
$+\ 47$
262

22. $28 \div 7 = $ **4**

23. $5\overline{)35}$ — 7

24. $6\overline{)54}$ — 9

25. $\dfrac{63}{7} = 9$

26. A 4-in. by 2-in. rectangle has an area of 8 sq. in.
D. 4 in. by 2 in.

27. **Associative property of multiplication**

LESSON 49, WARM-UP

a. **412**

b. **$632**

c. **941**

d. **183**

e. **775**

f. **365**

Problem Solving

$\dfrac{3}{4}$ **inch**

LESSON 49, LESSON PRACTICE

a. Pattern:
Number of groups × number in each group
= total

Problem: **6 flocks × 8 birds in each flock**
= N birds
48 birds

b. Pattern:
Number of groups × number in each group
= total

Problem: **9 cars × 6 people in each car**
= N people
54 people

c. Pattern:
Number of groups × number in each group
= total

Problem:
4 dozen × 12 doughnuts in each dozen
= N doughnuts
48 doughnuts

LESSON 49, MIXED PRACTICE

1. Pattern:
Number of groups × number in each group
= total

Problem: **4 rows × 8 boys in each row**
= N boys
32 boys

2. Pattern:
Number of groups × number in each group
= total

Problem: **9 rows × 7 girls in each row**
= N girls
63 girls

3. $5 \times 6 = 30$
$6 \times 5 = 30$
$30 \div 6 = 5$
$30 \div 5 = 6$

4. Pattern: Larger
                − Smaller
                Difference

Problem: 475 pounds
         − 111 pounds
         364 pounds

5.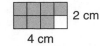

6. There are ten segments between 1 and 2, so each segment equals $\frac{1}{10}$. The arrow points to $1\frac{3}{10}$.

7. Step 1: Counting forward 30 minutes from 8:35 p.m. makes it 9:05 p.m.
Step 2: Counting forward 2 hours from 9:05 p.m. makes it **11:05 p.m.**

8. **One possibility:**

2 cm
4 cm

9. 3,750,000; 7

10. 1.4 + 0.7 = 2.1
0.7 + 1.4 = 2.1
2.1 − 1.4 = 0.7
2.1 − 0.7 = 1.4

11. 9 cm − 5 cm = **4 cm**

12.
$$\begin{array}{r} \overset{29}{\$3.\cancel{0}^17} \\ -\ \$2.2\,8 \\ \hline \$0.7\,9 \end{array}$$

13.
$$\begin{array}{r} \overset{3}{\cancel{4}.^17\,8} \\ -\ 3.9\,0 \\ \hline 0.8\,8 \end{array}$$

14.
$$\begin{array}{r} 7.07 \\ -\quad N \\ \hline 4.85 \end{array}$$

$$\begin{array}{r} \overset{6}{\cancel{7}.^10\,7} \\ -\ 4.8\,5 \\ \hline 2.2\,2 \end{array}$$

$N = $ **2.22**

15.
$$\begin{array}{r} C \\ -\ 2.3 \\ \hline 4.8 \end{array}$$

$$\begin{array}{r} \overset{1}{4.8} \\ +\ 2.3 \\ \hline 7.1 \end{array}$$

$C = $ **7.1**

16. 403 − (5 × 80)
403 − 400
3

17. (4 + 3) × $\sqrt{64}$
7 × 8
56

18. 6 × 5 = 30
$N = $ **5**

19. (587 − 238) + 415
349 + 415
764

20.
$$\begin{array}{r} \overset{3}{45} \\ \times\quad 6 \\ \hline 270 \end{array}$$

21.
$$\begin{array}{r} \overset{2}{23} \\ \times\quad 7 \\ \hline 161 \end{array}$$

22.
$$\begin{array}{r} \overset{3}{\$34} \\ \times\quad 8 \\ \hline \$272 \end{array}$$

23. 56 ÷ 7 = **8**

24. 64 ÷ 8 = **8**

25. $\dfrac{45}{9} = $ **5**

26. The diameter of a circle equals two radii.
3 ft + 3 ft = 6 ft, so the diameter of the circle is 6 ft.
12 in. = 1 ft
36 in. = 3 ft
A. 36 in.

27. B.

LESSON 50, WARM-UP

a. 635

b. 800

c. $7.10

d. $584

e. 95

f. 807

Problem Solving

The sequence consists of currency for the money amounts $100.00, $10.00, $1.00, $0.10, and $0.01. The names of the currency for $0.10 and $0.01 are **dime** and **penny,** respectively.

LESSON 50, LESSON PRACTICE

a. 5

b. 7

c. 5

d.
$$\begin{array}{r} 4.35 \\ +\ 2.6 \\ \hline 6.95 \end{array}$$

e.
$$\begin{array}{r} \overset{3}{\cancel{4}}.\overset{1}{3}\,5 \\ -\ 2.\,6 \\ \hline 1.\,7\,5 \end{array}$$

f.
$$\begin{array}{r} 12.1 \\ +\ \ 3.25 \\ \hline 15.35 \end{array}$$

g.
$$\begin{array}{r} 1\,\overset{4}{\cancel{5}}.\overset{1}{2}\,5 \\ -\ \ 2.\,5 \\ \hline 1\,2.\,7\,5 \end{array}$$

h.
$$\begin{array}{r} \overset{1}{0}.75 \\ +\ 0.5 \\ \hline 1.25 \end{array}$$

i.
$$\begin{array}{r} 0.75 \\ -\ 0.7 \\ \hline 0.05 \end{array}$$

LESSON 50, MIXED PRACTICE

1. Pattern:
Number of groups × number in each group
= total
Problem:
**3 lifeboats × 12 people in each lifeboat
= N people
36 people**

2.
$$\begin{array}{r} \overset{1}{\$}\overset{1}{6}.98 \\ +\ \$0.42 \\ \hline \mathbf{\$7.40} \end{array}$$

3. Pattern:
$$\begin{array}{r} \text{Larger} \\ -\ \text{Smaller} \\ \hline \text{Difference} \end{array}$$

Problem:
$$\begin{array}{r} 6\ \overset{1}{\cancel{2}}\overset{1}{0}\ \text{sit-ups} \\ -\ 4\ 1\ 7\ \text{sit-ups} \\ \hline \mathbf{2\ 0\ 3\ \text{sit-ups}} \end{array}$$

4.
$4 \times 12 = 48$
$12 \times 4 = 48$
$48 \div 4 = 12$
$48 \div 12 = 4$

5. 100 yd + 50 yd + 100 yd + 50 yd
= 300 yd

6. Fifty-eight and seven tenths

7. 12,750,000; 7

8. 783 rounds to 800 and 217 rounds to 200.
800 − 200 = 600

9. Step 1: Counting forward 30 minutes from
10:35 p.m. makes it 11:05 p.m.
Step 2: Counting forward 9 hours from 11:05 p.m.
makes it **8:05 a.m.**

10. $3.75 + $3.75 + $3.75 + $3.75
= 4 × $3.75

11. $(4 \times 50) - \sqrt{36}$
$200 - 6$
194

12.
$$\begin{array}{r} \overset{1}{3}.6 \\ 4.35 \\ +\ 4.2 \\ \hline 12.15 \end{array}$$

13.
$$\begin{array}{r} \overset{1}{\$}\overset{1}{4}.63 \\ \$2.00 \\ \$0.47 \\ +\ \$0.65 \\ \hline \$7.75 \end{array}$$

14.
$$\begin{array}{r} \overset{1}{4}3 \\ \times\ \ 6 \\ \hline 258 \end{array}$$

15.
$$
\begin{array}{r}
\overset{3}{54} \\
\times\ \ 8 \\
\hline
432
\end{array}
$$

16.
$$
\begin{array}{r}
\overset{2}{37} \\
\times\ \ 3 \\
\hline
111
\end{array}
$$

17.
$$
\begin{array}{r}
\$40 \\
\times\ \ \ 4 \\
\hline
\$160
\end{array}
$$

18.
$$
\begin{array}{r}
\overset{3}{4}.7 \\
5.5 \\
8.4 \\
6.3 \\
2.4 \\
+\ 2.7 \\
\hline
30.0\ \ \text{or}\ \ 30
\end{array}
$$

19.
$$
\begin{array}{r}
\$\overset{4}{5}.\overset{9}{\cancel{0}}{}^{1}0 \\
-\ \$4.\,2\,9 \\
\hline
\$0.\,7\,1
\end{array}
$$

20.
$$
\begin{array}{r}
\overset{6}{7}.^{1}0\,3 \\
-\ \ 4.\,2\ \ \\
\hline
2.\,8\,3
\end{array}
$$

21.
$$
\begin{array}{r}
N \\
-\ 27.9 \\
\hline
48.4
\end{array}
$$

$$
\begin{array}{r}
\overset{1\ 1}{48.4} \\
+\ 27.9 \\
\hline
76.3
\end{array}
$$

$$
N = \mathbf{76.3}
$$

22.
$$
\begin{array}{r}
46.2 \\
+\ \ \ C \\
\hline
52.9
\end{array}
$$

$$
\begin{array}{r}
\overset{4}{5}{}^{1}2.\,9 \\
-\ 4\,6.\,2 \\
\hline
6.\,7
\end{array}
$$

$$
C = \mathbf{6.7}
$$

23. $\dfrac{24}{3} = \mathbf{8}$

24. $\dfrac{36}{9} = \mathbf{4}$

25. $5\text{ cm} + 4\text{ cm} = \mathbf{9\text{ cm}}$

26.

27. 1 minute $\bigcirc\!\!>$ 58.7 seconds
60 seconds

INVESTIGATION 5

1. $\dfrac{25}{100} = \dfrac{1}{4}$

2. **25%**

3. $\dfrac{10}{100} = \dfrac{1}{10}$

4. **10%**

5. $\dfrac{1}{100}$

6. **1%**

7. $\dfrac{5}{100} = \dfrac{1}{20}$

8. **5%**

9. A. **20%**

10. C. **75%**

11. B. **40%**

12. D. **80%**

13. $100\% - 40\% = \mathbf{60\%}$

14. $100\% - 75\% =$ **25%**

15. $100\% - 80\% =$ **20%**

16. $100\% - 10\% =$ **90%**

17. $48\% < \frac{1}{2}$; **50% equals** $\frac{1}{2}$. **48% is less than 50%.**

18. $52\% > \frac{1}{2}$; **50% equals** $\frac{1}{2}$. **52% is greater than 50%.**

19. $50\% > \frac{1}{3}$; **50% equals** $\frac{1}{2}$, **and** $\frac{1}{2}$ **is greater than** $\frac{1}{3}$.

20. **more girls; Forty percent is less than** $\frac{1}{2}$, **so less than half of the children were boys. Therefore, more than half the children were girls.**

21. **6 eggs; A dozen is 12. Find half of 12, which is 6.**

22. **30 minutes; Find half of 60 minutes, which is 30 minutes.**

23. **$5; Find half of $10, which is $5.**

24. **12 hours; Find half of 24 hours, which is 12 hours.**

Activity

1. (a)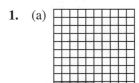

(b) **95%**

2. (a)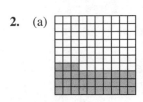

(b) **67%**

3. (a)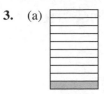

(b) **90%**

4. (a)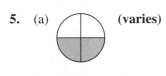

(b) **30%**

5. (a) 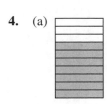 (varies)

(b) **50%**

6. (a) 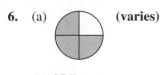 (varies)

(b) **25%**

LESSON 51, WARM-UP

a. 60

b. 30

c. 80

d. 1000, 2000, 3000, 4000, 5000, 6000, 7000, 8000, 9000, 10,000

e. 640

f. $2.15

g. 375

h. 3, 5, 2, 8

Problem Solving

Twelve months before Christmas Day, 1995, would be December 25, 1994. Count back six more months to get **June 25, 1994.**

LESSON 51, LESSON PRACTICE

a.
$$\begin{array}{r} \overset{1\ 1\ 1}{4356} \\ +\ 5644 \\ \hline 10{,}000 \end{array}$$

b.
$$\begin{array}{r} \overset{1\quad 1}{46{,}027} \\ +\ 39{,}682 \\ \hline 85{,}709 \end{array}$$

c.
$$\begin{array}{r} \overset{1\ \ 1\ 11}{360{,}147} \\ +\quad 96{,}894 \\ \hline 457{,}041 \end{array}$$

d.
$$\begin{array}{r} \overset{1\ 1}{_{1}436} \\ 5714 \\ +\quad 88 \\ \hline 6238 \end{array}$$

e.
$$\begin{array}{r} \overset{11\ 11}{43{,}284} \\ 572 \\ +\quad 7\ 635 \\ \hline 51{,}491 \end{array}$$

f.
$$\begin{array}{r} 7 \\ 3\overline{)21} \end{array}$$
$3 \times 7 = 21$

g.
$$\begin{array}{r} 6 \\ 7\overline{)42} \end{array}$$
$7 \times 6 = 42$

h.
$$\begin{array}{r} 8 \\ 6\overline{)48} \end{array}$$
$8 \times 6 = 48$

LESSON 51, MIXED PRACTICE

1. Pattern:
   Number of groups $\times$ number in each group
   $\qquad\qquad = $ total
   Problem:
   4 teams $\times$ 8 players on each team
   $\qquad = N$ **players**
   32 players

2. Pattern:
   Number of groups $\times$ number in each group
   $\qquad\qquad = $ total
   Problem:
   6 stacks $\times$ 7 pancakes in each stack
   $\qquad = N$ **pancakes**
   42 pancakes

3. Pattern:
   $$\begin{array}{r} \text{Larger} \\ -\ \text{Smaller} \\ \hline \text{Difference} \end{array}$$
   Problem:
   $$\begin{array}{r} 6\ \overset{4}{\cancel{5}}.{}^{1}3 \text{ seconds} \\ -\ 6\ 3.\ 4 \text{ seconds} \\ \hline 1.\ 9 \text{ seconds} \end{array}$$

4. $6 \times 7 = 42$
   $7 \times 6 = 42$
   $42 \div 6 = 7$
   $42 \div 7 = 6$

5. $1 + 3 + 5 + 7 + 9 \;\boxed{=}\; 5^2$
   $\qquad\qquad 25 \qquad\qquad 5 \times 5 = 25$

6. (a) 367 is closer to 400 than 300, so it rounds to
   400.

   (b) 367 is closer to 370 than 360, so it rounds to
   370.

7. $50\% = \dfrac{1}{2}$

8. (a) **Acute angle**
   (b) **Right angle**
   (c) **Obtuse angle**

9. (a) **4 ft**
   (b) **2 ft**
   (c) 4 ft $+$ 2 ft $+$ 4 ft $+$ 2 ft $=$ **12 ft**
   (d) 4 ft $\times$ 2 ft $=$ **8 sq. ft**

10. **Two and eighty-four hundredths liters**

11.
$$\begin{array}{r} \overset{1}{15.24} \\ +\ 18.5 \\ \hline 33.74 \end{array}$$

12.
$$\begin{array}{r} \overset{1\ \ 1}{63{,}285} \\ +\ 97{,}642 \\ \hline 160{,}927 \end{array}$$

13.
$$\begin{array}{r} {\scriptstyle 49} \\ \$5.\overset{1}{0}0 \\ -\ \$4.8\ 1 \\ \hline \$0.1\ 9 \end{array}$$

14.
$$\begin{array}{r} N \\ +\ 39.8 \\ \hline 61.4 \end{array}$$

$$\begin{array}{r} {\scriptstyle 5}\ {\scriptstyle 0} \\ \cancel{6}\ \cancel{1}.\overset{1}{4} \\ -\ 3\ 9.\ 8 \\ \hline 2\ 1.\ 6 \end{array}$$

$$N = \mathbf{21.6}$$

15.
$$\begin{array}{r} {\scriptstyle 2} \\ 85 \\ \times\ \ 5 \\ \hline \mathbf{425} \end{array}$$

16.
$$\begin{array}{r} {\scriptstyle 4} \\ 37 \\ \times\ \ 7 \\ \hline \mathbf{259} \end{array}$$

17.
$$\begin{array}{r} 40 \\ \times\ \ 8 \\ \hline \mathbf{320} \end{array}$$

18.
$$\begin{array}{r} 9 \\ \times\ 8 \\ \hline 72 \end{array}$$

$$F = \mathbf{9}$$

19.
$$\begin{array}{r} 47.8 \\ -\ \ \ C \\ \hline 20.3 \end{array}$$

$$\begin{array}{r} 47.8 \\ -\ 20.3 \\ \hline 27.5 \end{array}$$

$$C = \mathbf{27.5}$$

20.
$$\begin{array}{r} {\scriptstyle 1\ 1\ 1\ \ 1\ 1} \\ 462{,}586 \\ +\ \ \ 39{,}728 \\ \hline \mathbf{502{,}314} \end{array}$$

21.
$$\begin{array}{r} Z \\ -\ 4.78 \\ \hline 2.63 \end{array}$$

$$\begin{array}{r} {\scriptstyle 1\ \ 1} \\ 2.63 \\ +\ 4.78 \\ \hline 7.41 \end{array}$$

$$Z = \mathbf{7.41}$$

22.
$$\begin{array}{r} 9 \\ 2\overline{)18} \end{array}$$
$$9 \times 2 = 18$$

23.
$$\begin{array}{r} 3 \\ 7\overline{)21} \end{array}$$
$$3 \times 7 = 21$$

24.
$$\frac{56}{8} = \mathbf{7}$$
$$8 \times 7 = 56$$

25. $12\text{ cm} - 7\text{ cm} = \mathbf{5\ cm}$

26. If half of the students are boys, then half of the students are girls. One half $= 50\%$, so **50%** of the students are girls.

27. $5 \times 0 = 0$, so $N = 0$
$$6 \times 0 = 0$$
$$6N = \mathbf{0}$$

28. (a) The square is divided into 100 equal parts and 11 parts are shaded.
$$\frac{\mathbf{11}}{\mathbf{100}}$$

(b) **11%**

(c) **0.11**

LESSON 52, WARM-UP

a. **170**

b. **240**

c. **310**

d. **$5.47**

e. **551**

f. **83**

g. **1, 7, 4, 6**

Problem Solving

Sal's loop points: $4 \times 5 = 20$ points
Sal's tip points: $2 \times 3 = 6$ points
Cheryl's loop points: $3 \times 5 = 15$ points
Cheryl's tip points: $5 \times 3 = 15$ points
Sal earned 26 points. Cheryl earned 30 points.

LESSON 52, LESSON PRACTICE

a.
$$\begin{array}{r} 4\,\overset{6}{\cancel{7}}\,\overset{17}{\cancel{8}}3 \\ -\ 2\,4\,9\,7 \\ \hline 2\,2\,8\,6 \end{array}$$

b.
$$\begin{array}{r} \overset{399}{\cancel{400}}\!{}^{1}0 \\ -\ \ 5\,2\,7 \\ \hline 3\,4\,7\,3 \end{array}$$

c.
$$\begin{array}{r} \$\overset{199}{\cancel{20.0}}{}^{1}0 \\ -\ \$12.2\,5 \\ \hline \$7.7\,5 \end{array}$$

d. Pattern:
Number of groups $\times$ number in each group
$$=\ \text{total}$$
Problem:
7 cars $\times$ N people in each car $=$ 35 people
$$35 \div 7 = 5$$
5 people

e. Pattern:
Number of groups $\times$ number in each group
$$=\ \text{total}$$
Problem:
N rows $\times$ 6 cars in each row
$$=\ \textbf{30 cars}$$
$$30 \div 6 = 5$$
5 rows

LESSON 52, MIXED PRACTICE

1. Pattern:
Number of groups $\times$ number in each group
$$=\ \text{total}$$
Problem:
8 buses $\times$ 60 people in each bus
$$=\ N \textbf{ people}$$
480 people

2. Pattern:
Number of groups $\times$ number in each group
$$=\ \text{total}$$
Problem:
N vans $\times$ 9 children in each van
$$=\ \textbf{63 children}$$
$$63 \div 9 = 7$$
7 vans

3. Pattern:
Number of groups $\times$ number in each group
$$=\ \text{total}$$
Problem:
4 teams $\times$ N players on each team
$$=\ \textbf{28 players}$$
$$28 \div 4 = 7$$
7 players

4. Pattern:
Some $-$ some went away $=$ what is left
Problem:
10 swimmers $-$ 3 swimmers $=$ **7 swimmers**

5. Pattern:
$$\begin{array}{r} \text{Larger} \\ -\ \text{Smaller} \\ \hline \text{Difference} \end{array}$$
Problem:
$$\begin{array}{r} 5\,8.\,\overset{1}{\cancel{2}}{}^{1}6 \text{ seconds} \\ -\ 5\,7.\,1\,8 \text{ seconds} \\ \hline 1.\,0\,8 \text{ seconds} \end{array}$$

6.
$$7 \times 8 = 56$$
$$8 \times 7 = 56$$
$$56 \div 7 = 8$$
$$56 \div 8 = 7$$

7. $1 + 2 + 3 + 4\ \boxed{=}\ \sqrt{100}$
     $10 10$

8. The rule is "Count up by thousands."
9000, 10,000, 11,000

9. Pattern:
$$\begin{array}{r} \text{Larger} \\ -\ \text{Smaller} \\ \hline \text{Difference} \end{array}$$
Problem:
$$\begin{array}{r} \overset{3}{\cancel{4}}\,\overset{15}{\cancel{6}}5 \text{ apples} \\ -\ 2\,6\,7 \text{ apples} \\ \hline 1\,9\,8 \text{ apples} \end{array}$$

10.
$$\begin{array}{r} \overset{1}{8}.49 \\ 7.3 \\ +\ 6.15 \\ \hline 21.94 \end{array}$$

11. $6 \times 7 = 42$
$$N = 7$$

12.
$$\begin{array}{r} \overset{1\,1\,1}{47{,}586} \\ +\ 23{,}491 \\ \hline 71{,}077 \end{array}$$

13. $\overset{\;\;\overset{49}{}}{\$5.0^1 0}$
     $-\ \$3.2\ 6$
     $\overline{\;\;\$1.7\ 4}$

14. $\quad\quad N$
     $+\ 25.8$
     $\overline{\;\;60.4}$

     $\overset{\overset{5\;9}{}}{6\ \emptyset.^1 4}$
     $-\ 2\ 5.\ 8$
     $\overline{\;\;3\ 4.\ 6}$

     $N\ =\ \mathbf{34.6}$

15. $\overset{5}{\;\;49}$
     $\times\quad 6$
     $\overline{\;\;\mathbf{294}}$

16. $\overset{2}{\;\;84}$
     $\times\quad 5$
     $\overline{\;\;\mathbf{420}}$

17. $\quad 70$
     $\times\quad 8$
     $\overline{\;\;\mathbf{560}}$

18. $\overset{4}{\;\;35}$
     $\times\quad 9$
     $\overline{\;\;\mathbf{315}}$

19. $\quad 400$
     $-\quad N$
     $\overline{\;\;256}$

     $\overset{\overset{3\;9}{}}{4\ \emptyset^1 0}$
     $-\ 2\ 5\ 6$
     $\overline{\;\;1\ 4\ 4}$

     $N\ =\ \mathbf{144}$

20. $\overset{\;\;\overset{399}{}}{\$40.0^1 0}$
     $-\ \$24.6\ 8$
     $\overline{\;\;\$15.3\ 2}$

21. (a) 639 is closer to 600 than 700, so it rounds to **600.**
     (b) 639 is closer to 640 than 630, so it rounds to **640.**

22. $\overline{RQ}$ (or $\overline{QR}$)

23. 49% $\bigcirc\!\!\!<$ $\dfrac{1}{2}$

     $\dfrac{1}{2}\ =\ 50\%$

24. (a) $3\overline{)27}$ with quotient 9
         $3\ \times\ 9\ =\ 27$

     (b) $7\overline{)28}$ with quotient 4
         $7\ \times\ 4\ =\ 28$

     (c) $8\overline{)72}$ with quotient 9
         $8\ \times\ 9\ =\ 72$

25. $\overset{2}{\;\;18}$ m
     $\quad 17$ m
     $\quad 14$ m
     $+\ \ 16$ m
     $\overline{\;\;\mathbf{65\ m}}$

26. (a) 10 cents is less than 50, so $24.10 is closer to **$24.**

     (b) **24**

27. **D.** $2\ \times\ \triangle\ =\ \square\ +\ 2$

28. (a) The square is divided into 100 equal parts and 23 parts are shaded.
         $\dfrac{\mathbf{23}}{\mathbf{100}}$

     (b) **23%**

     (c) **0.23**

LESSON 53, WARM-UP

a. **54¢**

b. **36¢**

c. **72¢**

d. **48¢**

e. **83¢**

f. **15¢**

Problem Solving
$1\frac{3}{4}$ **inches**

LESSON 53, LESSON PRACTICE

a. ⟨••••⟩ ⟨••••⟩ ⟨••••⟩ ••
3 R 2

b.
$$\begin{array}{r} 5\ R\ 2 \\ 3\overline{)17} \\ -15 \\ \hline 2 \end{array}$$

c.
$$\begin{array}{r} 2\ R\ 2 \\ 5\overline{)12} \\ -10 \\ \hline 2 \end{array}$$

d.
$$\begin{array}{r} 5\ R\ 3 \\ 4\overline{)23} \\ -20 \\ \hline 3 \end{array}$$

e.
$$\begin{array}{r} 7\ R\ 1 \\ 2\overline{)15} \\ -14 \\ \hline 1 \end{array}$$

f.
$$\begin{array}{r} 4\ R\ 1 \\ 6\overline{)25} \\ -24 \\ \hline 1 \end{array}$$

g.
$$\begin{array}{r} 8\ R\ 1 \\ 3\overline{)25} \\ -24 \\ \hline 1 \end{array}$$

LESSON 53, MIXED PRACTICE

1. Pattern:
Number of groups $\times$ number in each group
         = total
Problem:
N piles $\times$ 8 washers in each pile
     = 56 washers
       $56 \div 8 = 7$
7 piles

2. Pattern:
Number of groups $\times$ number in each group
         = total
Problem:
7 cars $\times$ N children in each car
     = 42 children
       $42 \div 7 = 6$
6 children

3.
$$4 \times 7 = 28$$
$$7 \times 4 = 28$$
$$28 \div 4 = 7$$
$$28 \div 7 = 4$$

4. **September, April, June, November**

5. The rule is "Count up by thousands."
19,000, 20,000, 21,000

6. (a) 4728 is closer to 4700 than 4800, so it rounds
     to **4700.**
(b) 4728 is closer to 4730 than 4720, so it rounds
     to **4730.**

7. **4:15 p.m.**

8. 4 ft + 4 ft + 4 ft + 4 ft = **16 ft**

9. $4\frac{1}{8}$ **circles**

10. $\sqrt{64} + (42 \div 6)$
     8 + 7
       15

11.
$$\begin{array}{r} \overset{1\ 1}{} \\ \$6.35 \\ \$12.49 \\ +\ \ \$0.42 \\ \hline \mathbf{\$19.26} \end{array}$$

12.
$$\begin{array}{r} \overset{999}{\$1\cancel{00.0}{}^{1}0} \\ -\ \ \$59.8\ 8 \\ \hline \mathbf{\$40.1\ 2} \end{array}$$

13.
$$\begin{array}{r} {}^{4}\cancel{5}\,{}^{1}\cancel{1},\cancel{4}{}^{1}3\,8 \\ -\ 4\,7,4\,9\,5 \\ \hline \mathbf{3\ 9\ 4\ 3} \end{array}$$

14.
$$\begin{array}{r} 60 \\ \times\ \ 9 \\ \hline \mathbf{540} \end{array}$$

15.
$$\begin{array}{r} \overset{2}{5}7 \\ \times\ \ 4 \\ \hline \mathbf{228} \end{array}$$

16.
$$\begin{array}{r} \mathbf{6\ R\ 1} \\ 4\overline{)25} \\ -24 \\ \hline 1 \end{array}$$

17.
$$\begin{array}{r} \mathbf{4\ R\ 2} \\ 5\overline{)22} \\ -20 \\ \hline 2 \end{array}$$

18.
$$\begin{array}{r} \mathbf{6\ R\ 3} \\ 6\overline{)39} \\ -36 \\ \hline 3 \end{array}$$

19.
$$\begin{array}{r} \mathbf{4\ R\ 2} \\ 7\overline{)30} \\ -28 \\ \hline 2 \end{array}$$

20.
$$\begin{array}{r} \overset{4}{46} \\ \times\ \ 8 \\ \hline \mathbf{368} \end{array}$$

21.
$$\begin{array}{r} \overset{5}{38} \\ \times\ \ 7 \\ \hline \mathbf{266} \end{array}$$

22.
$$\begin{array}{r} Z \\ -\ 16.5 \\ \hline 40.2 \end{array}$$

$$\begin{array}{r} 40.2 \\ +\ 16.5 \\ \hline 56.7 \end{array}$$

$$Z = \mathbf{56.7}$$

23.
$$\begin{array}{r} \overset{1}{6.75} \\ \overset{}{4.5} \\ +\ \overset{1}{12.5} \\ \hline \mathbf{23.75} \end{array}$$

24. **7,260,000**

25. **One and eighty-nine hundredths liters**

26. $6 \times 0 = 0$

   D. 6 and 0

Saxon Math 5/4—Homeschool

27. (a) A quarter is $\frac{1}{4}$ or $\frac{25}{100}$ of a dollar.
   25%

   (b) 4 quarts = 1 gallon

   A quart is $\frac{1}{4}$ of a gallon

   $\frac{1}{4}$ = **25%**

28. (a) The square is divided into 100 equal parts and 51 parts are shaded.
   $$\frac{\mathbf{51}}{\mathbf{100}}$$

   (b) **51%**

   (c) **0.51**

LESSON 54, WARM-UP

a. **59¢**

b. **11¢**

c. **81¢**

d. **66¢**

e. **38¢**

f. **656**

g. **$9.60**

h. **833**

Problem Solving

$$\begin{array}{r} 3Q = 75¢ \\ 1D = 10¢ \\ +\ 3N = 15¢ \\ \hline 100¢ \text{ or } \$1.00 \end{array}$$

$$\begin{array}{r} 1HD = 50¢ \\ 1Q = 25¢ \\ +\ 5N = 25¢ \\ \hline 100¢ \text{ or } \$1.00 \end{array}$$

$$\begin{array}{r} 1HD = 50¢ \\ 4D = 40¢ \\ +\ 2N = 10¢ \\ \hline 100¢ \text{ or } \$1.00 \end{array}$$

3 quarters, 1 dime, and 3 nickels;
1 half-dollar, 1 quarter, and 5 nickels;
1 half-dollar, 4 dimes, and 2 nickels

LESSON 54, LESSON PRACTICE

a. **366 days**

b. **May 23, 2014**

c.
$$\begin{array}{r} 1\ 9\ \overset{3}{\cancel{4}}{}^{1}3 \\ -\ 1\ 9\ 1\ 8 \\ \hline \mathbf{2\ 5\ years} \end{array}$$

d. 1 decade = 10 years
    1 century = 100 years
    1 century = **10 decades**

e. 6746
    7 is 5 or more
    6746 rounds up to **7000**

f. 5280
    2 is 4 or less
    5280 rounds down to **5000**

g. 12,327
    3 is 4 or less
    12,327 rounds down to **12,000**

h. 21,694
    6 is 5 or more
    21,694 rounds up to **22,000**

i. 9870
    8 is 5 or more
    9870 rounds up to **10,000**

j. 27,462
    4 is 4 or less
    27,462 rounds down to **27,000**

k. 6472
    4 is 4 or less
    6472 rounds down to **6000**
    6472
    7 is 5 or more
    6472 rounds up to **6500**
    6472
    2 is 4 or less
    6472 rounds down to **6470**

LESSON 54, MIXED PRACTICE

1. Pattern:
    Number of groups × number in each group
    $\qquad\qquad = $ total
    Problem:
    N rows × 7 chairs in each row = 56 chairs
    $\qquad\quad 56 \div 7 = 8$
    8 rows

2. Pattern:
    Number of groups × number in each group
    $\qquad\qquad = $ total
    Problem:
    42 boards × 7 nails in each board = N nails

$$\begin{array}{r} \overset{1}{4}2 \\ \times\ \ 7 \\ \hline \mathbf{294\ nails} \end{array}$$

3. 1 decade = 10 years
    5 decades = **50 years**

4.
$$\begin{array}{r} 1938 \\ -\ 1921 \\ \hline \mathbf{17\ years} \end{array}$$

5. **Wednesday**

6. 5236
    2 is 4 or less
    5236 rounds down to 5000
    6929
    9 is 5 or more
    6929 rounds up to 7000
    5000 + 7000 = 12,000

7. **20 km**
    ⬜ **10 km**
    20 km + 10 km + 20 km + 10 km
    $\qquad = $ **60 km**

8. (a) There are four equal parts and one part is
        shaded.
        $$\frac{1}{4}$$

    (b) **25%**

9. **−15**

10.
$$17¢$$
$$20¢$$
$$50¢$$
$$+ \ 50¢$$
$$\overline{137¢ \ \text{or} \ \$1.37}$$

11.
$$\overset{111 \ \ 1}{794,150}$$
$$+ \quad 9,863$$
$$\overline{804,013}$$

12.
$$\overset{1 \ 11}{\$51,786}$$
$$+ \ \$36,357$$
$$\overline{\$88,143}$$

13.
$$\overset{2 \ 2}{87.6}$$
$$4.0$$
$$31.7$$
$$5.5$$
$$1.1$$
$$+ \quad 0.5$$
$$\overline{130.4}$$

14.
$$\$\overset{1 \ 9 \ 9}{2}\cancel{0}.\cancel{0}^{1}0$$
$$- \ \$1 \ 8 . \ 4 \ 7$$
$$\overline{\$1. \ 5 \ 3}$$

15.
$$\overset{3 \ {}^{0} \ {}^{12} \ {}^{0}}{\cancel{4} \ \cancel{1}, \ \cancel{3} \ \cancel{1}{}^{1}5}$$
$$- \ 2 \ 9, \ 4 \ 1 \ 8$$
$$\overline{1 \ 1, \ 8 \ 9 \ 7}$$

16.
$$\overset{4}{46}$$
$$\times \quad 7$$
$$\overline{322}$$

17.
$$\overset{3}{54}$$
$$\times \quad 8$$
$$\overline{432}$$

18.
$$\overset{8}{39}$$
$$\times \quad 9$$
$$\overline{351}$$

19.
$$40$$
$$\times \quad 9$$
$$\overline{360}$$

20.
$$\overset{1}{3.68}$$
$$\overset{1}{2.4}$$
$$+ \ 15.2$$
$$\overline{21.28}$$

21. $4 \times 8 = 32$
$$Y = \mathbf{8}$$

22.
$$\overset{\mathbf{6 \ R \ 1}}{7)\overline{43}}$$
$$\underline{-42}$$
$$1$$

23.
$$\overset{\mathbf{7 \ R \ 1}}{9)\overline{64}}$$
$$\underline{-63}$$
$$1$$

24. **Two and fifty-four hundredths centimeters**

25. (a) $8 \text{ cm} + 8 \text{ cm} + 8 \text{ cm} + 8 \text{ cm} = \mathbf{32 \ cm}$

(b) $8 \text{ cm} \times 8 \text{ cm} = \mathbf{64 \ sq. \ cm}$

26. **C.** $\overline{RT}$

27. (a) 80 cents is more than 50, so $136.80 is closer to **$137.**

(b) **137**

28. (a) The square is divided into 100 equal parts and 99 parts are shaded.
$$\mathbf{\frac{99}{100}}$$

(b) **99%**

(c) **0.99**

LESSON 55, WARM-UP

a. **74¢**

b. **8¢**

c. **86¢**

d. **24¢**

e. **69¢**

f. $3.50

g. 287

h. 600

Problem Solving

Counting by 3's: 3, 6, 9, 12, 15, 18, 21, 24, 27, 30, 33, 36, 39, 42, 45, 48, 51, 54, 57, 60, …

Counting by 4's: 4, 8, 12, 16, 20, 24, 28, 32, 36, 40, 44, 48, 52, 56, 60, 64, 68, 72, …

Counting by 5's: 5, 10, 15, 20, 25, 30, 35, 40, 45, 50, 55, 60, 65, 70, 75, 80, …

60

LESSON 55, LESSON PRACTICE

a. $1 \times 6 = 6$
$2 \times 6 = 12$
$3 \times 6 = 18$
$4 \times 6 = 24$
$5 \times 6 = 30$
6, 12, 18, 24, 30

b. $3 \times 9 = 27$
$4 \times 9 = 36$
$5 \times 9 = 45$
27, 36, 45

c. $7 \times 8 = $ **56**

d. **0, 2, 4, 6, 8**

e. **1, 2, 5, 10**

f. $1 \times 8, 8 \times 1, 2 \times 4, 4 \times 2$
1, 2, 4, 8

g. $2 \times 9, 9 \times 2, 3 \times 6, 6 \times 3$
2, 3, 6, 9

h. $1 \times 5, 5 \times 1$
1, 5

LESSON 55, MIXED PRACTICE

1. $\overset{1\ 1}{}$
$\$1.85$
$+\ \ \$0.75$
$\overline{\$2.60}$

2. $\overset{1\ 9\ 9}{2\ \cancel{0}\ \cancel{0}\ {}^1 0}$
$-\ \ \ \ \ \ 7$
$\overline{\textbf{1 9 9 3 people}}$

3. 60% are boys.
60% is more than half of the children at the pool. There were **more boys** at the pool.

4. Pattern: Larger
$-$ Smaller
$\overline{\text{Difference}}$

Problem: $\overset{26}{2}7,{}^1000$ people
$-\ \ \ 8, 400$ people
$\overline{\textbf{18, 600 people}}$

5.
```
    4 cm
 ┌──────┐
 │      │ 3 cm
 └──────┘
```

(a) $4\,\text{cm} + 3\,\text{cm} + 4\,\text{cm} + 3\,\text{cm} = $ **14 cm**

(b) $4\,\text{cm} \times 3\,\text{cm} = $ **12 sq. cm**

6. $3 \times 4 = 12$
$12 - 2 = $ **10**

7. $3 \times 5 = 15$
3, 5

8. Counting back 30 minutes from 3:49 p.m. makes it **3:19 p.m.**

9. 1789
-1776
$\overline{\textbf{13 years}}$

10. **5 cm**

11. $\overset{39}{\cancel{4}.0{}^1 0}$
$-\ 2.2\ 2$
$\overline{\textbf{1.7 8}}$

Saxon Math 5/4—Homeschool

12.
$$\begin{array}{r} \overset{6}{\cancel{7}}{}^{1}0.5 \\ -\ 4\,2.3 \\ \hline 2\,8.2 \end{array}$$

13.
$$\begin{array}{r} \overset{1\ 1}{\$45.87} \\ +\ \$23.64 \\ \hline \$69.51 \end{array}$$

14.
$$\begin{array}{r} \$\overset{1}{\cancel{2}}{}^{1}5.\overset{3}{\cancel{4}}{}^{1}2 \\ -\quad \$7.\,2\,5 \\ \hline \$1\,8.\,1\,7 \end{array}$$

15.
$$\begin{array}{r} \overset{2}{6}4 \\ \times\quad 5 \\ \hline 320 \end{array}$$

16.
$$\begin{array}{r} 70 \\ \times\quad 6 \\ \hline 420 \end{array}$$

17.
$$\begin{array}{r} \overset{3}{8}9 \\ \times\quad 4 \\ \hline 356 \end{array}$$

18.
$$\begin{array}{r} \overset{2}{6}3 \\ \times\quad 7 \\ \hline 441 \end{array}$$

19. $\dfrac{63}{7} = 9$

20.
$$\begin{array}{r} 1\ \text{R}\ 7 \\ 8\overline{)15} \\ \underline{-8} \\ 7 \end{array}$$

21.
$$\begin{array}{r} \overset{1\ 1}{{}_{1}4.68} \\ 12.2 \\ +\quad 3.75 \\ \hline 20.63 \end{array}$$

22.
$$\begin{array}{r} 5\ \text{R}\ 3 \\ 6\overline{)33} \\ \underline{-30} \\ 3 \end{array}$$

23. $\sqrt{64} \div (4 + 4)$
$8 \div 8$
$\qquad 1$

24. $0.75 + \$0.75 + \$0.75 + \$0.75$
$= 4 \times \$0.75$

25. **B. 35; Whole numbers ending in either zero or five are multiples of 5. Therefore, they can be divided by 5 without leaving a remainder.**

26.
$$\begin{array}{r} \overset{1}{2}.54\ \text{cm} \\ +\ 2.54\ \text{cm} \\ \hline 5.08\ \text{cm} \end{array}$$

27. (a) 54 cents is more than 50, so $2.54 is closer to **\$3.**

(b) **3**

28. (a) The square is divided into 100 equal parts and 1 part is shaded.
$\dfrac{1}{100}$

(b) **1%**

(c) **0.01**

LESSON 56, WARM-UP

a. **58¢**

b. **33¢**

c. **\$1.75**

d. **\$8.10**

e. **235**

f. **195**

Problem Solving

$\$1.00 - \$0.63 = \$0.37$

$$\begin{array}{r} 1\text{Q} = 25¢ \\ 1\text{D} = 10¢ \\ +\ 2\text{P} = \underline{\ 2¢} \\ 37¢ \end{array}$$

1 quarter, 1 dime, and 2 pennies

LESSON 56, LESSON PRACTICE

a. One possibility:

$$\frac{1}{2} \;\text{<}\; \frac{2}{3}$$

b. One possibility:

$$\frac{1}{2} \;\text{>}\; \frac{1}{4}$$

c. One possibility:

$$\frac{2}{5} \;\text{>}\; \frac{1}{3}$$

LESSON 56, MIXED PRACTICE

1. Pattern:
Number of groups $\times$ number in each group
= total
Problem:
N trays $\times$ 7 pies on each tray = 56 pies
$56 \div 7 = 8$
8 trays

2.
$$\begin{array}{r} \overset{1\ 1}{3.78}\text{ L} \\ +\ 3.78\text{ L} \\ \hline 7.56\text{ L} \end{array}$$
Seven and fifty-six hundredths liters

3. $6.87 rounds to $7 and $5.92 rounds to $6
$7 + $6 = $13

4.
$3 \times 8 = 24$
$8 \times 3 = 24$
$24 \div 3 = 8$
$24 \div 8 = 3$

5. **January, March, May, July, August, October, December**

6. $8 \times 6 = \mathbf{48}$
$48 + 1 = \mathbf{49}$
$\sqrt{49} = \mathbf{7}$

7. One possibility:

$$\frac{1}{4} \;\text{>}\; \frac{1}{6}$$

8. (a) 4873
8 is 5 or more
4873 rounds up to **5000**

(b) 4873
7 is 5 or more
4873 rounds up to **4900**

(c) 4873
3 is 4 or less
4873 rounds down to **4870**

9. (a) 7 mi + 4 mi + 7 mi + 4 mi = **22 mi**

(b) 7 mi $\times$ 4 mi = **28 sq. mi**

10.
$$\begin{array}{r} \overset{99}{\$10.\cancel{0}^{1}0} \\ -\ \$\ 5.4\ 6 \\ \hline \$4.5\ 4 \end{array}$$

11.
$$\begin{array}{r} 3\,\overset{5}{\cancel{6}},\overset{9}{\cancel{0}}\,\overset{1}{\cancel{2}}{}^{1}4 \\ -\ 1\,5,5\,3\,9 \\ \hline 2\,0,4\,8\,5 \end{array}$$

12.
$$\begin{array}{r} \overset{1\,1}{43,675} \\ +\ 52,059 \\ \hline 95,734 \end{array}$$

13.
$$\begin{array}{r} \overset{2}{73} \\ \times\ \ 9 \\ \hline 657 \end{array}$$

14.
$$\begin{array}{r} \overset{4}{46} \\ \times\ \ 7 \\ \hline 322 \end{array}$$

15.
$$\begin{array}{r} \overset{2}{84} \\ \times\ \ 6 \\ \hline 504 \end{array}$$

16.
$$\begin{array}{r} 40 \\ \times\ \ 5 \\ \hline 200 \end{array}$$

17.
$$\begin{array}{r} \mathbf{6\ R\ 6} \\ 7\overline{)48} \\ -\ 42 \\ \hline 6 \end{array}$$

18. $\dfrac{63}{7} = \mathbf{9}$

19.
$$\begin{array}{r} \overset{1}{3.75} \\ 2.5 \\ +\ 0.4 \\ \hline 6.65 \end{array}$$

20.
$$\begin{array}{r} \overset{3}{\cancel{4}}\overset{1}{2}.\overset{1}{2}\,5 \\ -\ \ \ 7.\ 5 \\ \hline 3\,4.\,7\,5 \end{array}$$

21. **B. 40; Multiples of 10 end with zero.**

22. (a) $\dfrac{1}{10}$

(b) **10%**

23. **Twelve million, three hundred fifty thousand dollars**

24. $2 \times 8 = 16,\ 4 \times 4 = 16$
2, 4, 8

25. (a) $\overline{DC}$ (or $\overline{CD}$)

(b) **Obtuse angle**

26. $2 \times 6 - 12$
B. 6

27. $2 \times 12 = 24$
D. 24

28. (a) $\dfrac{1}{100}$

(b) **1%**

(c) **$0.01**

LESSON 57, WARM-UP

a. **15¢**

b. **$1.37**

c. **$3.75**

d. **872**

e. **$7.42**

f. **370**

Problem Solving
First think, "3 plus what number equals 10?" (7). Then think, "8 plus 1 (from regrouping) plus what number equals 15?" (6). Next find the sum in the hundreds column (8).

$$\begin{array}{r} 5\underline{6}3 \\ +\ 2\underline{8}7 \\ \hline \underline{8}50 \end{array}$$

LESSON 57, LESSON PRACTICE

a. Pattern:
Number of groups $\times$ number in each group
$$= \text{total}$$
Problem:
6 hours $\times$ 55 miles per hour $=$ N miles

$$\begin{array}{r} \overset{3}{55} \\ \times\ \ 6 \\ \hline 330\ \text{miles} \end{array}$$

b. Pattern:
Number of groups $\times$ number in each group
$$= \text{total}$$
Problem:
7 days $\times$ 20 laps per day $=$ N laps

$$\begin{array}{r} 20 \\ \times\ \ 7 \\ \hline 140\ \text{laps} \end{array}$$

LESSON 57, MIXED PRACTICE

1. Pattern:
Number of groups $\times$ number in each group
$$= \text{total}$$
Problem:
8 minutes $\times$ 42 times per minute $=$ N times

$$\begin{array}{r} \overset{1}{42} \\ \times\ \ 8 \\ \hline 336\ \text{times} \end{array}$$

2. Pattern:
Number of groups $\times$ number in each group
$$= \text{total}$$
Problem:
3 hours $\times$ 7 miles per hour $=$ N miles
21 miles

3. $8 \times 9 = 72$
$9 \times 8 = 72$
$72 \div 8 = 9$
$72 \div 9 = 8$

4. $\sqrt{36} = 6$ and $\sqrt{64} = 8$
$6 + 8 = \mathbf{14}$

5. $\dfrac{1}{3}$ $\bigcirc\!\!\!<$ 50%

6. (a) 5280
$\underline{2}$ is 4 or less
5280 rounds down to **5000**

   (b) 5280
$\underline{8}$ is 5 or more
5280 rounds up to **5300**

7. Step 1: Counting back 5 minutes from 3:05 p.m. makes it 3:00 p.m.
Step 2: Counting back 6 hours from 3:00 p.m. makes it **9:00 a.m.**

8. $4 \times 6 = \mathbf{24}$
$3 \times 8 = \mathbf{24}$

$\begin{array}{r} 24 \\ + \ 24 \\ \hline \mathbf{48} \end{array}$

9. $\begin{array}{r} 1\,{}^{7}\!8\,{}^{9}\!\cancel{0}{}^{1}0 \\ - \ 1\ 4\ 9\ 2 \\ \hline \mathbf{3\ 0\ 8}\ \textbf{years} \end{array}$

10. (a) 7 in. + 7 in. + 7 in. + 7 in. = **28 inches**

   (b) 7 in. $\times$ 7 in. = **49 sq. in.**

11. $\begin{array}{r} {}^{6999}\!\!\!\!\!\!\!\!\!\! \\ 7\cancel{0},\!\cancel{0}\cancel{0}{}^{1}3 \\ - \ 36,\!41\ 8 \\ \hline \mathbf{33,\!58\ 5} \end{array}$

12. $\begin{array}{r} N \\ - \ 4.32 \\ \hline 2.57 \end{array}$

$\begin{array}{r} 2.57 \\ + \ 4.32 \\ \hline 6.89 \end{array}$
$N = \mathbf{6.89}$

13. $\begin{array}{r} {}^{1}\ {}^{1}\ {}^{1} \\ \$861.34 \\ + \ \$764.87 \\ \hline \mathbf{\$1626.21} \end{array}$

14. $\begin{array}{r} {}^{1} \\ 93 \\ \times \ \ 5 \\ \hline \mathbf{465} \end{array}$

15. $\begin{array}{r} {}^{2} \\ 84 \\ \times \ \ 6 \\ \hline \mathbf{504} \end{array}$

16. $\begin{array}{r} {}^{4} \\ 77 \\ \times \ \ 7 \\ \hline \mathbf{539} \end{array}$

17. $\begin{array}{r} 80 \\ \times \ \ 8 \\ \hline \mathbf{640} \end{array}$

18. $\dfrac{56}{8} = \mathbf{7}$

19. $\begin{array}{r} \mathbf{9\ R\ 2} \\ 7\overline{)65} \\ -63 \\ \hline 2 \end{array}$

20. $\begin{array}{r} \mathbf{7\ R\ 3} \\ 6\overline{)45} \\ -42 \\ \hline 3 \end{array}$

21. $7 \times 6 = 42$
$N = \mathbf{6}$

22. $\begin{array}{r} {}^{1} \\ 1.75 \\ + \ 17.5 \\ \hline \mathbf{19.25} \end{array}$

23. (a) $\overline{WY}$ (or $\overline{YW}$)

   (b) **Acute angle**

24. **One possibility:**

$\dfrac{2}{3}$ $\bigcirc\!\!\!<$ $\dfrac{3}{4}$

25. There are ten segments between 6 and 7, so each segment equals $\frac{1}{10}$. The arrow points to $6\frac{3}{10}$.

26.
$$\overset{1\ 1}{}$$
```
    2.54 cm
    2.54 cm
 +  2.54 cm
 ─────────
    7.62 cm
```

27. $2.54 + 2.54 + 2.54 = \mathbf{3 \times 2.54}$

28. (a) $\dfrac{3}{100}$

(b) **3%**

(c) **$0.03**

d. Sample answer: First multiply the pennies. Five times five pennies is 25 pennies, which equals two dimes and five pennies. Write the 5 below the bar and the 2 above the dimes.

Next multiply the dimes. Five times two dimes is 10 dimes, and we add the two dimes we carried to get a total of 12 dimes. Since 12 dimes equals one dollar and two dimes, write 2 below the bar and 1 above the dollars.

Finally, we multiply the dollars. Five times four dollars is 20 dollars. Add the one dollar we carried to get a total of 21 dollars. The product is $21.25.

LESSON 58, WARM-UP

a. **$2.75**

b. **$3.37**

c. **$1.65**

d. **360**

e. **194**

f. **$7.30**

Problem Solving
$1\frac{1}{2}$ in.

LESSON 58, LESSON PRACTICE

a.
$$\overset{1\ 1}{}$$
```
    234
 ×    3
 ──────
    702
```

b.
$$\overset{1}{}$$
```
   $340
 ×    4
 ──────
  $1360
```

c.
$$\overset{1\ 2}{}$$
```
   $4.25
 ×     5
 ───────
  $21.25
```

LESSON 58, MIXED PRACTICE

1. Pattern:
Number of groups $\times$ number in each group
$\qquad\qquad\qquad = $ total

Problem:
4 weeks $\times$ $7.50 per week $= N$ total cost

$$\overset{2}{}$$
```
     $7.50
  ×      4
  ────────
   $30.00 total cost
```

2. Pattern:
Number of groups $\times$ number in each group
$\qquad\qquad\qquad = $ total

Problem:
5 pies $\times$ 4 apples in each pie $=$ **20 apples**

3. Counting back 8 hours from 6:00 a.m. makes it **10:00 p.m.**

4. (a) **Two and eleven hundredths quarts**

(b) 2L $\gtrdot$ 2 qt
2.11 qt

5. $8000 + 400 + 2$
Eight thousand, four hundred two

6. $4 \times 7 = \mathbf{28}$
$6 \times 6 = \mathbf{36}$

$$\overset{1}{}$$
```
    28
 +  36
 ─────
    64
```
$\sqrt{64} = \mathbf{8}$

7. **September 9, 2042**

8.
$$\begin{array}{r} 5 \\ + \; N \\ \hline 23 \end{array}$$

$$\begin{array}{r} \overset{1}{2}{}^{1}3 \\ - \quad 5 \\ \hline 1\;8 \end{array}$$

$$N = 18$$

$$\begin{array}{r} 18 \\ - \; 5 \\ \hline 13 \end{array}$$

9. 7 ft + 9 ft + 3 ft + 2 ft + 5 ft = **26 ft**

10.

$$\frac{1}{2} \; \overset{\small ?}{=} \; \frac{2}{4}$$

11. There are ten segments between 7 and 8, so each segment equals $\frac{1}{10}$. The arrow points to **$7\frac{7}{10}$**.

12. One possibility:

13.
$$\begin{array}{r} {}^{2}\;{}^{1} \\ 0.47 \\ 3.62 \\ 0.85 \\ + \; 4.54 \\ \hline \mathbf{9.48} \end{array}$$

14.
$$\begin{array}{r} {}^{1}\;{}^{1} \\ \$ \; 3.00 \\ \$_1\,4.39 \\ + \; \$12.62 \\ \hline \mathbf{\$20.01} \end{array}$$

15. 36.47 − (3.5 + 12.6)
   36.47 − 16.1
      20.37

16. $20.00 − (29¢ + $7)
   $20.00 − $7.29
      $12.71

17.
$$\begin{array}{r} {}^{3}\;{}^{1}0\;{}^{9} \\ \cancel{4}\,\cancel{1},\cancel{0}{}^{1}5\,9 \\ - \; 3\,6,2\,7\,5 \\ \hline 4\;\;7\;8\;4 \end{array}$$

18.
$$\begin{array}{r} {}^{2}\;{}^{2} \\ 768 \\ \times \quad 3 \\ \hline \mathbf{2304} \end{array}$$

19.
$$\begin{array}{r} {}^{3} \\ \$2.80 \\ \times \quad 4 \\ \hline \mathbf{\$11.20} \end{array}$$

20.
$$\begin{array}{r} 436 \\ - \quad Z \\ \hline 252 \end{array}$$

$$\begin{array}{r} {}^{3} \\ \cancel{4}{}^{1}3\,6 \\ - \; 2\,5\,2 \\ \hline 1\,8\,4 \end{array}$$

$$Z = \mathbf{184}$$

21.
$$\begin{array}{r} \mathbf{7\;R\;1} \\ 5\overline{)36} \\ - \; 35 \\ \hline 1 \end{array}$$

22.
$$\begin{array}{r} \mathbf{6\;R\;3} \\ 7\overline{)45} \\ -42 \\ \hline 3 \end{array}$$

23.
$$\begin{array}{r} \mathbf{8\;R\;3} \\ 4\overline{)35} \\ -32 \\ \hline 3 \end{array}$$

24. (a) $\dfrac{1}{4}$

   (b) **25%**

25. 2 × 10 = 20
    4 × 5 = 20
   2, 4, 5, 10

26. (a) 6$\underline{7}$81
      7 is 5 or more
      6781 rounds up to **7000**

   (b) 67$\underline{8}$1
      8 is 5 or more
      6781 rounds up to **6800**

27. 4 × 6 = 24, so N = 6
    4 × 6 ≠ 6
    D. 4N = 6

28. (a) $\dfrac{7}{100}$

(b) **7%**

(c) **$0.07**

LESSON 59, WARM-UP

a. **$1.05**

b. **$3.61**

c. **$1.25**

d. **$7.34**

e. **350**

f. **566**

Problem Solving

$$
\begin{array}{r}
1HD = 50¢ \\
3D = 30¢ \\
+ \quad 4N = 20¢ \\
\hline
100¢ \text{ or } \$1.00
\end{array}
$$

$$
\begin{array}{r}
2Q = 50¢ \\
4D = 40¢ \\
+ \quad 2N = 10¢ \\
\hline
100¢ \text{ or } \$1.00
\end{array}
$$

$$
\begin{array}{r}
3Q = 75¢ \\
+ \quad 5N = 25¢ \\
\hline
100¢ \text{ or } \$1.00
\end{array}
$$

1 half-dollar, 3 dimes, and 4 nickels;
2 quarters, 4 dimes, and 2 nickels;
3 quarters and 5 nickels

LESSON 59, LESSON PRACTICE

a. 59 rounds to 60, 68 rounds to 70, and 81 rounds to 80.

$$
\begin{array}{r}
60 \\
70 \\
+ \quad 80 \\
\hline
210
\end{array}
$$

$$
\begin{array}{r}
\overset{1}{5}9 \\
68 \\
+ \quad 81 \\
\hline
208
\end{array}
$$

b. 607 rounds to 600 and 891 rounds to 900.

$$
\begin{array}{r}
600 \\
+ \quad 900 \\
\hline
1500
\end{array}
$$

$$
\begin{array}{r}
607 \\
+ \quad 891 \\
\hline
1498
\end{array}
$$

c. 585 rounds to 600 and 294 rounds to 300.

$$
\begin{array}{r}
600 \\
- \quad 300 \\
\hline
300
\end{array}
$$

$$
\begin{array}{r}
\overset{4}{\cancel{5}}{}^{1}8\,5 \\
- \quad 2\,9\,4 \\
\hline
\mathbf{2\,9\,1}
\end{array}
$$

d. 82 rounds to 80 and 39 rounds to 40.

$$
\begin{array}{r}
80 \\
- \quad 40 \\
\hline
40
\end{array}
$$

$$
\begin{array}{r}
\overset{7}{\cancel{8}}{}^{1}2 \\
- \quad 3\,9 \\
\hline
\mathbf{4\,3}
\end{array}
$$

e. 59 rounds to 60.

$$
\begin{array}{r}
60 \\
\times \quad 6 \\
\hline
360
\end{array}
$$

$$
\begin{array}{r}
\overset{5}{5}9 \\
\times \quad 6 \\
\hline
354
\end{array}
$$

f. 397 rounds to 400.

$$
\begin{array}{r}
400 \\
\times \quad 4 \\
\hline
1600
\end{array}
$$

$$
\begin{array}{r}
\overset{3\,2}{3}97 \\
\times \quad 4 \\
\hline
1588
\end{array}
$$

g. 42 is close to 40.
$40 \div 5 = \mathbf{8}$

$$
\begin{array}{r}
\mathbf{8\ R\ 2} \\
5\overline{)42} \\
-40 \\
\hline
2
\end{array}
$$

h. 29 is close to 28.

$28 \div 7 = \mathbf{4}$

$$\begin{array}{r} \mathbf{4\,R\,1} \\ 7\overline{)29} \\ -28 \\ \hline 1 \end{array}$$

i. **Dixie's estimate was less than the actual product because she rounded 5280 down to 5000 before multiplying.**

———————————

LESSON 59, MIXED PRACTICE

1. Pattern:
Number of groups $\times$ number in each group
= total
Problem:
7 big baskets $\times$ 42 apples in each big basket
= N apples

$$\begin{array}{r} \overset{1}{42} \\ \times\ 7 \\ \hline \mathbf{294\ apples} \end{array}$$

2. Pattern:
Number of groups $\times$ number in each group
= total
Problem: N boxes $\times$ 6 pears in each box
= 48 pears

8 boxes

3. (a) **One and sixty-one hundredths kilometers**

(b) 1 mi $\overset{\textstyle\bigcirc}{>}$ 1 km
1.61 km

4. 193 rounds to 200

$$\begin{array}{r} 200 \\ \times\ \ \ 5 \\ \hline \mathbf{1000} \end{array}$$

5. 50% of 16 $\overset{\textstyle\bigcirc}{>}$ $\sqrt{16}$
Half of 16 is 8 $\qquad$ $4 \times 4 = 16$
$\qquad$ 8 $\qquad\qquad$ 4

6. $3 \times 4 = 12$
$2 \times 6 = 12$
$12 - 12 = \mathbf{0}$

7. $\begin{array}{r} 1\,{\overset{6}{7}}{\overset{9}{0}}{}^{1}1 \\ -1\,4\,9\,2 \\ \hline \mathbf{2\,0\,9\ years} \end{array}$

8. (a) **Angle B**

(b) $\overline{AC}$ (or $\overline{CA}$)

9. One possibility:

$>$

$\dfrac{2}{5}$ $\overset{\textstyle\bigcirc}{>}$ $\dfrac{1}{4}$

10. Pattern:
Number of groups $\times$ number in each group
= total
Problem:
5 hours $\times$ 40 packages per hour = N packages

$$\begin{array}{r} 40 \\ \times\ \ 5 \\ \hline \mathbf{200\ packages} \end{array}$$

11. **15,210,000**

12.

	3 mi	
2 mi		2 mi
	3 mi	

3 mi + 2 mi + 3 mi + 2 mi = **10 mi**

13. $\begin{array}{r} \overset{1\,1\ 1}{\$37.75} \\ +\ \$45.95 \\ \hline \mathbf{\$83.70} \end{array}$

14. $\begin{array}{r} \overset{1\,1\ 1}{43{,}793} \\ +\ 76{,}860 \\ \hline \mathbf{120{,}653} \end{array}$

15. $\begin{array}{r} \overset{3\,1}{48.0} \\ 9.7 \\ 12.6 \\ \overset{1}{5.3} \\ +\ 236.2 \\ \hline \mathbf{311.8} \end{array}$

16. $\begin{array}{r} \overset{499}{\$5\cancel{0}.0^{1}0} \\ -\ \$42.8\ 7 \\ \hline \mathbf{\$7.1\ 3} \end{array}$

17. $\begin{array}{r} {\overset{3}{\cancel{4}}}\,{\overset{\overset{1}{2}}{\cancel{3}}}{,}7\,9\,3 \\ -\ 2\,6{,}8\,6\,0 \\ \hline \mathbf{1\,6{,}9\,3\,3} \end{array}$

18.
$$\overset{\scriptstyle 3\ 1}{483}$$
$$\underline{\times\quad 4}$$
$$\mathbf{1932}$$

19.
$$\overset{\scriptstyle 2}{360}$$
$$\underline{\times\quad 4}$$
$$\mathbf{1440}$$

20.
$$\overset{\scriptstyle 5}{207}$$
$$\underline{\times\quad 8}$$
$$\mathbf{1656}$$

21.
$$\mathbf{5\ R\ 3}$$
$$8\overline{)43}$$
$$\underline{-40}$$
$$3$$

22.
$$\mathbf{8\ R\ 3}$$
$$5\overline{)43}$$
$$\underline{-40}$$
$$3$$

23.
$$\mathbf{6\ R\ 1}$$
$$7\overline{)43}$$
$$\underline{-42}$$
$$1$$

24. (a) Count up by twos from 0: 0, 2
   2°C

   (b) Count down by twos from 2: 2, 0, –2
   −2°C

25. **See student work.**

26.
$$\overset{\scriptstyle 2\ 1}{2.54}\text{ cm}$$
$$2.54\text{ cm}$$
$$2.54\text{ cm}$$
$$\underline{+\ 2.54\text{ cm}}$$
$$\mathbf{10.16\ cm}$$

27. From Problem 26, we know that
   4 inches = 10.16 centimeters.
   10.16 centimeters $\bigcirc\!\!\!>$ 10 centimeters
   4 inches $\bigcirc\!\!\!>$ 10 centimeters

28. (a) $\dfrac{\mathbf{9}}{\mathbf{100}}$

   (b) **9%**

   (c) **$0.09**

LESSON 60, WARM-UP

a. $1.11

b. $2.75

c. $1.33

d. 175

e. $9.71

f. 255

Problem Solving

What day is this?;
Coded answer varies:
Monday: 13-15-14-4-1-25
Tuesday: 20-21-5-19-4-1-25
Wednesday: 23-5-4-14-5-19-4-1-25
Thursday: 20-8-21-18-19-4-1-25
Friday: 6-18-9-4-1-25
Saturday: 19-1-20-21-18-4-1-25
Sunday: 19-21-14-4-1-25

LESSON 60, LESSON PRACTICE

a. Pattern:
   Number of groups $\times$ number in each group
   $$= \text{total}$$
   Problem:
   N minutes $\times$ 5 pencils per minute
   $$= 40 \text{ pencils}$$
   $$40 \div 5 = 8$$
   8 minutes

b. Pattern:
   Number of groups $\times$ number in each group
   $$= \text{total}$$
   Problem:
   4 hours $\times$ N miles per hour = 12 miles
   $$12 \div 4 = 3$$
   3 miles per hour

c. Pattern:
   Number of groups $\times$ number in each group
   $$= \text{total}$$
   Problem: 5 hours $\times$ N dollars per hour = $40
   $$40 \div 5 = 8$$
   $8 per hour

LESSON 60, MIXED PRACTICE

1. Half of an hour is **30 minutes.**

2. Pattern:
    Some
    Some more
   + Some more
    Total

Problem:
$$\begin{array}{r} \overset{1}{2}14 \text{ parrots} \\ {}_1752 \text{ crows} \\ + 2042 \text{ blue jays} \\ \hline 3008 \text{ birds} \end{array}$$

3. 1 dozen = 12
Pattern:
Number of groups × number in each group
          = total
Problem:
N pounds of beans × 4 burritos per pound
          = 12 burritos
12 ÷ 4 = 3
3 pounds

4. Pattern:
Number of groups × number in each group
          = total
Problem:
3 hours × 12 signs per hour = **36 signs**

5. 286 rounds to 300 and 415 rounds to 400
300 + 400 = 700

6. **A. 23; All multiples of 2 end in 0, 2, 4, 6, or 8.**

7. **6:45 a.m.**

8. 3 × 5 = 15
     N = **5**

9. 6 × 7 = 42
6 + 7 = 13

$$\begin{array}{r} \overset{3}{\cancel{4}}{}^{1}2 \\ - 1\ 3 \\ \hline 2\ 9 \end{array}$$

10. **3 cm**

11. (32 ÷ 8) ÷ 2 ⊘ 32 ÷ (8 ÷ 2)
    4 ÷ 2   32 ÷ 4
      2   8

12.
$$\begin{array}{r} \$\overset{1}{{}_1}6.\overset{2}{4}9 \\ \$12.00 \\ \$7.59 \\ + \$0.08 \\ \hline \$26.16 \end{array}$$

13.
$$\begin{array}{r} \overset{1}{6}.5 \\ {}_14.75 \\ + \ 11.3 \\ \hline 22.55 \end{array}$$

14.
$$\begin{array}{r} \overset{0}{\cancel{1}}{}^{1}2.56 \\ - \ 4.3 \\ \hline 8.26 \end{array}$$

15.
$$\begin{array}{r} \overset{2}{3}50 \\ \times \quad 5 \\ \hline 1750 \end{array}$$

16.
$$\begin{array}{r} \overset{2}{2}04 \\ \times \quad 7 \\ \hline 1428 \end{array}$$

17.
$$\begin{array}{r} \overset{3\ 1}{4}63 \\ \times \quad 6 \\ \hline 2778 \end{array}$$

18.
$$\begin{array}{r} \textbf{9 R 1} \\ 4\overline{)37} \\ - 36 \\ \hline 1 \end{array}$$

19.
$$\begin{array}{r} \textbf{6 R 3} \\ 6\overline{)39} \\ - 36 \\ \hline 3 \end{array}$$

20.
$$\begin{array}{r} \textbf{9 R 1} \\ 3\overline{)28} \\ - 27 \\ \hline 1 \end{array}$$

21. (a) $\dfrac{5}{100} = \dfrac{1}{20}$

    (b) **5%**

22. 1 × 25 = 25
    5 × 5 = 25
    1, 5, and 25

23. 5% $\bigcirc$ $\frac{1}{2}$
    50%

24. (a) 5 yd + 4 yd + 5 yd + 4 yd = **18 yd**

    (b) 5 yd × 4 yd = **20 sq. yd**

25. (a) **Obtuse angle**

    (b) **Acute angle**

26. 15 + 10 = 25, so N = 15
    15 − 5 ≠ 20
    C. N − 5 = 20

27. (a) 8 ÷ (4 ÷ 2) $\bigcirc$ (8 ÷ 4) ÷ 2
          8 ÷ 2   2 ÷ 2
            4   1

    (b) **No, the associative property does not apply to division.**

28. (a) $\frac{19}{100}$

    (b) **19%**

    (c) **$0.19**

INVESTIGATION 6

1. **Favorite Lunches**

2. **4 different types**

3. **The legend tells us that each whole picture represents the favorite choice of two students. We count the pictures and multiply by 2 to find the number.**

4. **6 students; There are three whole pictures of corn dogs, and 3 times 2 is 6.**

5. **9 students; There are four whole pictures, and 4 × 2 = 8. The half picture represents one more student, and 8 + 1 = 9.**

6. **30 students; There are fourteen whole pictures, and 14 × 2 = 28. There are two half pictures that show the choices of two more students, and 28 + 2 = 30.**

7. **Number of Students**

8. **8 students**

9. **The longest bar is the bar for tacos. That means that tacos are the favorite lunch of more students than any other lunch on the graph.**

10. **Four more students named pizza than named hamburgers. One way to tell is to subtract the number who named hamburgers (5) from the number who named pizza (9). Another way to find the answer is to notice that the bar for pizza is 4 students longer than the bar for hamburgers.**

11. **The 8 represents the day Robert turned eight years old. (The distance from 8 to 9 is the year during which Robert was 8 years old.)**

12. **Robert was 40 inches tall. Find 4 on the age scale. Go straight up from 4 to the graph of Robert's height. Then look left to the vertical scale, and see that the height is 40 inches.**

13. **Robert became 45 inches tall while he was 7. Find 45 inches on the vertical scale (the mark halfway between 40 and 50). Go straight to the right to the graph of Robert's height. From there look straight down to the age scale and see that Robert was between his 7th and 8th birthdays.**

14. **Robert's height was increasing rapidly during his early years. As Robert became older, his height changed more gradually.**

15. **The largest slice is the one labeled "sleeping." Since it is the biggest slice, it means that Vanessa spends more time sleeping than on any other single activity.**

16. 5 hr + 2 hr + 2 hr = **9 hours**

17. **24 hours**

18. **15 hours; The quickest way to find the answer is to subtract 9 hours from 24 hours.**

Activity

Pictograph: **See student work.**

Bar Graph:

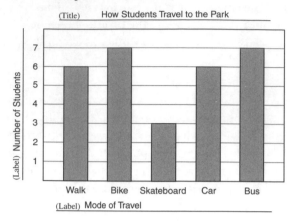

Line Graph:

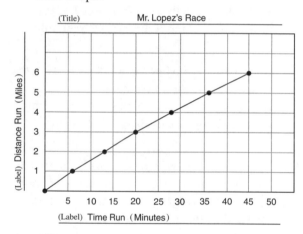

Circle Graph:

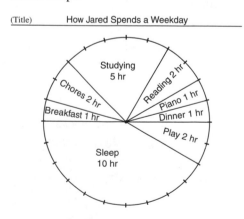

Extensions

a. **See student work.**

b. **See student work.**

LESSON 61, WARM-UP

a. **120**

b. **1200**

c. **100**

d. **298**

e. **$0.64**

f. **$7.50**

Problem Solving

$\frac{1}{4} = \textbf{25\%}; \frac{3}{4} = \textbf{75\%}$

LESSON 61, LESSON PRACTICE

a. Five of the six equal parts are not shaded.
$$\frac{\textbf{5}}{\textbf{6}}$$

b. Two out of five parts are left.
$$\frac{\textbf{2}}{\textbf{5}}$$

c. $2N = 10$
$N = \textbf{5}$

d. $2 + N = 16$
$N = \textbf{14}$

LESSON 61, MIXED PRACTICE

1. The diameter is equal to two radii.
12 inches $+$ 12 inches $=$ 24 inches
12 inches

2. Pattern:
Number of groups $\times$ number in each group
$= $ total
Problem:
35 birdhouses $\times$ 5 sparrows in each birdhouse
$= N$ sparrows

$$\begin{array}{r} \overset{2}{35} \\ \times\ \ 5 \\ \hline \textbf{175 sparrows} \end{array}$$

3. (a) One nickel is $\frac{1}{20}$ of a dollar, so two nickels are $\frac{2}{20}$ of a dollar; or two nickels have the same value as one dime, which is $\frac{1}{10}$ of a dollar.

   (b) **0.1 or 0.10**

4. Pattern:
   Number of groups $\times$ number in each group
   $\qquad = $ total
   Problem: 3 days $\times$ N quarts per day
   $\qquad = 39$ quarts of milk
   $\qquad 39 \div 3 = 13$
   13 quarts each day

5. Pattern: $\quad$ Some
   $\qquad\quad +$ Some More
   $\qquad\qquad$ Total

   Problem: $\quad\overset{1}{8}8$ pounds
   $\qquad\qquad\ 2$ pounds
   $\qquad\qquad\ 2$ pounds
   $\qquad\quad +\ 3$ pounds
   $\qquad\qquad$ **95 pounds**

6. Seven of the ten equal parts are not shaded.
   $$\frac{7}{10}$$

7. Multiples of 10 end with zero, so 50 is a multiple of 10.
   D. 50

8. Five out of six parts are left.
   $$\frac{5}{6}$$

9. **One possibility:**

   $$\frac{2}{3} \,\textcircled{<}\, \frac{3}{4}$$

10. 5070 rounds to 5000 and 3840 rounds to 4000
    $5000 + 4000 = $ **9000**

11. $60\% + 40\% = 100\%$
    60% of the answers were true and 40% of the answers were false
    $60\% > 40\%$, so there were **more true answers**

12. (a) $8\,\text{cm} + 4\,\text{cm} + 8\,\text{cm} + 4\,\text{cm} = $ **24 cm**

    (b) $8\,\text{cm} \times 4\,\text{cm} = $ **32 sq. cm**

13. $$\begin{array}{r} \overset{1\ 1\ \ 1}{\$62.59} \\ +\ \$17.47 \\ \hline \$80.06 \end{array}$$

14. $$\begin{array}{r} Z \\ -\ 417 \\ \hline 268 \end{array}$$

    $$\begin{array}{r} \overset{1}{2}68 \\ +\ 417 \\ \hline 685 \end{array}$$

    $Z = $ **685**

15. $$\begin{array}{r} 110 \\ \times\quad 9 \\ \hline 990 \end{array}$$

    $$\begin{array}{r} \overset{9}{1}0^10\ 0 \\ -\ \ 9\ 9\ 0 \\ \hline 1\ 0 \end{array}$$

16. $$\begin{array}{r} \overset{2}{3}.^16\ 7\ 5 \\ -\ 1.\ 7\ 6 \\ \hline 1.\ 9\ 1\ 5 \end{array}$$

17. $$\begin{array}{r} \overset{2}{\$6}.70 \\ \times\qquad 4 \\ \hline \$26.80 \end{array}$$

18. $$\begin{array}{r} \overset{1}{7}03 \\ \times\quad 6 \\ \hline 4218 \end{array}$$

19. $$\begin{array}{r} \overset{4\ 5}{\$346} \\ \times\quad\ 9 \\ \hline \$3114 \end{array}$$

20. $$\begin{array}{r} 7\ \text{R}\ 4 \\ 5\overline{)39} \\ -35 \\ \hline 4 \end{array}$$

21. $$\begin{array}{r} 5\ \text{R}\ 4 \\ 7\overline{)39} \\ -35 \\ \hline 4 \end{array}$$

22.
$$4\overline{)39} \quad \mathbf{9\,R\,3}$$
$$\underline{-36}$$
$$3$$

23.
$$3\overline{)16} \quad \mathbf{5\,R\,1}$$
$$\underline{-15}$$
$$1$$

24.
$$6\overline{)26} \quad \mathbf{4\,R\,2}$$
$$\underline{-24}$$
$$2$$

25. $36 \div 6 = \mathbf{6}$

26. $\mathbf{-3}$

27. $-5 \;\boxed{<}\; -3$

28. (a) The square is divided into 100 equal parts and 67 parts are not shaded.
$$\frac{\mathbf{67}}{\mathbf{100}}$$

(b) **67%**

(c) **0.67**

LESSON 62, WARM-UP

a. **420**

b. **400**

c. **1200**

d. **$8.32**

e. **647**

f. **$3.13**

Patterns

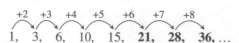

$$1, \; 3, \; 6, \; 10, \; 15, \; \mathbf{21}, \; \mathbf{28}, \; \mathbf{36}, \ldots$$

with increments +2 +3 +4 +5 +6 +7 +8

LESSON 62, LESSON PRACTICE

a. $2 \times 3 = 6$
$6 \times 4 = \mathbf{24}$

b. $3 \times 4 = 12$
$12 \times 10 = \mathbf{120}$

c. $8 \times 8 = \mathbf{64}$

d. $3 \times 3 \times 3$
$3 \times 3 = 9$
$9 \times 3 = \mathbf{27}$

e. $10 \times 10 = 100$
$6 \times 6 = 36$

$$\overset{9}{\cancel{10}}{}^{1}0$$
$$\underline{-\quad 3\,6}$$
$$\mathbf{6\,4}$$

f. $3 \times 3 = 9$
$2 \times 2 \times 2 = 8$
$9 - 8 = \mathbf{1}$

g. Four is a factor three times, so the exponent is 3.
$\mathbf{4^3}$

LESSON 62, MIXED PRACTICE

1. Pattern:
Number of groups $\times$ number in each group
$= $ total
Problem:
N peahens $\times$ 2 peacocks for each peahen
$= 12$ peacocks
$12 \div 2 = 6$
6 peahens

2. Counting back 20 minutes from 6:00 p.m. makes it **5:40 p.m.**

3.
$$\overset{2\ 3}{\$1.98}$$
$$\$0.49$$
$$\$0.49$$
$$\underline{+\ \ \$0.18}$$
$$\mathbf{\$3.14}$$

4.
$$\begin{array}{r} 52 \\ \times\ \$3 \\ \hline \$156 \end{array}$$

5. One out of three parts is left.
$$\frac{1}{3}$$

6. 887 rounds to 900 and 291 rounds to 300
$$900 - 300 = \textbf{600}$$

7. $99 = 100 - y$
$$y = \textbf{1}$$

8. =
$$\frac{2}{4} \ominus \frac{4}{8}$$

9. $8 \times 5 = 40$
$4 \times 10 = 40$
$40 + 40 = \textbf{80}$

10. Friday

11. $6 \times 3 = \textbf{18}$
$3 \times 6 = \textbf{18}$
$18 \div 3 = \textbf{6}$
$18 \div 6 = \textbf{3}$

12. $5 \times 6 = 30$
$30 \times 7 = \textbf{210}$

13. $4 \times 4 \times 4$
$4 \times 4 = 16$

$$\begin{array}{r} {}^{2}16 \\ \times\ \ 4 \\ \hline \textbf{64} \end{array}$$

14.
$$\begin{array}{r} {}^{11}\ {}^{11}\ \\ 476{,}385 \\ +\ 259{,}518 \\ \hline \textbf{735,903} \end{array}$$

15.
$$\begin{array}{r} \$\overset{1}{\cancel{2}}\,\overset{9}{\cancel{0}}.\,\overset{9}{\cancel{0}}{}^{1}0 \\ -\ \$1\,7.\,8\,4 \\ \hline \$2.\,1\,6 \end{array}$$

16.
$$\begin{array}{r} C \\ -\ 19{,}434 \\ \hline 45{,}579 \end{array}$$

$$\begin{array}{r} {}^{11}\ {}^{11}\ \\ 45{,}579 \\ +\ 19{,}434 \\ \hline 65{,}013 \end{array}$$

$$C = \textbf{65,013}$$

17.
$$\begin{array}{r} {}^{1}\ {}^{5}\ \\ \$4.17 \\ \times\ \ \ \ \ 8 \\ \hline \$33.36 \end{array}$$

18.
$$\begin{array}{r} {}^{4}\ \\ \$470 \\ \times\ \ \ \ 7 \\ \hline \$3290 \end{array}$$

19.
$$\begin{array}{r} {}^{3}\ \\ 608 \\ \times\ \ \ \ 4 \\ \hline 2432 \end{array}$$

20.
$$\begin{array}{r} \textbf{7 R 1} \\ 4\overline{)29} \\ -28 \\ \hline 1 \end{array}$$

21.
$$\begin{array}{r} \textbf{8 R 1} \\ 8\overline{)65} \\ -64 \\ \hline 1 \end{array}$$

22.
$$\begin{array}{r} \textbf{5 R 4} \\ 5\overline{)29} \\ -25 \\ \hline 4 \end{array}$$

23.
$$\begin{array}{r} \textbf{9 R 2} \\ 7\overline{)65} \\ -63 \\ \hline 2 \end{array}$$

24.
$$\begin{array}{r} \textbf{7 R 4} \\ 5\overline{)39} \\ -35 \\ \hline 4 \end{array}$$

25.
$$\begin{array}{r} \textbf{7 R 2} \\ 9\overline{)65} \\ -63 \\ \hline 2 \end{array}$$

26. $100\% - 40\% = $ **60%**

27. (a) 6 in. + 6 in. + 6 in. + 6 in. = **24 in.**

(b) 6 in. $\times$ 6 in. = **36 sq. in.**

28. B. Right

29. (a) **3 red candies**

(b) 9 green candies $-$ 5 orange candies
       = **4 more green candies**

LESSON 63, WARM-UP

a. 2400

b. 1200

c. 2000

d. $1.75

e. $8.54

f. 440

Problem Solving

Do we know the number of kittens in the litter?
No

Do we know whether there are more males or
females? **Yes** (same number)

Do we know whether the number of kittens is
even or odd? **Yes** (even)

LESSON 63, LESSON PRACTICE

a. Sample drawing:

b. Sample drawing:

c. Sample drawing:

d. Sample drawing:

e. Sample drawing:

f. Sample drawing:

g. The polygon has four sides, so it is a **quadrilateral.**

h. The polygon has six sides, so it is a **hexagon.**

i. The polygon has five sides, so it is a **pentagon.**

j. The polygon has eight sides, so it is an **octagon.**

k. g. Quadrilateral and j. Octagon

l. Stop sign

m. A decagon has ten sides and ten vertices.
A hexagon has six vertices.
10 vertices $-$ 6 vertices = 4 vertices
4 more vertices

LESSON 63, MIXED PRACTICE

1. Pattern:
Number of groups $\times$ number in each group
          = total
Problem: N yards $\times$ 3 feet per yard = 15 feet
          $15 \div 3 = 5$
5 yards

2. $3 \times 10 = 30$
      $10 \times 3 = 30$
      $30 \div 3 = 10$
      $30 \div 10 = 3$

3. $1.50 + $0.30 + $0.14 = **$1.94**

4.
$$\begin{array}{r} \overset{2}{12} \\ 14 \\ 16 \\ + \ 18 \\ \hline \mathbf{60} \end{array}$$

5. 715 rounds to 700 and 594 rounds to 600
700 + 600 = **1300**

6. 1 glass = 1 cup
2 cups = 1 pint
2 pints = 1 quart
4 quarts = 1 gallon

$$\begin{array}{c} 2 \times 2 \times 4 = N \text{ glasses} \\ 2 \times 2 = 4 \\ 4 \times 4 = 16 \end{array}$$

16 glasses

7. There are eight segments between 6 and 7, so each segment equals $\frac{1}{8}$. The arrow points to **$6\frac{5}{8}$.**

8. 5 out of 12 pieces were left.

$$\mathbf{\frac{5}{12}}$$

9. $4 \times 3 = 12$
$4 + 3 = 7$
$12 - 7 = \mathbf{5}$

10. $9 \times 9 = 81$
$\sqrt{9} = 3$
$81 + 3 = \mathbf{84}$

11. (a) The polygon has six sides, so it is a **hexagon.**

(b) 1 cm + 1 cm + 1 cm + 1 cm + 1 cm + 1 cm = **6 cm**

12. Pattern:
Number of groups $\times$ number in each group
= total
Problem: 8 minutes $\times$ N flowers per minute
= 56 flowers
$56 \div 8 = \mathbf{7 \text{ flowers}}$

13. Pattern:
Number of groups $\times$ number in each group
= total
Problem: 5 minutes $\times$ 11 flowers per minute
= **55 flowers**

14. $40.00 − D = $2.43

$$\begin{array}{r} \overset{3\ \ 9\ \ 9}{\$4\ \cancel{0}.\ \cancel{0}{}^1 0} \\ - \ \ \$2.\ 4\ 3 \\ \hline \$3\ 7.\ 5\ 7 \end{array}$$

D = **$37.57**

15. $5 \times 5 = 25$
$5 \times N = 15 + 5$
$5 \times N = 20$
$N = \mathbf{4}$

16. $6 \times 4 = 24$
$24 \times 10 = \mathbf{240}$

17. $5 \times 5 \times 5$
$5 \times 5 = 25$

$$\begin{array}{r} \overset{2}{25} \\ \times \ \ 5 \\ \hline 125 \end{array}$$

18.
$$\begin{array}{r} 3.5 \\ + \ 2.45 \\ \hline \mathbf{5.95} \end{array}$$

19.
$$\begin{array}{r} 1.95 \\ - \ 0.4 \\ \hline \mathbf{1.55} \end{array}$$

20.
$$\begin{array}{r} \overset{1}{\$0.36} \\ + \ \$0.57 \\ \hline \$0.93 \end{array}$$

$$\begin{array}{r} \overset{0\ \ 9}{\$\cancel{1}.\ \cancel{0}{}^1 0} \\ - \ \$0.\ 9\ 3 \\ \hline \mathbf{\$0.\ 0\ 7} \end{array}$$

21.
$$\begin{array}{r} \overset{3\ 7}{349} \\ \times \ \ \ \ 8 \\ \hline \mathbf{2792} \end{array}$$

22.
$$\begin{array}{r} \overset{4}{\$7.60} \\ \times \ \ \ \ \ \ 7 \\ \hline \mathbf{\$53.20} \end{array}$$

23.
$$\begin{array}{r} \mathbf{5\ R\ 4} \\ 6\overline{)34} \\ -30 \\ \hline 4 \end{array}$$

24.
$$\begin{array}{r} 7\text{ R }6 \\ 8\overline{)62} \\ -56 \\ \hline 6 \end{array}$$

25.
$$\begin{array}{r} 4\text{ R }4 \\ 5\overline{)24} \\ -20 \\ \hline 4 \end{array}$$

26. 9

27. $100\% - 60\% = \mathbf{40\%}$

28. $2 \times 5 = 10$
     B. 5

29. (a) The square is divided into 100 equal parts
         and 21 parts are shaded.
         $$\frac{\mathbf{21}}{\mathbf{100}}$$

     (b) 21% of the large square is shaded.
         $21\% + N = 100\%$

         $$\begin{array}{r} {\overset{9}{\cancel{10}}}{}^{1}0\% \\ - \quad 2\ 1\% \\ \hline \mathbf{7\ 9\%} \end{array}$$

LESSON 64, WARM-UP

a. **1600**

b. **1500**

c. **4200**

d. **2000**

Problem Solving
   Which coin equals half of a dime?
   14-9-3-11-5-12 (nickel)

LESSON 64, LESSON PRACTICE

a.
$$\begin{array}{r} 17 \\ 3\overline{)51} \\ 3 \\ \hline 21 \\ 21 \\ \hline 0 \end{array}$$

b.
$$\begin{array}{r} 13 \\ 4\overline{)52} \\ 4 \\ \hline 12 \\ 12 \\ \hline 0 \end{array}$$

c.
$$\begin{array}{r} 15 \\ 5\overline{)75} \\ 5 \\ \hline 25 \\ 25 \\ \hline 0 \end{array}$$

d.
$$\begin{array}{r} 24 \\ 3\overline{)72} \\ 6 \\ \hline 12 \\ 12 \\ \hline 0 \end{array}$$

e.
$$\begin{array}{r} 24 \\ 4\overline{)96} \\ 8 \\ \hline 16 \\ 16 \\ \hline 0 \end{array}$$

f.
$$\begin{array}{r} 37 \\ 2\overline{)74} \\ 6 \\ \hline 14 \\ 14 \\ \hline 0 \end{array}$$

g. **A. 75; The sum of 7 and 5 is 12, which is a
     multiple of 3.**

LESSON 64, MIXED PRACTICE

1. **27,878,400 square feet**

2. Pattern:
    Number of groups $\times$ number in each group
    $= $ total
    Problem: 113 paces $\times$ 3 feet in each pace
    $= N$ feet

    $$\begin{array}{r} 113 \\ \times \quad 3 \\ \hline \mathbf{339 \text{ feet}} \end{array}$$

3. Pattern: Some
             − Some went away
             What is left

      Problem: $\overset{4}{\cancel{5}}\,\overset{9}{\cancel{0}}{}^{1}0$ cards
             − 3 8 4 cards
             1 1 6 cards

4. Pattern:

    Number of groups × number in each group
                 = total

    Problem: N weeks × 7 days per week
                = 21 days

        21 ÷ 7 = 3 weeks

    3 weeks

5. An octagon has 8 sides.

    Pattern:

    Number of groups × number in each group
                = total

    Problem:

    7 stop signs × 8 sides in each stop sign
                  = 56 sides

6. $1\dfrac{3}{4}$ inches

7. **400,000 + 6000 + 900 + 10 + 2; four hundred six thousand, nine hundred twelve**

8.   12 in.
     12 in.
     12 in.
   + 12 in.
     48 in.

9. 586 rounds to 600 and 797 rounds to 800.

    600 + 800 = **1400**

10. $\dfrac{3}{6} = \dfrac{1}{2}$

11. Half of 100 is 50.

    $\sqrt{100}$ = 10

    50 $\gt$ 10

12. Pattern: Some
             + Some more
             Total

      Problem: N birds
            + 47 birds
           112 birds

          $\overset{0}{\cancel{1}}\,\overset{1}{\cancel{1}}{}^{1}2$
         − 4 7
         6 5 birds

13. $\overset{1\ 1\ 1}{\$32.47}$
   + $67.54
    $100.01

14. $\overset{4}{\cancel{5}}\,\overset{1}{}\overset{0}{\cancel{1}},\overset{9}{\cancel{0}}\,\overset{1}{}\overset{2}{\cancel{3}}{}^{1}6$
   − 7, 6 4 8
    4 3, 3 8 8

15. $\overset{2\ 2}{53.6}$
     2.9
    97.4
    $\overset{1}{}\ 8.8$
  + 436.1
    598.8

16.
$$
\begin{array}{r}
15 \\
5\overline{)75} \\
\underline{5} \\
25 \\
\underline{25} \\
0
\end{array}
$$

17.
$$
\begin{array}{r}
28 \\
3\overline{)84} \\
\underline{6} \\
24 \\
\underline{24} \\
0
\end{array}
$$

18.
$$
\begin{array}{r}
23 \\
4\overline{)92} \\
\underline{8} \\
12 \\
\underline{12} \\
0
\end{array}
$$

19.
$$
\begin{array}{r}
9\ \text{R}\ 4 \\
6\overline{)58} \\
\underline{-54} \\
4
\end{array}
$$

20.
$$\overset{2\ 3}{257}$$
$$\times\quad 5$$
$$\overline{1285}$$

21.
$$\overset{2}{\$7.09}$$
$$\times\qquad 3$$
$$\overline{\$21.27}$$

22.
$$\overset{3\ 3}{\$334}$$
$$\times\qquad 9$$
$$\overline{\$3006}$$

23.
$$\begin{array}{r} 18 \\ 2\overline{)36} \\ 2 \\ \overline{16} \\ 16 \\ \overline{0} \end{array}$$

24. $4 \times 9 = 36$
$N = \mathbf{9}$

25. $\quad\ 4 \times 4 = 16$
$2 \times 2 \times 2 = 8$

$$\begin{array}{r} \overset{1}{16} \\ +\ \ 8 \\ \overline{24} \end{array}$$

26.
$$\begin{array}{r} 2.4 \\ -\ 1.3 \\ \overline{1.1} \end{array}$$

$$\begin{array}{r} 3.5 \\ -\ 1.1 \\ \overline{2.4} \end{array}$$

27. $1\frac{1}{2}$ in. $+\ 1\frac{1}{2}$ in. $=$ **3 in.**

28. One out of four parts were left.
$$\frac{\mathbf{1}}{\mathbf{4}}$$

29. (a) The square is divided into 100 equal parts and 49 parts are shaded.
$$\frac{\mathbf{49}}{\mathbf{100}}$$

(b) 49% of the large square is shaded
$100\% - 49\% = \mathbf{51\%}$

(c) **0.49**

LESSON 65, WARM-UP

a. **600**

b. **2400**

c. **1200**

d. **$0.76**

e. **$8.87**

f. **355**

Problem Solving

$$\$1.00 - \$0.44 = \$0.56$$
$$\begin{array}{r} 2Q = 50¢ \\ 1N = 5¢ \\ +\ 1P = 1¢ \\ \overline{56¢} \end{array}$$

2 quarters, 1 nickel, and 1 penny

LESSON 65, LESSON PRACTICE

a. **8**

b. **32**

c. **4**

d.
$$\begin{array}{r} 48 \\ 3\overline{)144} \\ 12 \\ \overline{24} \\ 24 \\ \overline{0} \end{array}$$

e.
$$\begin{array}{r} 36 \\ 4\overline{)144} \\ 12 \\ \overline{24} \\ 24 \\ \overline{0} \end{array}$$

f.
$$\begin{array}{r} 24 \\ 6\overline{)144} \\ 12 \\ \overline{24} \\ 24 \\ \overline{0} \end{array}$$

g.
$$\begin{array}{r} 45 \\ 5\overline{)225} \\ \underline{20} \\ 25 \\ \underline{25} \\ 0 \end{array}$$

h.
$$\begin{array}{r} 65 \\ 7\overline{)455} \\ \underline{42} \\ 35 \\ \underline{35} \\ 0 \end{array}$$

i.
$$\begin{array}{r} 25 \\ 8\overline{)200} \\ \underline{16} \\ 40 \\ \underline{40} \\ 0 \end{array}$$

j. **A. 288; The sum of the digits 2, 8, and 8 is 18, which is a multiple of 9. So 288 is a multiple of 9.**

LESSON 65, MIXED PRACTICE

1. Pattern:
    Number of groups $\times$ number in each group
    $= $ total
    Problem:
    N omelettes $\times$ 3 eggs in each omelette
    $= 24$ eggs
    $24 \div 3 = $ **8 omelettes**

2. Pattern: Some
    $-$ Some went away
    What is left

    Problem: 72 young knights
    $-$ N young knights
    27 young knights

    $$\begin{array}{r} \overset{6}{\cancel{7}}{}^{1}2 \text{ young knights} \\ - 2\,7 \text{ young knights} \\ \hline \mathbf{4\,5} \text{ young knights} \end{array}$$

3. Pattern:
    Number of groups $\times$ number in each group
    $= $ total
    Problem: 3 years $\times$ 12 months in each year
    $3 \times 12 = $ **36 months**

4. $12.89 rounds to $13 and $3.95 rounds to $4
    $13 + $4 = **$17**

5. Pattern:
    Number of groups $\times$ number in each group
    $= $ total
    Problem: 4 hours $\times$ N miles per hour
    $= 28$ miles
    $28 \div 4 = $ **7 miles per hour**

6. Pattern: Some
    $+$ Some more
    Total

    Problem: $\overset{1\ 1\ \ 1}{}$
    $18.95
    $+$ $42.85
    $61.80

7. Five out of six parts are not shaded.
    $$\frac{5}{6}$$

8. A hexagon has six sides.
    1 cm + 1 cm + 1 cm + 1 cm + 1 cm
    + 1 cm = **6 cm**

9. There are **2 hours and 20 minutes** between 8:05 a.m. and 10:25 a.m.

10. $\overset{1}{}$
    18 in.
    $+$ 18 in.
    36 in.

11. **6 cm**

12. **D. Quotient**

13. $3 \times 3 = 9$
    $27 \div 9 \;\textcircled{<}\; 27 \div 3$
    $3 \phantom{\div 9 \textcircled{<} 27}9$

14. $\overset{1\ 1\ \ 1}{}$
    $97.56
    $+$ $8.49
    $106.05

15. $\overset{\ \ 5\ \ 9\ \ 9}{\$\cancel{6}\,\cancel{0}.\,\cancel{0}^{1}0}$
    $-$ $5\,4\,.\,7\,8$
    $5.\,2\,2

16.

$$
\begin{array}{r}
\overset{2\,1\ 1}{37.64} \\
29.45 \\
3.01 \\
+\ 75.38 \\
\hline
\mathbf{145.48}
\end{array}
$$

17.

$$
\begin{array}{r}
\mathbf{56} \\
3\overline{)168} \\
\underline{15} \\
18 \\
\underline{18} \\
0
\end{array}
$$

18.

$$
\begin{array}{r}
\mathbf{54} \\
7\overline{)378} \\
\underline{35} \\
28 \\
\underline{28} \\
0
\end{array}
$$

19.

$$
\begin{array}{r}
\overset{1}{840} \\
\times\quad 3 \\
\hline
\mathbf{2520}
\end{array}
$$

20.

$$
\begin{array}{r}
\overset{2\,1}{564} \\
\times\quad 4 \\
\hline
\mathbf{2256}
\end{array}
$$

21.

$$
\begin{array}{r}
\overset{2}{304} \\
\times\quad 6 \\
\hline
\mathbf{1824}
\end{array}
$$

22.

$$
\begin{array}{r}
\mathbf{34} \\
4\overline{)136} \\
\underline{12} \\
16 \\
\underline{16} \\
0
\end{array}
$$

23.

$$
\begin{array}{r}
\mathbf{66} \\
2\overline{)132} \\
\underline{12} \\
12 \\
\underline{12} \\
0
\end{array}
$$

24.

$$
\begin{array}{r}
\mathbf{32} \\
6\overline{)192} \\
\underline{18} \\
12 \\
\underline{12} \\
0
\end{array}
$$

25. $7 \times 8 = 56$
    $N = \mathbf{8}$

26. $12 \times 10 = 120$

$$
\begin{array}{r}
\overset{1}{120} \\
\times\quad 7 \\
\hline
\mathbf{840}
\end{array}
$$

27. Count up by twos from -10: $-10, -8, -6, -4$
    $\mathbf{-4°F}$

28. (a) One quarter is $\frac{1}{4}$ of a dollar, so three quarters
    are $\frac{3}{4}$ of a dollar.

    (b) $\frac{1}{4} = 25\%$

$$
\begin{array}{r}
\overset{1}{25\%} \\
\times\quad 3 \\
\hline
\mathbf{75\%}
\end{array}
$$

29. **Sample answer:**

4 vertices

LESSON 66, WARM-UP

a. **2000**

b. **6000**

c. **20,000**

d. **2500**

e. **$1.29**

f. **$7.42**

Patterns
   2, 7, 12, 17, 22, 27, …

LESSON 66, LESSON PRACTICE

a. **Triangles *A*, *C*, and *D***

b. **Triangles *A* and *C***

LESSON 66, MIXED PRACTICE

1. Each apple has 2 halves.
Pattern:
Number of groups $\times$ number in each group
$\qquad = $ total
Problem: 10 apples $\times$ 2 halves in each apple
$\qquad = $ **20 apple halves**

2.
$$\begin{array}{r} \textbf{47 beads} \\ 3\overline{)141} \\ \underline{12} \\ 21 \\ \underline{21} \\ 0 \end{array}$$

3. $100\% - 25\% = \textbf{75\%}$

4.

20 km   ↑ | 15 km ↓
5 km

5. Step 1: Counting forward 30 minutes from 11:45 a.m. makes it 12:15 p.m.
Step 2: Counting forward 2 hours from 12:15 p.m. makes it **2:15 p.m.**

6. (a) 6 units $+$ 3 units $+$ 6 units $+$ 3 units
$\qquad = $ **18 units**

(b) 6 units $\times$ 3 units $=$ **18 square units**

7. Pattern:
Number of groups $\times$ number in each group
$\qquad = $ total
Problem: 8 gallons $\times$ 30 miles per gallons
$\qquad = $ **240 miles**

8. Five out of seven parts stood silently.
$$\frac{5}{7}$$

9. 4 quarts $=$ 1 gallon
Pattern:
Number of groups $\times$ number in each group
$\qquad = $ total
Problem: 4 quarts $\times$ 42 glops per quart
$\qquad = N$ glops
$$\begin{array}{r} 42 \\ \times \quad 4 \\ \hline \textbf{168 glops} \end{array}$$

10. One possibility:

▨□ $>$ ▥▥▥

$$\frac{1}{2} \; \boxed{>} \; \frac{2}{5}$$

11. $N + 2 = 36$
$36 - 2 = 34$
$\qquad N = \textbf{34}$

12.
$$\begin{array}{r} \overset{2}{\cancel{3}}.{}^{1}3 \\ -\; 1.5 \\ \hline 1.8 \end{array}$$

$$\begin{array}{r} \overset{5}{\cancel{6}}.{}^{1}4\,2 \\ -\quad 1.8 \\ \hline \textbf{4.6 2} \end{array}$$

13. $\sqrt{81} = 9$
$3 \times 3 = 9$
$$\begin{array}{r} 9 \\ 82 \\ +\quad 9 \\ \hline \textbf{100} \end{array}$$

14.
$$\begin{array}{r} \overset{0\;9}{\$\cancel{10}}.{}^{1}0\,0 \\ -\;\$0.1\,0 \\ \hline \textbf{\$9.9 0} \end{array}$$

15.
$$\begin{array}{r} \overset{3\;\;{}^{1}2\;\;9\;\;{}^{1}0}{\cancel{4}\,\cancel{3},\cancel{0}\,\cancel{1}\,6} \\ 5,987 \\ \hline \textbf{3 7,0 2 9} \end{array}$$

16.
$$\begin{array}{r} \overset{1}{2}4 \\ \times \quad 3 \\ \hline 72 \end{array}$$
$72 \times 10 = \textbf{720}$

17.
$$\begin{array}{r} \overset{6\;4}{\$4.86} \\ \times \qquad 7 \\ \hline \textbf{\$34.02} \end{array}$$

18.
$$\begin{array}{r} \overset{5}{3}07 \\ \times \quad 8 \\ \hline \textbf{2456} \end{array}$$

19.
$$\begin{array}{r} {\scriptstyle 5} \\ \$460 \\ \times \quad 9 \\ \hline \$4140 \end{array}$$

20.
$$\begin{array}{r} 76 \\ 2\overline{)152} \\ \underline{14} \\ 12 \\ \underline{12} \\ 0 \end{array}$$

21.
$$\begin{array}{r} 44 \\ 6\overline{)264} \\ \underline{24} \\ 24 \\ \underline{24} \\ 0 \end{array}$$

22.
$$\begin{array}{r} 14 \\ 4\overline{)56} \\ \underline{4} \\ 16 \\ \underline{16} \\ 0 \end{array}$$

23.
$$\begin{array}{r} 46 \\ 5\overline{)230} \\ \underline{20} \\ 30 \\ \underline{30} \\ 0 \end{array}$$

24.
$$\begin{array}{r} 13 \\ 7\overline{)91} \\ \underline{7} \\ 21 \\ \underline{21} \\ 0 \end{array}$$

25.
$$\begin{array}{r} 45 \\ 3\overline{)135} \\ \underline{12} \\ 15 \\ \underline{15} \\ 0 \end{array}$$

26. (a) **$0.17**

   (b) **$0.08**

   (c) **$3.45**

27. (a) **Two and three tenths**

   (b) **Two and three tenths**

28. A. 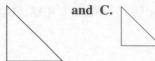 and C.

29. Sample answer:

   5 vertices

LESSON 67, WARM-UP

a. **1000**

b. **6000**

c. **24,000**

d. **$9.32**

e. **580**

f. **$1.21**

Problem Solving

$50\% = \frac{1}{2}$

$\frac{1}{2}$ of 60 minutes is **30 minutes.**

$\frac{1}{2}$ of 30 minutes is **15 minutes.**

LESSON 67, LESSON PRACTICE

a. **750**

b. **320**

c. **$5.30**

d.
$$\begin{array}{r} {\scriptstyle 1} \\ 26 \\ \times \quad 20 \\ \hline 520 \end{array}$$

e.
$$\begin{array}{r} {\scriptstyle 1\ 1} \\ \$1.64 \\ \times \quad 30 \\ \hline \$49.20 \end{array}$$

f.
$$\begin{array}{r} \overset{2}{4}5 \\ \times\ \ 50 \\ \hline 2250 \end{array}$$

g. **12 × 3 × 10**
12 × 3 = 36
36 × 10 = **360**

LESSON 67, MIXED PRACTICE

1.
$$\begin{array}{r} \text{15 beans} \\ 5\overline{)75} \\ \underline{5} \\ 25 \\ \underline{25} \\ 0 \end{array}$$

2. Perimeter: 8 units + 3 units + 8 units
+ 3 units = **22 units**
Area: 3 units × 8 units = **24 square units**

3. **B. 2 quarts**

4. 1 × 12 = 12
2 × 6 − 12
3 × 4 = 12
Factors of 12: 1, 2, 3, 4, 6, 12
B. 5

5. There are **12 hours 5 minutes** between 4:05 a.m. and 4:10 p.m.

6. **Three million, ninety-seven thousand, six hundred square yards**

7. Three out of five parts are not shaded.
$$\frac{3}{5}$$

8. 50% is one half. Less than half of the pentagon is shaded, so **less than 50%** of the pentagon is shaded.

9. **July 27, 2019**

10. There are eight segments between 7 and 8, so each segment equals $\frac{1}{8}$. The arrow points to **$7\frac{5}{8}$**.

11. 78 rounds to 80
$$\begin{array}{r} 80 \\ \times\ \ 4 \\ \hline 320 \end{array}$$

12. 2 × 2 × 2
2 × 2 = 4
4 × 2 = 8
8 ⊘ 6

13.
$$\begin{array}{r} \overset{1\ \ 1}{\$6.25} \\ \$_14.00 \\ +\ \$12.78 \\ \hline \$23.03 \end{array}$$

14.
$$\begin{array}{r} \overset{1}{3}.6 \\ 12.4 \\ +\ \ 0.84 \\ \hline 16.84 \end{array}$$

15.
$$\begin{array}{r} \$30.25 \\ -\ \ \ \ \ B \\ \hline \$13.06 \end{array}$$

$$\begin{array}{r} \$\overset{2}{3}{}^{1}0.\ \overset{1}{2}{}^{1}5 \\ -\ \$1\ 3.\ 0\ 6 \\ \hline \$1\ 7.\ 1\ 9 \end{array}$$
B = **$17.19**

16.
$$\begin{array}{r} \overset{0}{1}{}^{1}4\ \overset{8}{9},{}^{1}3\ \overset{7}{8}{}^{1}4 \\ -\ \ \ 9\ 8,7\ 6\ 5 \\ \hline 5\ 0,6\ 1\ 9 \end{array}$$

17.
$$\begin{array}{r} \overset{6}{4}09 \\ \times\ \ \ 70 \\ \hline 28,630 \end{array}$$

18.
$$\begin{array}{r} \overset{2\ 3}{\$3.46} \\ \times\ \ \ \ \ 5 \\ \hline \$17.30 \end{array}$$

19.
$$\begin{array}{r} \overset{5}{\$0.79} \\ \times\ \ \ \ \ 6 \\ \hline \$4.74 \end{array}$$

20.
$$\begin{array}{r} \$0.39 \\ \times\ \ \ \ 10 \\ \hline \$3.90 \end{array}$$

21.
$$\begin{array}{r} 15 \\ 6\overline{)90} \\ \underline{6} \\ 30 \\ \underline{30} \\ 0 \end{array}$$

22.
$$\begin{array}{r} 24 \\ 4\overline{)96} \\ \underline{8} \\ 16 \\ \underline{16} \\ 0 \end{array}$$

23.
$$\begin{array}{r} 57 \\ 8\overline{)456} \\ \underline{40} \\ 56 \\ \underline{56} \\ 0 \end{array}$$

24.
$$\begin{array}{r} 19 \\ 5\overline{)95} \\ \underline{5} \\ 45 \\ \underline{45} \\ 0 \end{array}$$

25.
$$\begin{array}{r} 78 \\ 3\overline{)234} \\ \underline{21} \\ 24 \\ \underline{24} \\ 0 \end{array}$$

26. Three out of ten parts are shaded.

$$\frac{3}{10}; 0.3$$

27. C.  and D.

28. 25¢

29. Sample answer:

6 vertices

30. 17 correct answers

128

LESSON 68, WARM-UP

a. 1200

b. 240

c. 360

d. 8000

e. $7.34

f. 185

Problem Solving

First think, "6 plus what number equals 15?" (9). Then think, "4 plus 1 (from regrouping) plus what number equals 14?" (9). Next think, "7 plus 1 (from regrouping) plus what number equals a single-digit number?" (1). Then find the sum in the hundreds column (9).

$$\begin{array}{r} 7\underline{9}6 \\ + \ 1\underline{4}9 \\ \hline \underline{9}45 \end{array}$$

LESSON 68, LESSON PRACTICE

a.
$$\begin{array}{r} 44 \text{ R } 2 \\ 3\overline{)134} \\ \underline{12} \\ 14 \\ \underline{12} \\ 2 \end{array}$$

b.
$$\begin{array}{r} 34 \text{ R } 2 \\ 7\overline{)240} \\ \underline{21} \\ 30 \\ \underline{28} \\ 2 \end{array}$$

c.
$$\begin{array}{r} 17 \text{ R } 3 \\ 5\overline{)88} \\ \underline{5} \\ 38 \\ \underline{35} \\ 3 \end{array}$$

d.
$$\begin{array}{r} 32 \text{ R } 3 \\ 8\overline{)259} \\ 24 \\ \hline 19 \\ 16 \\ \hline 3 \end{array}$$

e.
$$\begin{array}{r} 23 \text{ R } 3 \\ 4\overline{)95} \\ 8 \\ \hline 15 \\ 12 \\ \hline 3 \end{array}$$

f.
$$\begin{array}{r} 54 \text{ R } 1 \\ 6\overline{)325} \\ 30 \\ \hline 25 \\ 24 \\ \hline 1 \end{array}$$

g. **Multiply 58 by 4. Then add 3 to the product. The answer should be 235.**
($58 \times 4 = 232$; $232 + 3 = 235$)

LESSON 68, MIXED PRACTICE

1. Pattern:
Number of groups $\times$ number in each group
$\qquad = $ total
Problem:
40 batches $\times$ 4 spoonfuls in each batch
$\qquad = $ **160 spoonfuls**

2. Perimeter: 6 units + 4 units + 6 units
$\qquad + $ 4 units = **20 units**
Area: 6 units $\times$ 4 units = **24 square units**

3. **12.14 seconds**

4. Maura walked two out of five parts.
$\dfrac{2}{5}$

5. 50% is one half. Maura ran **more than 50%** of the course.

6. An octagon has eight sides and a pentagon has five sides.
$8 + 5 = $ **13 sides**

7. There are five segments between 7 and 8, so each segment equals $\frac{1}{5}$. The arrow points to **$7\frac{2}{5}$.**

8. **4392 meters**

9. Pattern:
Number of groups $\times$ number in each group
$\qquad = $ total
Problem: 7 minutes $\times$ N prizes per minute
$\qquad = $ 35 prizes
$35 \div 7 = $ **5 prizes**

10. 6810 rounds to 7000 and 9030 rounds to 9000
$$\begin{array}{r} 7000 \\ + 9000 \\ \hline \mathbf{16,000} \end{array}$$

11.
$$\begin{array}{r} \$8.95 \\ + \$0.75 \\ \hline \$9.70 \end{array}$$
$$\begin{array}{r} \$20.00 \\ - \$9.70 \\ \hline \$10.30 \end{array}$$

12. $5 \times 5 = 25$
$4 \times 4 = 16$
$$\begin{array}{r} 25 \\ - 16 \\ \hline 9 \end{array}$$

13.
$$\begin{array}{r} 23.64 \\ - 5.45 \\ \hline 18.19 \end{array}$$

14.
$$\begin{array}{r} \$0.43 \\ \times \quad 8 \\ \hline \$3.44 \end{array}$$

15.
$$\begin{array}{r} \$3.05 \\ \times \quad 5 \\ \hline \$15.25 \end{array}$$

16.
$$\begin{array}{r} \$2.63 \\ \times \quad 7 \\ \hline \$18.41 \end{array}$$

17.
$$
\begin{array}{r}
\overset{2}{64} \\
\times\ 5 \\
\hline
320
\end{array}
$$

$5 \times 64 = 320$

18.
$$
\begin{array}{r}
\overset{2}{47} \\
\times\ 30 \\
\hline
1410
\end{array}
$$

19.
$$
\begin{array}{r}
\overset{5}{39} \\
\times\ 60 \\
\hline
2340
\end{array}
$$

20.
$$
\begin{array}{r}
\overset{2}{85} \\
\times\ 40 \\
\hline
3400
\end{array}
$$

21.
$$
\begin{array}{r}
19\ R\ 1 \\
5\overline{)96} \\
5 \\
\hline
46 \\
45 \\
\hline
1
\end{array}
$$

22.
$$
\begin{array}{r}
22\ R\ 2 \\
7\overline{)156} \\
14 \\
\hline
16 \\
14 \\
\hline
2
\end{array}
$$

23.
$$
\begin{array}{r}
82 \\
3\overline{)246} \\
24 \\
\hline
06 \\
6 \\
\hline
0
\end{array}
$$

24.
$$
\begin{array}{r}
36 \\
6\overline{)216} \\
18 \\
\hline
36 \\
36 \\
\hline
0
\end{array}
$$

25.
$$
\begin{array}{r}
39 \\
4\overline{)156} \\
12 \\
\hline
36 \\
36 \\
\hline
0
\end{array}
$$

26.
$$
\begin{array}{r}
24\ R\ 3 \\
8\overline{)195} \\
16 \\
\hline
35 \\
32 \\
\hline
3
\end{array}
$$

27. $AB = 1\frac{1}{4}$ in.; $BC = 1\frac{1}{2}$ in.; $AC = 2\frac{3}{4}$ in.

28. **D. Congruent**

29. 1 gallon = 3.78 liters

2 liters is greater than half of 3.78

2 liters $\bigcirc\!\!\!\!>$ $\frac{1}{2}$ gallon

30. Sample answer:

8 vertices

LESSON 69, WARM-UP

a. **420**

b. **500**

c. **480**

d. **27,000**

e. **$7.01**

f. **$9.15**

Problem Solving

$1.50 + ($0.75 \times 3) = **$3.75**

LESSON 69, LESSON PRACTICE

a. 10 mm = 1 cm
   About 10 dimes

b. **45 mm; 4.5 cm**

c. $\quad$ 1 cm $=$ 10 mm

$\quad$ 10 mm $+$ 10 mm $+$ 10 mm

$\qquad\qquad + $ 10 mm $=$ **40 mm**

d. **About 1.9 cm**

e. $\quad$ **3.2** $-$ **1.5** $=$ **1.7**

LESSON 69, MIXED PRACTICE

1. **52 mm**

2. Pattern: $\qquad$ Larger

$\qquad\qquad\quad\underline{- \text{ Smaller}}$

$\qquad\qquad\qquad$ Difference

$\quad$ Problem: $\quad$ 384 cards

$\qquad\qquad\quad\underline{- \text{ 260 cards}}$

$\qquad\qquad\qquad$ **124 cards**

3. Pattern:

$\quad$ Number of groups $\times$ number in each group

$\qquad\qquad\qquad = $ total

$\quad$ Problem:

$\quad$ 30 baskets $\times$ 42 plums in each basket

$\qquad = N$ plums

$\qquad\qquad 42$

$\qquad\underline{\times\quad 30}$

$\qquad$ **1260 plums**

4. **-4**

5.

6. (a) **4 cm**

$\quad$ (b) **40 mm**

7. $100\% - 25\% = $ **75%**

8. 3 feet $+$ 3 feet $+$ 3 feet $+$ 3 feet $=$ **12 feet**

9. 412 rounds to 400, 695 rounds to 700, and 379 rounds to 400

$\qquad 400$

$\qquad 700$

$\qquad\underline{+\ \ 400}$

$\qquad \mathbf{1500}$

10. $\qquad 11.6$ cm

$\qquad\underline{-\ \ 3.5 \text{ cm}}$

$\qquad\quad\mathbf{8.1}$ **cm**

$\quad$ **11.6** $-$ **3.5** $=$ **8.1**

11. Pattern:

$\quad$ Number of groups $\times$ number in each group

$\qquad\qquad\qquad = $ total

$\quad$ Problem: 5 hours $\times\ N$ miles per hour

$\qquad\qquad\qquad = $ 125 miles

$$\begin{array}{r} \textbf{25 miles per hour} \\ 5\overline{)125} \\ \underline{10} \\ 25 \\ \underline{25} \\ 0 \end{array}$$

12. Pattern:

$\quad$ Number of groups $\times$ number in each group

$\qquad\qquad\qquad = $ total

$\quad$ Problem: 7 hours $\times$ 21 miles per hour

$\qquad\qquad\qquad = N$ miles

$\qquad\qquad 21$

$\qquad\underline{\times\quad 7}$

$\qquad$ **147 miles**

13. $7.95 rounds to $8 and $8.95 rounds to $9

$\quad$ $8 $+$ $9 $=$ **$17**

14.
$$\begin{array}{r} \textbf{41 R 4} \\ 6\overline{)250} \\ \underline{24} \\ 10 \\ \underline{6} \\ 4 \end{array}$$

15.
$$\begin{array}{r} \textbf{11 R 1} \\ 9\overline{)100} \\ \underline{9} \\ 10 \\ \underline{9} \\ 1 \end{array}$$

16.
$$\begin{array}{r} {}^{3\,2} \\ 36.2 \\ 4.7 \\ {}_{1}15.9 \\ 148.4 \\ 30.5 \\ \underline{+\quad 6.0} \\ \mathbf{241.7} \end{array}$$

17.
$$\begin{array}{r} 32 \\ 8\overline{)256} \\ \underline{24} \\ 16 \\ \underline{16} \\ 0 \end{array}$$

18. $4 \times W = 60$

$$\begin{array}{r} 15 \\ 4\overline{)60} \\ \underline{4} \\ 20 \\ \underline{20} \\ 0 \end{array}$$

$W = \mathbf{15}$

19.
$$\begin{array}{r} {}^{5\ 2}\ \ \\ \$4.63 \\ \times \qquad 9 \\ \hline \mathbf{\$41.67} \end{array}$$

20.
$$\begin{array}{r} {}^{7}\ \ \\ \$0.29 \\ \times \qquad 80 \\ \hline \mathbf{\$23.20} \end{array}$$

21.
$$\begin{array}{r} {}^{0\ 9\ 9}\\ \$\cancel{1}\,\cancel{0}.\,\cancel{0}{}^{1}0 \\ - \quad \$1.\,7\,3 \\ \hline \mathbf{\$8.\,2\,7} \end{array}$$

22.
$$\begin{array}{r} {}^{2\ \ \ 3}\\ \cancel{3}{}^{1}6,\cancel{4}{}^{1}2\,8 \\ -\ 2\,7,3\,3\,8 \\ \hline \mathbf{9\ 0\ 9\ 0} \end{array}$$

23.
$$\begin{array}{r} {}^{4}\ \\ 78 \\ \times \quad 60 \\ \hline \mathbf{4680} \end{array}$$

24.
$$\begin{array}{r} 82 \\ 4\overline{)328} \\ \underline{32} \\ 08 \\ \underline{8} \\ 0 \end{array}$$

25.
$$\begin{array}{r} 53\ R\ 4 \\ 7\overline{)375} \\ \underline{35} \\ 25 \\ \underline{21} \\ 4 \end{array}$$

26.
$$\begin{array}{r} 64 \\ 5\overline{)320} \\ \underline{30} \\ 20 \\ \underline{20} \\ 0 \end{array}$$

27. $A + 5 = 50$
$45 + 5 = 50$
$A = \mathbf{45}$

28.
$$\begin{array}{r} {}^{2}\ \\ \cancel{3}.{}^{1}6 \\ -\ 1.\ 7 \\ \hline 1.\ 9 \end{array}$$

$$\begin{array}{r} {}^{3}\ \\ \cancel{4}.{}^{1}7 \\ -\ 1.\ 9 \\ \hline \mathbf{2.\ 8} \end{array}$$

29. (a) 3 cm = 30 mm and 2 cm = 20 mm
30 mm + 20 mm + 30 mm + 20 mm
= **100 mm**

(b) 2 cm × 3 cm = **6 sq. cm**

30. **A. Acute**

LESSON 70, WARM-UP

a. **9000**

b. **2400**

c. **440**

d. **$0.72**

e. **$6.70**

f. **404**

Problem Solving

$1.00 − $0.10 = $0.90

$$\begin{array}{ll} 1HD = 50¢ & \quad 1HD = 50¢ \\ 1Q = 25¢ & \quad +\ 8N = 40¢ \\ 2N = 10¢ & \overline{\qquad 90¢\ \text{or}\ \$0.90} \\ +\ 5P = \ 5¢ & \\ \overline{\qquad 90¢\ \text{or}\ \$0.90} & \end{array}$$

**1 half-dollar, 1 quarter, 2 nickels, and
5 pennies; or 1 half-dollar and 8 nickels**

LESSON 70, LESSON PRACTICE

a.

$\frac{1}{3}$ of 60 $\left\{ \begin{array}{l} 20 \end{array} \right.$ 60

$\frac{2}{3}$ of 60 $\left\{ \begin{array}{l} 20 \\ 20 \end{array} \right.$

20

b.

$\frac{1}{2}$ of 60 $\left\{ \begin{array}{l} 30 \end{array} \right.$ 60

$\frac{1}{2}$ of 60 $\left\{ \begin{array}{l} 30 \end{array} \right.$

30

c.

$\frac{1}{4}$ of 60 $\left\{ \begin{array}{l} 15 \end{array} \right.$ 60

$\frac{3}{4}$ of 60 $\left\{ \begin{array}{l} 15 \\ 15 \\ 15 \end{array} \right.$

15

d.

$\frac{1}{5}$ of 60 $\left\{ \begin{array}{l} 12 \end{array} \right.$ 60

$\frac{4}{5}$ of 60 $\left\{ \begin{array}{l} 12 \\ 12 \\ 12 \\ 12 \end{array} \right.$

12

e.

$\frac{1}{2}$ were boys. $\left\{ \begin{array}{l} 16\ \text{children} \end{array} \right.$ 32 children

$\frac{1}{2}$ were girls. $\left\{ \begin{array}{l} 16\ \text{children} \end{array} \right.$

16 boys

f.

$\frac{1}{3}$ were quarters. $\left\{ \begin{array}{l} 8\ \text{coins} \end{array} \right.$ 24 coins

$\frac{2}{3}$ were not quarters. $\left\{ \begin{array}{l} 8\ \text{coins} \\ 8\ \text{coins} \end{array} \right.$

8 quarters

LESSON 70, MIXED PRACTICE

1. Pattern:

$$\begin{array}{r} \text{Larger} \\ -\ \text{Smaller} \\ \hline \text{Difference} \end{array}$$

Problem:

$$\begin{array}{r} \overset{3}{4}{}^{1}2,0\,0\,0,0\,0\,0 \\ -\ 2\,4,0\,0\,0,0\,0\,0 \\ \hline \mathbf{1\,8,0\,0\,0,0\,0\,0} \end{array}$$

2. Pattern:

$$\begin{array}{r} \text{Larger} \\ -\ \text{Smaller} \\ \hline \text{Difference} \end{array}$$

Problem:

$$\begin{array}{r} 1\,\overset{4}{5}{}^{1}0\ \text{seats} \\ -\ 1\,2\,8\ \text{seats} \\ \hline \mathbf{2\,2\ seats} \end{array}$$

3.

$$\begin{array}{r} 1\,\overset{1}{2}.\,\overset{1}{1}{}^{1}4\ \text{seconds} \\ -\ 1\,1.\,9\,8\ \text{seconds} \\ \hline \mathbf{0.\,1\,6\ second} \end{array}$$

4. Pattern:

Number of groups $\times$ number in each group $=$ total

Problem: 5 days $\times$ \$1.25 per day $= N$ dollars

$$\begin{array}{r} \overset{1\ 2}{\$1.25} \\ \times \qquad 5 \\ \hline \mathbf{\$6.25} \end{array}$$

5. Perimeter: 5 units $+$ 4 units $+$ 5 units $+$ 4 units $=$ **18 units**

Area: 5 units $\times$ 4 units $=$ **20 square units**

6. Pattern:

$$\begin{array}{r} \text{Some} \\ +\ \text{Some more} \\ \hline \text{Total} \end{array}$$

Problem:

$$\begin{array}{r} \overset{1}{3}0\ \text{pages} \\ 30\ \text{pages} \\ 30\ \text{pages} \\ 45\ \text{pages} \\ +\ 26\ \text{pages} \\ \hline \mathbf{161\ pages} \end{array}$$

7.

$\frac{1}{2}$ sprouted. $\left\{ \begin{array}{l} 37\ \text{seeds} \end{array} \right.$ 74 seeds

$\frac{1}{2}$ did not sprout. $\left\{ \begin{array}{l} 37\ \text{seeds} \end{array} \right.$

37 seeds

8. $\frac{1}{2} = \mathbf{50\%}$

9.

$\frac{1}{6}$ $\left\{ \begin{array}{l} 10 \end{array} \right.$ 60

$\frac{5}{6}$ $\left\{ \begin{array}{l} 10 \\ 10 \\ 10 \\ 10 \\ 10 \end{array} \right.$

10

10. Pattern:
Number of groups × number in each group
= total
Problem: 3 hours × 65 miles per hour
= N miles

$$\begin{array}{r} \overset{1}{65} \\ \times\ 3 \\ \hline 195 \text{ miles} \end{array}$$

11. Pattern:
Number of groups × number in each group
= total
Problem: 4 hours × N miles per hour
= 248 miles

$$\begin{array}{r} 62 \text{ miles} \\ 4\overline{)248} \\ 24 \\ \hline 08 \\ 8 \\ \hline 0 \end{array}$$

12. (a) **1 cm**

(b) 1 cm = 10 mm
A diameter equals two radii.
5 mm + 5 mm = 10 mm
5 mm

13.
$$\begin{array}{r} \overset{1}{2.7} \text{ cm} \\ +\ 4.8 \text{ cm} \\ \hline 7.5 \text{ cm} \end{array}$$
2.7 + 4.8 = 7.5

14.
$$\begin{array}{r} \overset{1\ 1}{\$8.00} \\ \$9.48 \\ +\ \$0.79 \\ \hline \$18.27 \end{array}$$

15.
$$\begin{array}{r} 5.36 \\ 2.1 \\ +\ 0.43 \\ \hline 7.89 \end{array}$$

16.
$$\begin{array}{r} \$\overset{0}{\cancel{1}}\overset{9}{\cancel{0}}\overset{9}{\cancel{0}}.\overset{9}{\cancel{0}}{}^{1}0 \\ -\ \ \$5\ 9.\ 4\ 7 \\ \hline \$4\ 0.\ 5\ 3 \end{array}$$

17.
$$\begin{array}{r} \overset{2}{\cancel{3}}\ \overset{6}{\cancel{7}},\ \overset{1}{\cancel{1}}{}^{1}0\ 2 \\ -\ 1\ 8,\ 5\ 9\ 0 \\ \hline 1\ 8,\ 5\ 1\ 2 \end{array}$$

18.
$$\sqrt{49} = 7$$
$$2 \times 2 \times 2 = 8$$
$$7 \times 8 = \mathbf{56}$$

19.
$$\begin{array}{r} \overset{2\ 1}{\$1.63} \\ \times\ \ \ \ 40 \\ \hline \$65.20 \end{array}$$

20.
$$\begin{array}{r} \overset{5}{39} \\ \times\ \ 60 \\ \hline 2340 \end{array}$$

21.
$$\begin{array}{r} \overset{3\ 4}{\$2.56} \\ \times\ \ \ \ 7 \\ \hline \$17.92 \end{array}$$

22.
$$\begin{array}{r} 29 \text{ R } 2 \\ 3\overline{)89} \\ 6 \\ \hline 29 \\ 27 \\ \hline 2 \end{array}$$

23.
$$\begin{array}{r} 26 \\ 9\overline{)234} \\ 18 \\ \hline 54 \\ 54 \\ \hline 0 \end{array}$$

24.
$$\begin{array}{r} 15 \\ 6\overline{)90} \\ 6 \\ \hline 30 \\ 30 \\ \hline 0 \end{array}$$

25.
$$\begin{array}{r} 34 \text{ R } 5 \\ 7\overline{)243} \\ 21 \\ \hline 33 \\ 28 \\ \hline 5 \end{array}$$

26.
$$\begin{array}{r} 71 \\ 5\overline{)355} \\ 35 \\ \hline 05 \\ 5 \\ \hline 0 \end{array}$$

27. $7 + N = 28$
$28 - 7 = 21$
$N = \mathbf{21}$

28. $12\dfrac{3}{10}$; **12.3**

29. $1 \times 12 = 12$
$2 \times 6 = 12$
$3 \times 4 = 12$
Factors of 12: 1, 2, 3, 4, 6, 12

$1 \times 20 = 20$
$2 \times 10 = 20$
$4 \times 5 = 20$
Factors of 20: 1, 2, 4, 5, 10, 20

B. **4** is a factor of both 12 and 20.

30. **One possibility:**

INVESTIGATION 7

1. **Sample answer:** **1) Which of these drinks is your favorite to have with lunch: milk, lemonade, juice, or water?**
2) What is your favorite drink to have with lunch?

2. **Words like "cool" and "sweet" bias the question to favor the choice of lemonade.**

3. **One possibility: Which drink do you prefer for lunch, lemonade or milk?**

4. **B. All children about your age in your community.**

   Reasons why/why not the larger population is like student's sample:

   A. The age span may be too great.
   B. This is the population most similar to the sample, since the age and regional differences will be small.
   C. Regional differences may be too great.
   D. Parents may make different choices than their children.

5. **9 students**

6. **Sample tally sheet:**

Question:	Which of these drinks is your favorite to have with lunch: milk, lemonade, juice, water, no opinion?
Tallies:	milk
	lemonade
	juice
	water
	no opinion

LESSON 71, WARM-UP

a. **240**

b. **360**

c. **480**

d. **295**

e. **$8.85**

f. **64¢**

Problem Solving

LESSON 71, LESSON PRACTICE

a.
$$\begin{array}{r} 40 \\ 3\overline{)120} \\ \underline{12} \\ 00 \\ \underline{0} \\ 0 \end{array}$$

b.
$$\begin{array}{r} 60 \\ 4\overline{)240} \\ \underline{24} \\ 00 \\ \underline{0} \\ 0 \end{array}$$

c. $\begin{array}{r} 30 \text{ R } 2 \\ 5\overline{)152} \\ \underline{15} \\ 02 \\ \underline{0} \\ 2 \end{array}$

d. $\begin{array}{r} 30 \text{ R } 1 \\ 4\overline{)121} \\ \underline{12} \\ 01 \\ \underline{0} \\ 1 \end{array}$

e. $\begin{array}{r} 30 \text{ R } 1 \\ 3\overline{)91} \\ \underline{9} \\ 01 \\ \underline{0} \\ 1 \end{array}$

f. $\begin{array}{r} 20 \text{ R } 1 \\ 2\overline{)41} \\ \underline{4} \\ 01 \\ \underline{0} \\ 1 \end{array}$

LESSON 71, MIXED PRACTICE

1. $\begin{array}{r} 30 \\ \times\ 40 \\ \hline \textbf{1200 tiles} \end{array}$

2. $\begin{array}{r} 2\ \overset{5}{\cancel{6}}{}^1 0 \text{ seats} \\ -\quad 4\ 3 \text{ seats} \\ \hline \textbf{2 1 7 seats} \end{array}$

3. Pattern:
Number of groups $\times$ number in each group
$\qquad\qquad$ = total
Problem:
115 cookies $\times$ 5 chocolate chips in each cookie
$\qquad$ = N chocolate chips

$\begin{array}{r} \overset{2}{1}15 \\ \times\quad 5 \\ \hline \textbf{575 chocolate chips} \end{array}$

4. 4 quarts = 1 gallon
4 cups

5. $\begin{array}{r} \overset{1}{\ }\$0.05 \\ \$0.30 \\ \$0.50 \\ +\ \$0.15 \\ \hline \textbf{\$1.00} \end{array}$

6.
$\frac{1}{4}$ were tired. $\Big\{$

$\frac{3}{4}$ were not tired. $\Big\{$

280 swimmers
| 70 swimmers |
| 70 swimmers |
| 70 swimmers |
| 70 swimmers |

70 swimmers

7. $\frac{1}{4}$ = **25%**

8.
$\frac{1}{2} \Big\{$
$\frac{1}{2} \Big\{$

560
| 280 |
| 280 |

280

9. (a) **3 cm**

   (b) **30 mm**

10. $\begin{array}{l} 1 \times 90 = \textbf{90} \\ 2 \times 90 = \textbf{180} \\ 3 \times 90 = \textbf{270} \\ 4 \times 90 = \textbf{360} \end{array}$

11. **One possibility:**

$\frac{2}{3} \gtrdot \frac{2}{5}$

12. Pattern:
Number of groups $\times$ number in each group
$\qquad\qquad$ = total
Problem: 19 minutes $\times$ 72 times per minute
$\qquad\qquad$ = N times

$\begin{array}{r} \overset{1}{7}2 \\ \times\quad 9 \\ \hline \textbf{648 times} \end{array}$

13. $\begin{array}{r} \overset{1\ 1\ 1}{\ }\$375.48 \\ +\ \$536.70 \\ \hline \textbf{\$912.18} \end{array}$

14.
$$\overset{1}{3}\overset{1}{6}\overset{1}{7},419$$
$$+\ \ 90,852$$
$$\overline{458,271}$$

15.
$$\overset{3\,1}{42.3}$$
$$57.1$$
$$28.9$$
$$96.4$$
$$+\ 38.0$$
$$\overline{262.7}$$

16.
$$\$\overset{1}{2}\,\overset{9}{\cancel{0}}.\,\overset{9}{\cancel{0}}\,{}^{1}0$$
$$-\ \$1\,9.\,3\,9$$
$$\overline{\$0.\ 6\ 1}$$

17.
$$\overset{2}{\cancel{3}}\,{}^{10}\overset{9}{\cancel{1}}\,\overset{9}{\cancel{0}},\overset{13}{\cancel{4}}\,{}^{1}9$$
$$-\ 2\,5\,0,5\,2\,7$$
$$\overline{5\,9,8\,9\,2}$$

18.
$$\$6.\overset{5}{0}8$$
$$\times\ \ \ \ \ 7$$
$$\overline{\$42.56}$$

19.
$$\overset{2}{8}6$$
$$\times\ \ 40$$
$$\overline{3440}$$

20.
$$\$0.\overset{7}{5}9$$
$$\times\ \ \ \ \ 8$$
$$\overline{\$4.72}$$

21.
$$\begin{array}{r}60 \\ 3\overline{)180} \\ \underline{18} \\ 00 \\ \underline{0} \\ \overline{0}\end{array}$$

22.
$$\begin{array}{r}30\ R\ 1 \\ 8\overline{)241} \\ \underline{24} \\ 01 \\ \underline{0} \\ \overline{1}\end{array}$$

23.
$$\begin{array}{r}64\ R\ 3 \\ 5\overline{)323} \\ \underline{30} \\ 23 \\ \underline{20} \\ \overline{3}\end{array}$$

24.
$$\begin{array}{r}30\ R\ 4 \\ 6\overline{)184} \\ \underline{18} \\ 04 \\ \underline{0} \\ \overline{4}\end{array}$$

25.
$$\begin{array}{r}60\ R\ 3 \\ 7\overline{)423} \\ \underline{42} \\ 03 \\ \underline{0} \\ \overline{3}\end{array}$$

26.
$$\sqrt{36}\ =\ 6$$
$$4\ \times\ 4\ =\ 16$$
$$10\ \times\ 10\ =\ 100$$

$$\overset{1}{1}\overset{6}{6}$$
$$+\ 100$$
$$\overline{122}$$

27.
$$9\ +\ M\ =\ 99$$
$$99\ -\ 9\ =\ 90$$
$$M\ =\ \mathbf{90}$$

28.
$$\begin{array}{r}15 \\ 6\overline{)90} \\ \underline{6} \\ 30 \\ \underline{30} \\ \overline{0}\end{array}$$

$$N\ =\ \mathbf{15}$$

29. $AB\ =\ 1\frac{3}{4}$ in.; $BC\ =\ 1\frac{1}{2}$ in.; $AC\ =\ 3\frac{1}{4}$ in.

30. The radius is 1 cm, and 1 cm = 10 mm.

The radius is **10 mm.**

LESSON 72, WARM-UP

a. **420**

b. **750**

c. **500**

d. **$7.02**

e. **377**

f. **365**

Problem Solving

$50\% = \frac{1}{2}$

$\frac{1}{2}$ of 2000 is 1000, so John consumes **about 1000 calories** each day from carbohydrates.

LESSON 72, LESSON PRACTICE

a. 3 hours + 4 hours = **7 hours**

b. 3 hours × $6 per hour = **$18**

c. 7 hours × $6 per hour = **$42**

LESSON 72, MIXED PRACTICE

1. Pattern:
    Number of groups × number in each group
    = total
    Problem: 10 gallons × 18 miles per gallon
    = **180 miles**

2. 12 inches = 1 ft
    24 inches = 2 ft
    Pattern:
    Number of groups × number in each group
    = total
    Problem: *N* times × 2 ft per time = 50 feet

    $$\begin{array}{r} \textbf{25 times} \\ 2\overline{)50} \\ \underline{4} \\ 10 \\ \underline{10} \\ 0 \end{array}$$

3. Pattern:
    Number of groups × number in each group
    = total
    Problem: 8 pieces × *N* pounds in each piece
    = 160 pounds

    $$\begin{array}{r} \textbf{20 pounds} \\ 8\overline{)160} \\ \underline{16} \\ 00 \\ \underline{0} \\ 0 \end{array}$$

4. Step 1: Counting forward 30 minutes from
    3:15 p.m. makes it 3:45 p.m.
    Step 2: Counting forward 1 hour from 3:45 p.m.
    makes it **4:45 p.m.**

5. $\frac{1}{3}$ scored by Lucy   { 36 points
    $\frac{2}{3}$ not scored by Lucy { 12 points / 12 points / 12 points

    12 points

6. 4 units + 3 units + 4 units
    + 3 units = **14 units**
    4 units × 3 units = **12 square units**

7. 10 mm = 1 cm
    6 cm

8. **Wednesday**

9. 100% − 30% = **70%**

10. 100 cm + 100 cm + 100 cm
    + 100 cm = **400 cm**

11. 1 × 90 = **90**
    2 × 90 = **180**
    3 × 90 = **270**
    4 × 90 = **360**

12. $$\begin{array}{r} {}^{1}{}^{1} \\ \$1.68 \\ \$0.32 \\ \$6.37 \\ + \ \$5.00 \\ \hline \textbf{\$13.37} \end{array}$$

13.
$$\overset{2}{4.3}$$
$$2.4$$
$$0.8$$
$$+\ 6.7$$
$$\overline{14.2}$$

14.
$$\overset{1}{\$6.46}$$
$$+\ \$2.17$$
$$\overline{\$8.63}$$

$$\$\overset{0}{1}\overset{9}{0}.\overset{9}{0}{}^{1}0$$
$$-\ \ \$8.\ 6\ 3$$
$$\overline{\$1.\ 3\ 7}$$

15. $5 \times 4 = 20$
$20 \times 5 = \textbf{100}$

16.
$$\overset{4\ 6}{359}$$
$$\times\ \ \ \ 70$$
$$\overline{25,130}$$

17.
$$\overset{2}{74}$$
$$\times\ \ \ 50$$
$$\overline{3700}$$

18.
$$\overset{\textbf{80 R 1}}{2\overline{)161}}$$
$$\underline{16}$$
$$01$$
$$\underline{\ 0}$$
$$1$$

19.
$$\overset{\textbf{80}}{5\overline{)400}}$$
$$\underline{40}$$
$$00$$
$$\underline{\ 0}$$
$$0$$

20.
$$\overset{\textbf{51 R 3}}{9\overline{)462}}$$
$$\underline{45}$$
$$12$$
$$\underline{\ 9}$$
$$3$$

21.
$$\overset{\textbf{72}}{3\overline{)216}}$$
$$\underline{21}$$
$$06$$
$$\underline{\ 6}$$
$$0$$

22.
$$\overset{\textbf{39 R 3}}{4\overline{)159}}$$
$$\underline{12}$$
$$39$$
$$\underline{36}$$
$$3$$

23.
$$\overset{\textbf{70}}{7\overline{)490}}$$
$$\underline{49}$$
$$00$$
$$\underline{\ 0}$$
$$0$$

24.
$$\overset{\textbf{42}}{3\overline{)126}}$$
$$\underline{12}$$
$$06$$
$$\underline{\ 6}$$
$$0$$

25. $\sqrt{36} = 6$

$$\overset{\textbf{60}}{6\overline{)360}}$$
$$\underline{36}$$
$$00$$
$$\underline{\ 0}$$
$$0$$

26. $5 \times N = 120$

$$\overset{24}{5\overline{)120}}$$
$$\underline{10}$$
$$20$$
$$\underline{20}$$
$$0$$

$N = \textbf{24}$

27. 5 goals $-$ 2 goals = **3 goals**

28. 4 games $+$ 2 games = **6 games**

29. **Angle** *B*

30.

LESSON 73, WARM-UP

a. 2500

b. 4000

c. 3600

d. 337

e. $7.50

f. $9.48

Problem Solving

$2.50 + ($1.50 × 7) = **$13.00**

LESSON 73, LESSON PRACTICE

a. **Slide, translation; flip, reflection; turn, rotation**

b. **Possibilities include:**
   • **Rotation and reflection**
   • **Reflection only**

Note: various combinations of transformations can be equivalent.

LESSON 73, MIXED PRACTICE

1. Pattern:
   Number of groups × number in each group
                     = total
   Problem: *N* cars × 5 visitors in each car
                     = 30 visitors
            30 ÷ 5 = **6 cars**

2. 30 × $5 = **$150**

3.
$$\begin{array}{r} \$\overset{4}{5}.\overset{9}{\cancel{0}}{}^{1}0 \\ -\ \$3.25 \\ \hline \$1.75 \end{array}$$

4. Step 1: Counting forward 30 minutes from 6:30 a.m. makes it 7:00 a.m.
   Step 2: Counting forward 3 hours from 7:00 a.m. makes it **10:00 a.m.**

5.
$$\frac{1}{2} \text{ sharpened}$$
$$\frac{1}{2} \text{ not sharpened}$$
48 pencils
{ 24 pencils
  24 pencils }

24 pencils
One half of the pencils were not sharpened.
One half = **50%.**

6.
$$\frac{1}{4} \Big\{ \begin{array}{|c|} \hline 15 \\ \hline 15 \\ \hline \end{array}$$
$$\frac{3}{4} \Big\{ \begin{array}{|c|} \hline 15 \\ \hline 15 \\ \hline \end{array}$$
60

15

7. 1 gallon = 4 quarts
   4 1-quart bottles

8. 1 cm = 10 mm
   10 mm + 10 mm + 10 mm + 10 mm
            + 10 mm + 10 mm = **60 mm**

9.
$$\begin{array}{r} 5{,}280 \text{ ft} \\ -\ 4{,}200 \text{ ft} \\ \hline 1{,}080 \text{ ft} \end{array}$$

10. The smallest multiple of 90 is 90.
    A. 45

11. **350**

12.
$$\begin{array}{r} 37.56 \\ -\ 4.2 \\ \hline 33.36 \end{array}$$

13.
$$\begin{array}{r} 4.2 \\ 3.5 \\ 0.25 \\ +\ 4.0 \\ \hline 11.95 \end{array}$$

14.
$$\begin{array}{r} \$\overset{0}{1}\overset{9}{0}\overset{9}{0}.\overset{9}{\cancel{0}}{}^{1}0 \\ -\ \$31.53 \\ \hline \$68.47 \end{array}$$

15.
$$\begin{array}{r} 2\,\overset{4}{5}{}^{1}1, \overset{4}{5}{}^{1}4\,6 \\ -\ 37{,}156 \\ \hline 214{,}390 \end{array}$$

16.
$$\begin{array}{r} N \\ +\ 423 \\ \hline 618 \end{array}$$

$$\begin{array}{r} \overset{5}{\cancel{6}}{}^{1}1\ 8 \\ -\ 4\ 2\ 3 \\ \hline 1\ 9\ 5 \end{array}$$

$N = \mathbf{195}$

17.
$$\begin{array}{r} \overset{3\ 4}{\$3.46} \\ \times\quad 7 \\ \hline \mathbf{\$24.22} \end{array}$$

18.
$$\begin{array}{r} \overset{1}{96} \\ \times\ \ 30 \\ \hline \mathbf{2880} \end{array}$$

19.
$$\begin{array}{r} \overset{7}{\$0.59} \\ \times\qquad 8 \\ \hline \mathbf{\$4.72} \end{array}$$

20.
$$\begin{array}{r} \mathbf{90\ R\ 3} \\ 7\overline{)633} \\ \underline{63} \\ 03 \\ \underline{\ 0} \\ 3 \end{array}$$

21.
$$\begin{array}{r} \mathbf{19\ R\ 3} \\ 5\overline{)98} \\ \underline{\mathbf{5}} \\ 48 \\ \underline{45} \\ 3 \end{array}$$

22.
$$\begin{array}{r} \mathbf{50} \\ 3\overline{)150} \\ \underline{15} \\ 00 \\ \underline{\ 0} \\ 0 \end{array}$$

23.
$$\begin{array}{r} \mathbf{54\ R\ 5} \\ 6\overline{)329} \\ \underline{30} \\ 29 \\ \underline{24} \\ 5 \end{array}$$

24.
$$\begin{array}{r} \mathbf{68\ R\ 2} \\ 4\overline{)274} \\ \underline{24} \\ 34 \\ \underline{32} \\ 2 \end{array}$$

25.
$$\begin{array}{r} \mathbf{30\ R\ 7} \\ 8\overline{)247} \\ \underline{24} \\ 07 \\ \underline{\ 0} \\ 7 \end{array}$$

26.
$$\sqrt{25}\ =\ 5$$
$$5\ \times\ M\ =\ 135$$

$$\begin{array}{r} 27 \\ 5\overline{)135} \\ \underline{10} \\ 35 \\ \underline{35} \\ 0 \end{array}$$

$M\ =\ \mathbf{27}$

27. $Z\ -\ 476\ =\ 325$

$$\begin{array}{r} \overset{1\ 1}{325} \\ +\ 476 \\ \hline 801 \end{array}$$

$Z\ =\ \mathbf{801}$

28.
$$6A\ =\ 18$$
$$6\ \times\ 3\ =\ 18$$
$$A\ =\ \mathbf{3}$$

29.
$$\begin{array}{r} 2.3\ cm \\ +\ 3.5\ cm \\ \hline \mathbf{5.8\ cm} \end{array}$$
$$\mathbf{2.3\ +\ 3.5\ =\ 5.8}$$

30. **Reflection**

LESSON 74, WARM-UP

a. **65**

b. **72**

c. **91**

d. **83**

Problem Solving

First think, "13 (after regrouping) minus what number equals 4?" (9). Then think, "11 (after regrouping) minus 4 equals what number?" (7).

$$\begin{array}{r} 123 \\ -\ \ 49 \\ \hline 74 \end{array}$$

LESSON 74, LESSON PRACTICE

a. Five of the twelve circles are shaded.

$$\frac{5}{12}$$

b. Six of the seven hexagons are not shaded.

$$\frac{6}{7}$$

c. Thirteen of the 27 members are boys.

$$\frac{13}{27}$$

d. Four of the seven letters are A's.

$$\frac{4}{7}$$

LESSON 74, MIXED PRACTICE

1. Pattern:
```
        Some
    +   Some more
        Total
```

Problem:
```
          1
        62 crawfish
         7 crawfish
        12 crawfish
    +   12 crawfish
        93 crawfish
```

2. Pattern:
```
        Larger
    −   Smaller
        Difference
```

Problem:
$$\begin{array}{r} 1\ 5,\overset{6}{\cancel{7}}\ {}^{1}7\ 1\ \text{feet} \\ -\ 1\ 4,6\ \ 9\ 1\ \text{feet} \\ \hline 1\ 0\ 8\ 0\ \text{feet} \end{array}$$

3. Pattern:
Number of groups × number in each group
= total

Problem:
4 bingo cards × 25 squares on each bingo card
= **100 squares**

4.
$$\begin{array}{r} 24 \text{ books} \\ 4\overline{)96} \\ \underline{8} \\ 16 \\ \underline{16} \\ 0 \end{array}$$

5.
$$\begin{array}{r} 390 \text{ fans} \\ 2\overline{)780} \\ \underline{6} \\ 18 \\ \underline{18} \\ 00 \\ \underline{0} \\ 0 \end{array}$$

One half is **50%**

6. 1 century = 100 years
10 centuries = **1000 years**

7. 493 rounds to 500 and 387 rounds to 400
500 + 400 = **900**

8. Four of the seven circles are not shaded.

$$\frac{4}{7}$$

9. 1 L = 1000 mL
2 L = **2000 mL**

10. 6 in. + 4 in. + 6 in. + 4 in. = **20 in.**

11. **24 squares**

12. **Possibilities include:**
- **Rotation and translation**
- **Rotation only**

13.
$$\begin{array}{r} \$0.57 \\ +\ \$1.20 \\ \hline \$1.77 \end{array}$$

$$\begin{array}{r} \overset{5}{\cancel{\$6}}.\overset{1}{\cancel{1}}{}^{0}5 \\ -\ \$1.7\ 7 \\ \hline \$4.3\ 8 \end{array}$$

14.
$$\begin{array}{r} \overset{3}{\cancel{4}}\overset{12}{\cancel{3}},\overset{1}{1}\overset{5}{\cancel{6}}{}^{}0 \\ -\ \ 8,4\ 5\ 9 \\ \hline 3\ 4,7\ 0\ 1 \end{array}$$

Saxon Math 5/4—Homeschool

15. $8 \times 8 = 64$

$$\begin{array}{r} \overset{3}{64} \\ \times\ \ 8 \\ \hline 512 \end{array}$$

16.
$$\begin{array}{r} \overset{3\ 2}{\$3.54} \\ \times\ \ \ \ \ 6 \\ \hline \$21.24 \end{array}$$

17.
$$\begin{array}{r} \overset{5}{57} \\ \times\ \ 80 \\ \hline 4560 \end{array}$$

18.
$$\begin{array}{r} \overset{3}{704} \\ \times\ \ \ 9 \\ \hline 6336 \end{array}$$

19.
$$\begin{array}{r} 39\text{ R }3 \\ 9\overline{)354} \\ 27\ \ \\ \hline 84 \\ 81 \\ \hline 3 \end{array}$$

20.
$$\begin{array}{r} 40\text{ R }5 \\ 7\overline{)285} \\ 28\ \ \\ \hline 05 \\ 0 \\ \hline 5 \end{array}$$

21.
$$\begin{array}{r} 87\text{ R }4 \\ 5\overline{)439} \\ 40\ \ \\ \hline 39 \\ 35 \\ \hline 4 \end{array}$$

22.
$$\begin{array}{r} 85\text{ R }5 \\ 6\overline{)515} \\ 48\ \ \\ \hline 35 \\ 30 \\ \hline 5 \end{array}$$

23.
$$\begin{array}{r} 90 \\ 4\overline{)360} \\ 36\ \ \\ \hline 00 \\ 0 \\ \hline 0 \end{array}$$

24.
$$\begin{array}{r} 98 \\ 8\overline{)784} \\ 72\ \ \\ \hline 64 \\ 64 \\ \hline 0 \end{array}$$

25. $\sqrt{36} = 6$

$6 \times 6 = 36$

$6 + N = 36$

$6 + 30 = 36$

$N = \mathbf{30}$

26.
$$\begin{array}{r} 462 \\ -\ \ Y \\ \hline 205 \end{array}$$

$$\begin{array}{r} 4\ \overset{5}{\cancel{6}}{}^{1}2 \\ -\ 2\ 0\ 5 \\ \hline 2\ 5\ 7 \end{array}$$

$Y = \mathbf{257}$

27. $\qquad 50 = 5 \times R$

$50 \div 5 = 10$

$R = \mathbf{10}$

28. The rule is "Count up by nineties."
360

29. A diameter equals two radii.
20 in. $+\ 20$ in. $= \mathbf{40\ in.}$

30. $6 \times 8 = 48$
C. 48

LESSON 75, WARM-UP

a. **84**

b. **92**

c. **72**

d. **12,000**

e. **$5.02**

f. **597**

Problem Solving

7 days after Monday is **Monday.**

70 days after Monday is Monday, so 71 days after Monday is **Tuesday.**

700 days after Monday is Monday, so 699 days after Monday is **Sunday.**

LESSON 75, LESSON PRACTICE

a. **90° counterclockwise or 270° clockwise**

b. **West**

LESSON 75, MIXED PRACTICE

1. Pattern:
   Number of groups $\times$ number in each group
   $\qquad\qquad$ = total
   Problem: 4 pounds $\times$ 59 cents per pound
   $\qquad\qquad$ = N pounds

$$\begin{array}{r} \overset{3}{} \\ \$0.59 \\ \times \quad 4 \\ \hline \mathbf{\$2.36} \end{array}$$

2. Perimeter: 6 units + 4 units + 6 units
   $\qquad\qquad$ + 4 units = **20 units**
   Area: 6 units $\times$ 4 units = **24 square units**

3. (a) $\begin{array}{r} \mathbf{90\ books} \\ 4\overline{)360} \\ \underline{36} \\ 00 \\ \underline{0} \\ 0 \end{array}$

   (b) $\begin{array}{r} \overset{2}{\cancel{3}}{}^{1}6\ 0\ \text{books} \\ -\quad 9\ 0\ \text{books} \\ \hline \mathbf{2\ 7\ 0\ books} \end{array}$

4. One fourth of the books were on the table.
   One fourth = 25%.
   100% − 25% = **75%**

5. There are nine segments between 201 and 202, so each segment equals $\frac{1}{9}$.
   The arrow points to **201 $\frac{4}{9}$.**

6. 272 rounds to 300 and 483 rounds to 500
   300 + 500 = **800**

7. Four of the nine triangles are shaded.
   $$\frac{4}{9}$$

8. 1 qt = 2 pt
   1 pt = 2 c
   1 c = 8 oz
   1 qt = 4 $\times$ 8 oz = **32 oz**

9. One quart is **25%** of a gallon.

10. **5 shoppers**

11. 7 shoppers + 5 shoppers = **12 shoppers**

12. **90° clockwise or 270° counterclockwise**

13. $\begin{array}{r} {}^{1\ 1\ \ 1} \\ \$86.47 \\ +\ \$47.98 \\ \hline \mathbf{\$134.45} \end{array}$

14. $\begin{array}{r} \overset{2}{\cancel{3}}{}^{1}6.\ 7 \\ -\ 1\ 8.\ 5 \\ \hline \mathbf{1\ 8.\ 2} \end{array}$

15. $\begin{array}{r} {}^{1\ 1\ 2} \\ 2358 \\ 4715 \\ 317 \\ 2103 \\ +\quad 62 \\ \hline \mathbf{9555} \end{array}$

16. $\begin{array}{r} \mathbf{89\ R\ 4} \\ 8\overline{)716} \\ \underline{64} \\ 76 \\ \underline{72} \\ 4 \end{array}$

17. $\begin{array}{r} \mathbf{80\ R\ 1} \\ 2\overline{)161} \\ \underline{16} \\ 01 \\ \underline{0} \\ 1 \end{array}$

18.
$$\begin{array}{r} 62 \\ 7\overline{)434} \\ \underline{42} \\ 14 \\ \underline{14} \\ 0 \end{array}$$

19.
$$\begin{array}{r} 85\ R\ 3 \\ 6\overline{)513} \\ \underline{48} \\ 33 \\ \underline{30} \\ 3 \end{array}$$

20.
$$\begin{array}{r} 30 \\ 9\overline{)270} \\ \underline{27} \\ 00 \\ \underline{0} \\ 0 \end{array}$$

21.
$$\begin{array}{r} 89 \\ 3\overline{)267} \\ \underline{24} \\ 27 \\ \underline{27} \\ 0 \end{array}$$

22. $N - 7.5 = 21.4$

$$\begin{array}{r} 21.4 \\ + \ \ 7.5 \\ \hline 28.9 \end{array}$$

$$N = 28.9$$

23.
$$\begin{array}{r} {\scriptstyle 7\ 4} \\ \$6.95 \\ \times \quad\ 8 \\ \hline \$55.60 \end{array}$$

24.
$$\begin{array}{r} {\scriptstyle 4} \\ 46 \\ \times\ \ 70 \\ \hline 3220 \end{array}$$

25.
$$\begin{array}{r} {\scriptstyle 5} \\ 460 \\ \times\ \ \ 9 \\ \hline 4140 \end{array}$$

26. $3 \times A = 60$

$60 \div 3 = 20$

$A = 20$

27. $3 \times 3 = 9$

$2 \times 2 \times 2 = 8$

$9 - 8 = 1$

28. $39 + N = 43$

$43 - 39 = 4$

$N = 4$

29. Segment *AC* (or segment *CA*)

30.

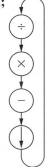

Triangle may be larger or smaller than $\triangle ABC$.

LESSON 76, WARM-UP

a. **102**

b. **72**

c. **491**

d. **$8.20**

e. **450**

f. **$0.62**

Patterns

Counting by fives from four gives us this sequence: **4, 9, 14, 19, 24, 29, …**

The two digits that appear as final digits in the terms of the sequence are **4 and 9.**

LESSON 76, LESSON PRACTICE

a. See student work;

Divide, multiply, subtract, bring down

b.
$$\begin{array}{r} 243\ \text{R}\ 2 \\ 4\overline{)974} \\ \underline{8} \\ 17 \\ \underline{16} \\ 14 \\ \underline{12} \\ 2 \end{array}$$

c.
$$\begin{array}{r} \$1.59 \\ 5\overline{)\$7.95} \\ \underline{5} \\ 2\ 9 \\ \underline{2\ 5} \\ 45 \\ \underline{45} \\ 0 \end{array}$$

d.
$$\begin{array}{r} 252 \\ 6\overline{)1512} \\ \underline{12} \\ 31 \\ \underline{30} \\ 12 \\ \underline{12} \\ 0 \end{array}$$

e.
$$\begin{array}{r} \$6.25 \\ 8\overline{)\$50.00} \\ \underline{48} \\ 2\ 0 \\ \underline{1\ 6} \\ 40 \\ \underline{40} \\ 0 \end{array}$$

LESSON 76, MIXED PRACTICE

1. Pattern:
$$\begin{array}{r} \text{Larger} \\ -\ \text{Smaller} \\ \hline \text{Difference} \end{array}$$

Problem:
$$\begin{array}{r} {}^{0}\cancel{1}{}^{7}1,\ {}^{7}\cancel{8}\ {}^{16}\cancel{7}{}^{1}3 \\ -\quad 7,\ 3\ 9\ 6 \\ \hline 4\ 4\ 7\ 7 \end{array}$$

2. Pattern:
Number of groups $\times$ number in each group
$= \text{total}$
Problem: 5 pages $\times$ N pages per day
$= 200$ pages
$200 \div 5 = $ **40 pages**

3. Pattern:
$$\begin{array}{r} \text{Some} \\ +\ \text{Some more} \\ \hline \text{Total} \end{array}$$

Problem:
$$\begin{array}{r} {}^{2}\ {}^{2}\ \\ \$6.99 \\ \$8.99 \\ +\ \$5.99 \\ \hline \mathbf{\$21.97} \end{array}$$

4. Pattern:
Number of groups $\times$ number in each group
$= \text{total}$
Problem: 7 weeks $\times$ 7 days per week
$= $ **49 days**

5. Two out of three parts were not placed on the first shelf.
$$\frac{2}{3}$$

6. **2500**

7. Three of the eleven letters are P's.
$$\frac{3}{11}$$

8. 1 km $=$ 1000 m
5 km $=$ **5000 m**

9. $30 + N = 72$
$$\begin{array}{r} 72 \\ -\ 30 \\ \hline 42 \end{array}$$
$N = $ **42**

10.
$$\begin{array}{r} {}^{1}\quad \\ 25\ \text{mm} \\ 15\ \text{mm} \\ +\ 20\ \text{mm} \\ \hline \mathbf{60\ \text{mm}} \end{array}$$

11.
$$\begin{array}{r} 11.8\ \text{cm} \\ -\ 3.6\ \text{cm} \\ \hline \mathbf{8.2\ \text{cm}} \end{array}$$
$\mathbf{11.8 - 3.6 = 8.2}$

12.
$$\begin{array}{r} {}^{1}\ {}^{1}\ \\ \$19.71 \\ +\ \$0.98 \\ \hline \$20.69 \end{array}$$

$$\begin{array}{r} {}^{4}\ {}^{9}\ \\ \$2\ \cancel{5}.\ \cancel{0}{}^{1}0 \\ -\ \$2\ 0.\ 6\ 9 \\ \hline \mathbf{\$4.\ 3\ 1} \end{array}$$

13. $10 \times 10 = 100$
$3 \times 3 = 9$
$9 \times 3 = 27$

$$\begin{array}{r} \overset{1}{3}65 \\ 100 \\ +\ 27 \\ \hline \mathbf{492} \end{array}$$

14.
$$\begin{array}{r} \$5\overset{49}{.\cancel{0}}{}^{1}0 \\ -\ \$2.9\ 2 \\ \hline \mathbf{\$2.0\ 8} \end{array}$$

15.
$$\begin{array}{r} 3\overset{5}{\cancel{6}}.{}^{1}2\ 1 \\ -\ \ 5.\ 7 \\ \hline \mathbf{3\ 0.\ 5\ 1} \end{array}$$

16. $5 \times 6 = 30$
$30 \times 9 = \mathbf{270}$

17.
$$\begin{array}{r} \overset{1}{6}3 \\ \times\ \ 50 \\ \hline \mathbf{3150} \end{array}$$

18.
$$\begin{array}{r} \overset{4\ 4}{4}78 \\ \times\ \ \ \ 6 \\ \hline \mathbf{2868} \end{array}$$

19.
$$\begin{array}{r} \mathbf{145} \\ 3\overline{)435} \\ \underline{3} \\ 13 \\ \underline{12} \\ 15 \\ \underline{15} \\ 0 \end{array}$$

20.
$$\begin{array}{r} \mathbf{123\ R\ 6} \\ 7\overline{)867} \\ \underline{7} \\ 16 \\ \underline{14} \\ 27 \\ \underline{21} \\ 6 \end{array}$$

21.
$$\begin{array}{r} \mathbf{\$2.73} \\ 5\overline{)\$13.65} \\ \underline{10} \\ 3\ 6 \\ \underline{3\ 5} \\ 15 \\ \underline{15} \\ 0 \end{array}$$

22.
$$\begin{array}{r} \mathbf{75\ R\ 3} \\ 6\overline{)453} \\ \underline{42} \\ 33 \\ \underline{30} \\ 3 \end{array}$$

23.
$$\begin{array}{r} \mathbf{135\ R\ 3} \\ 4\overline{)543} \\ \underline{4} \\ 14 \\ \underline{12} \\ 23 \\ \underline{20} \\ 3 \end{array}$$

24.
$$\begin{array}{r} \mathbf{\$0.59} \\ 8\overline{)\$4.72} \\ \underline{4\ 0} \\ 72 \\ \underline{72} \\ 0 \end{array}$$

25. $N + 6 = 120$
$114 + 6 = 120$
$N = \mathbf{114}$

26. $4 \times W = 132$

$$\begin{array}{r} 33 \\ 4\overline{)132} \\ \underline{12} \\ 12 \\ \underline{12} \\ 0 \end{array}$$

$W = \mathbf{33}$

27. $43 + N = 55$

$$\begin{array}{r} 55 \\ -\ 43 \\ \hline 12 \end{array}$$

$N = \mathbf{12}$

28. **South**

29. A diameter equals two radii.
$1\ \text{ft} = 12\ \text{in.}$
$6\ \text{in.} + 6\ \text{in.} = 1\ \text{ft}$
6 in.

30. **Possibilities include:**
- **Reflection and translation**
- **Reflection only, if the reflection is made about a vertical line halfway between the triangles**

LESSON 77, WARM-UP

a. **91**

b. **47**

c. **3500**

d. **$0.64**

e. **$7.37**

f. **1200**

Problem Solving

$1\frac{1}{2}$ **in.**

LESSON 77, LESSON PRACTICE

a.   1 ton = 2000 lb
  a half ton = 2000 lb ÷ 2 = **1000 lb**

b. 1 lb = 16 oz
  8 lb = 7 lb 16 oz
  7 lb 12 oz is **4 oz** less than 7 lb 16 oz

c. **57 g**

d. **6 kg**

e. **7 kg**

f.   1 kg = 1000 g
  7 × 1000 g = **7000 g**

LESSON 77, MIXED PRACTICE

1.   2 cups = 1 pint
  14 cups ÷ 2 = **7 pints**

2. **Juice**

3. 4 cups = 1 quart
  Milk

4.
$$
\begin{array}{r}
14 \text{ guests} \\
4\overline{)56} \\
\underline{4} \\
16 \\
\underline{16} \\
0
\end{array}
$$

One fourth = **25%**

5. 2500 is halfway between 2000 and 3000.
  B

6. 682 rounds to 700, 437 rounds to 400,
  and 396 rounds to 400

$$
\begin{array}{r}
700 \\
400 \\
+\ \ 400 \\
\hline
1500
\end{array}
$$

7. One out of seven hexagons are shaded.
  $\dfrac{1}{7}$

8. 1 lb = 16 oz

$$
\begin{array}{r}
\overset{5}{16} \text{ oz} \\
\times\ \ \ 9 \\
\hline
144 \text{ oz}
\end{array}
$$

9. (a) **4 cm**

  (b) 1 cm = 10 mm
    4 cm = **40 mm**

10. **Seven million, four hundred fifty thousand dollars**

11. A hexagon has six sides.
  1 foot = 12 in.

$$
\begin{array}{r}
\overset{1}{12} \text{ in.} \\
12 \text{ in.} \\
12 \text{ in.} \\
12 \text{ in.} \\
12 \text{ in.} \\
+\ \ 12 \text{ in.} \\
\hline
72 \text{ in.}
\end{array}
$$

12.
$$
\begin{array}{r}
\overset{1\ 1}{9}\overset{\ \ 1}{3},417 \\
+\ \ \ 8,915 \\
\hline
102,332
\end{array}
$$

13.
```
  42,718
−      K
  26,054
```

```
     3   6
  4̶ 2, 7̶ ¹1 8
− 2 6, 0 5 4
  1 6, 6 6 4
```

$K = \textbf{16,664}$

14.
```
  ¹ ¹ ³
  1307
   638
  5219
   138
+   16
  7318
```

15.
```
    99    9
  $1̶0̶0̶. 0̶¹0
− $ 86. 3 2
  $ 13. 6 8
```

16.
```
  39  ¹4  ¹0
  4̶0̶ 5̶, 1̶¹5 8
− 3 9 6, 3 7 0
        8 7 8 8
```

17.
```
  5 5
  567
×   8
 4536
```

18.
```
    ¹
  $0.84
×    30
 $25.20
```

19.
```
    ³
  $2.08
×    4
 $8.32
```

20.
```
        $3.75
  4)$15.00
     12
      3 0
      2 8
        20
        20
         0
```

21.
```
      156
  6)936
     6
     33
     30
      36
      36
       0
```

22.
```
       567 R 1
  8)4537
     40
     53
     48
      57
      56
       1
```

23.
```
      90 R 2
  5)452
     45
     02
      0
      2
```

24.
```
       42
  9)378
     36
     18
     18
      0
```

25.
```
      137 R 1
  7)960
     7
     26
     21
      50
      49
       1
```

26.
$$\sqrt{16} = 4$$
$$4 \times N = 100$$
$$4 \times 25 = 100$$
$$N = \textbf{25}$$

27.
$$10 \times 10 = 100$$
$$5 \times B = 100$$
$$5 \times 20 = 100$$
$$B = \textbf{20}$$

28. **−7**

29.
$$
\begin{array}{r}
90° \\
90° \\
+\ 90° \\
\hline
270°
\end{array}
$$

30. Perimeter: 5 units + 3 units + 5 units
        + 3 units = **16 units**
   Area: 5 units × 3 units = **15 square units**

LESSON 78, WARM-UP

a. 85

b. 91

c. 182

d. 24,000

e. $7.50

f. $8.44

Problem Solving
   37¢ + (23¢ × 5) = $1.52
   $2.00 − $1.52 = $0.48 or **48¢**

LESSON 78, LESSON PRACTICE

a. **No. The sides would not meet to form a triangle:**

b. **Isosceles triangle**

c. 4 in. + 4 in. + 4 in. = **12 in.**

LESSON 78, MIXED PRACTICE

1. Pattern:
   Number of groups × number in each group
                = total
   Problem: *N* lb × 5 cents per lb = 85 cents

$$
\begin{array}{r}
\mathbf{19\ lb} \\
5)\overline{95} \\
5 \\
\hline
45 \\
45 \\
\hline
0
\end{array}
$$

2. Pattern:
   $$
   \begin{array}{l}
   \ \ \text{Some} \\
   -\ \text{Some went away} \\
   \hline
   \ \ \text{What is left}
   \end{array}
   $$

   Problem:
   $$
   \begin{array}{r}
   \overset{1}{2}\ \overset{13}{\cancel{4}}\,^{1}3 \text{ paint cans} \\
   -\ \ 9\ 5 \text{ paint cans} \\
   \hline
   \mathbf{1\ 4\ 8} \text{ paint cans}
   \end{array}
   $$

3.
$$
\begin{array}{r}
\mathbf{45\ minutes} \\
2)\overline{90} \\
8 \\
\hline
10 \\
10 \\
\hline
0
\end{array}
$$

4. Three out of four parts did not gather in the living room.
   $$\frac{3}{4}$$

   One fourth = 25%
   Three fourths = 3 × 25% = **75%**

5. (a) 3 cm + 3 cm + 3 cm = **9 cm**

   (b) 1 cm = 10 mm
       9 cm = **90 mm**

6. 2750 is greater than 2500, so it is more than halfway between 2000 and 3000.
   C

7. 2 cups = 1 pint
   2 pints = 1 quart
   2 × 2 = 4 cups
   4 cups = 1 quart, so a cup is $\frac{1}{4}$ of a quart.

8. 1 kg = 1000 g
   3 kg = **3000 g**

9. 396 rounds to 400.
   400 × 7 = **2800**

10. Counting back 12 hours from 4:56 p.m. makes it **4:56 a.m.**

11. **One possibility:**

$$\frac{3}{4} \ \textcircled{<}\ \frac{4}{5}$$

12.
$$\begin{array}{r} \overset{3}{\cancel{4}}.{}^{1}3\,2\,5 \\ -\ 2.\,5 \\ \hline 1.\,8\,2\,5 \end{array}$$

13.
$$\begin{array}{r} {}^{1}\;{}^{1}\;\\ 3.65 \\ 5.2 \\ +\ 0.18 \\ \hline 9.03 \end{array}$$

14.
$$\begin{array}{r} \overset{49}{\$5}\cancel{0}.{}^{1}0\,0 \\ -\ \$42.\,6\,0 \\ \hline \$7.\,4\,0 \end{array}$$

15.
$$\begin{array}{r} {}^{1}\;{}^{1}\;{}^{1}\\ \$17.54 \\ \$0.49 \\ +\ \$15.00 \\ \hline \$33.03 \end{array}$$

16.
$$\begin{array}{r} 283\ R\ 1 \\ 2\overline{)567} \\ 4 \\ \hline 16 \\ 16 \\ \hline 07 \\ 6 \\ \hline 1 \end{array}$$

17.
$$\begin{array}{r} \$5.76 \\ 6\overline{)\$34.56} \\ 30 \\ \hline 4\,5 \\ 4\,2 \\ \hline 36 \\ 36 \\ \hline 0 \end{array}$$

18.
$$\begin{array}{r} 244\ R\ 2 \\ 4\overline{)978} \\ 8 \\ \hline 17 \\ 16 \\ \hline 18 \\ 16 \\ \hline 2 \end{array}$$

19.
$$\begin{array}{r} {}^{5\,4}\\ 398 \\ \times\ \ 6 \\ \hline 2388 \end{array}$$

20.
$$\begin{array}{r} {}^{4}\\ 47 \\ \times\ 60 \\ \hline 2820 \end{array}$$

21.
$$\begin{array}{r} {}^{2}\;{}^{4}\\ \$6.25 \\ \times\ \ \ 8 \\ \hline \$50.00 \end{array}$$

22. $\sqrt{25} = 5$

$$\begin{array}{r} 194 \\ 5\overline{)970} \\ 5 \\ \hline 47 \\ 45 \\ \hline 20 \\ 20 \\ \hline 0 \end{array}$$

23.
$$\begin{array}{r} 124 \\ 3\overline{)372} \\ 3 \\ \hline 07 \\ 6 \\ \hline 12 \\ 12 \\ \hline 0 \end{array}$$

24.
$$\begin{array}{r} 70\ R\ 1 \\ 7\overline{)491} \\ 49 \\ \hline 01 \\ 0 \\ \hline 1 \end{array}$$

25. $8 \times N = 120$

$$\begin{array}{r} 15 \\ 8\overline{)120} \\ 8 \\ \hline 40 \\ 40 \\ \hline 0 \end{array}$$

$N = 15$

26. $3 \times 3 = 9$
$F \times 9 = 108$

$$\begin{array}{r} 12 \\ 9\overline{)108} \\ 9 \\ \hline 18 \\ 18 \\ \hline 0 \end{array}$$

$F = 12$

27. $36 + N = 54$

$$\begin{array}{r} \overset{4}{\cancel{5}}{}^{1}4 \\ -\ 3\ 6 \\ \hline 1\ 8 \end{array}$$

$N = \mathbf{18}$

28. Perimeter: 8 units + 4 units + 8 units
+ 4 units = **24 units**
Area: 8 units × 4 units = **32 square units**

29. Possibilities include:
- **Reflection and translation**
- **Reflection only, if the reflection is made about a vertical line drawn halfway between the triangles**

30. Since the first four multiples of 18 are 18, 36, 54, and 72, the first four multiples of 180 are **180, 360, 540,** and **720.**

LESSON 79, WARM-UP

a. **6**

b. **29**

c. **15**

d. **47**

LESSON 79, ACTIVITY

1.

2.

3.

4.

5.

6.

7.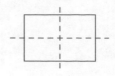

8.

9. no lines of symmetry

10.

11.

12.

13. A line of symmetry is a line that divides a figure into two halves that are mirror images of each other. (Drawings will vary.)

LESSON 79, LESSON PRACTICE

a.

b. None

c.

d.

e.

f. None

LESSON 79, MIXED PRACTICE

1. 6 hr + 9 hr + 2 hr + 1 hr + 3 hr
   + 2 hr + 1 hr = **24 hours**

2. 1 hr out of 24 hours was spent watching TV.

$$\frac{1}{24}$$

3. Counting forward 9 hours from 9:30 p.m. makes
it **6:30 a.m.**

4. Studying and sleeping combined take up more
than half of the graph.

C. Studying and sleeping

5.
$$\begin{array}{r} \textbf{12 eggs} \\ 5\overline{)60} \text{ eggs} \\ \underline{5} \\ 10 \\ \underline{10} \\ 0 \end{array}$$

6. 2250 is less than 2500, so it is less than halfway
between 2000 and 3000.

A

7. 427 rounds to 400, 533 rounds to 500, and 764
rounds to 800.

400 + 500 + 800 = **1700**

8. Seven out of ten circles are not shaded.

$$\frac{7}{10}$$

9. Pattern:

Number of groups $\times$ number in each group

= total

Problem: 30 boxes $\times$ 42 oranges in each box

= *N* oranges

$$\begin{array}{r} 42 \\ \times\ \ 30 \\ \hline \textbf{1260 oranges} \end{array}$$

10. Pattern:

Number of groups $\times$ number in each group
= total

Problem: *N* boxes $\times$ 5 apples in each box
= 145 apples

$$\begin{array}{r} \textbf{29 boxes} \\ 5\overline{)145} \text{ boxes} \\ \underline{10} \\ 45 \\ \underline{45} \\ 0 \end{array}$$

11. 5 in. + 5 in. + 5 in. + 5 in. = **20 in.**

12. **25 squares**

13. ⊓

14.
$$\begin{array}{r} {}^{1}2\!\!\!/\,{}^{9}0.\,\cancel{1}{}^{1}0 \\ -\ \$1\,6.\,4\,5 \\ \hline \$\ \ 3.\,6\,5 \end{array}$$

15.
$$\begin{array}{r} {}^{1}\ {}^{1} \\ \$98.54 \\ +\ \ \ \$9.85 \\ \hline \$108.39 \end{array}$$

16.
$$\begin{array}{r} {}^{3} \\ 380 \\ \times\ \ \ \ 4 \\ \hline 1520 \end{array}$$

17.
$$\begin{array}{r} {}^{5} \\ 97 \\ \times\ \ \ 80 \\ \hline 7760 \end{array}$$

18.
$$\begin{array}{r} \textbf{768} \\ 5\overline{)3840} \\ \underline{35} \\ 34 \\ \underline{30} \\ 40 \\ \underline{40} \\ 0 \end{array}$$

19.
$$\begin{array}{r} {}^{4}\ {}^{2} \\ \$8.63 \\ \times\ \ \ \ 7 \\ \hline \$60.41 \end{array}$$

20.
$$\begin{array}{r} \overset{3}{\cancel{4}}.{}^{1}25 \\ -\ 2.4 \\ \hline \textbf{1.85} \end{array}$$

21.
$$\begin{array}{r} \textbf{\$8.75} \\ 8)\overline{\$70.00} \\ \underline{64} \\ 6\,0 \\ \underline{5\,6} \\ 40 \\ \underline{40} \\ 0 \end{array}$$

22.
$$\begin{array}{r} \textbf{632 R 3} \\ 6)\overline{3795} \\ \underline{36} \\ 19 \\ \underline{18} \\ 15 \\ \underline{12} \\ 3 \end{array}$$

23. $4 \times 40 = 160$
$$P = \textbf{40}$$

24. $\sqrt{64} = 8$
$\sqrt{16} = 4$
$8 \div 4 = \textbf{2}$

25.
$$\begin{array}{r} \textbf{41} \\ 7)\overline{287} \\ \underline{28} \\ 07 \\ \underline{7} \\ 0 \end{array}$$

26. $6 \times 6 = 36$
$2 \times 2 \times 2 = 8$

$$\begin{array}{r} \overset{1}{3}6 \\ +\ 8 \\ \hline 44 \end{array}$$

$10 \times 44 = \textbf{440}$

27. (a)
$$\begin{array}{r} \overset{2}{1}.5\ \text{cm} \\ 0.8\ \text{cm} \\ 1.5\ \text{cm} \\ +\ 0.8\ \text{cm} \\ \hline \textbf{4.6 cm} \end{array}$$

(b) $4.6\ \text{cm} \times 10 = \textbf{46 mm}$

28. Count down from 0 or up from −5.
−3

29. One complete spin is 360°.
Two complete spins are

$$\begin{array}{r} \overset{1}{3}60° \\ \times\ \ \ 2 \\ \hline \textbf{720°} \end{array}$$

30. U

LESSON **80**, WARM-UP

a. 35

b. 28

c. 34

d. 82

e. 2400

f. **\$4.15**

Patterns

Counting by fives from two gives us this
sequence: 2, 7, 12, 17, 22, 27, …
The terms of the sequence end with either **2 or 7.**

Counting by fives from three gives us this
sequence: 3, 8, 13, 18, 23, 28, …
The terms of the sequence end with either **3 or 8.**

Counting by fives from four gives us this
sequence: 4, 9, 14, 19, 24, 29, …
The terms of the sequence end with either **4 or 9.**

LESSON **80**, LESSON PRACTICE

a. **Divide, multiply, subtract, bring down;**

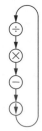

b.
$$\begin{array}{r} 203 \text{ R } 3 \\ 4\overline{)815} \\ \underline{8} \\ 01 \\ \underline{0} \\ 15 \\ \underline{12} \\ 3 \end{array}$$

c.
$$\begin{array}{r} 830 \text{ R } 2 \\ 5\overline{)4152} \\ \underline{40} \\ 15 \\ \underline{15} \\ 02 \\ \underline{0} \\ 2 \end{array}$$

d.
$$\begin{array}{r} 905 \text{ R } 2 \\ 6\overline{)5432} \\ \underline{54} \\ 03 \\ \underline{0} \\ 32 \\ \underline{30} \\ 2 \end{array}$$

e.
$$\begin{array}{r} 120 \text{ R } 5 \\ 7\overline{)845} \\ \underline{7} \\ 14 \\ \underline{14} \\ 05 \\ \underline{0} \\ 5 \end{array}$$

f. **300**

g. **500**

LESSON 80, MIXED PRACTICE

1. The chance of rain is less than 50%, so the chance **that it will not rain** is more likely.

2. Pattern:
$$\begin{array}{r} \text{Larger} \\ - \text{ Smaller} \\ \hline \text{Difference} \end{array}$$
Problem:
$$\begin{array}{r} N \text{ seconds} \\ - 58 \text{ seconds} \\ \hline 12 \text{ seconds} \end{array}$$
$$\begin{array}{r} \overset{1}{12} \text{ seconds} \\ + 58 \text{ seconds} \\ \hline \textbf{70 seconds} \end{array}$$

3. $100\% \div 5 = \textbf{20\%}$

4. $72 \div 6 = \textbf{12 sticks}$

5. Four out of every five leaves had not fallen.
$$\frac{4}{5}$$

6. 5263 is less than 5500, so it is less than halfway between 5000 and 6000.
A

7. Seven out of twelve months have 31 days.
$$\frac{7}{12}$$

8. **1000**

9. 393 rounds to 400, 589 rounds to 600, and 241 rounds to 200.
$400 + 600 + 200 = \textbf{1200}$

10. $2 \text{ cm} = 20 \text{ mm}$
$20 \text{ mm} + 20 \text{ mm} + 20 \text{ mm} = \textbf{60 mm}$

11. $1 \text{ L} = 1000 \text{ mL}$
$3 \text{ L} = \textbf{3000 mL}$

12. Pattern:
Number of groups $\times$ number in each group
$\qquad\qquad\qquad = \text{total}$
Problem:
N hours $\times$ 5 miles per hour
$\qquad\qquad\qquad = 40 \text{ miles}$
$\qquad\quad 40 \div 5 = \textbf{8 hours}$

13. $2 \times N = 150$
$$\begin{array}{r} 75 \\ 2\overline{)150} \\ \underline{14} \\ 10 \\ \underline{10} \\ 0 \end{array}$$
$N = \textbf{75}$

14.
$$\begin{array}{r} \overset{1}{6}.2 \\ + 4.8 \\ \hline 11.0 \end{array}$$
$$\begin{array}{r} 24.25 \\ - 11.0 \\ \hline \textbf{13.25} \end{array}$$

15.
$$\begin{array}{r} {}^{111}{}^{1} \\ 103,279 \\ +97,814 \\ \hline \mathbf{201,093} \end{array}$$

16.
$$\begin{array}{r} {}^{11} \\ \$36.14 \\ +\$27.95 \\ \hline \mathbf{\$64.09} \end{array}$$

17.
$$\begin{array}{r} {}^{2}{}^{8}{}^{1} \\ \cancel{3}\,\cancel{9},{}^{1}4\,\cancel{2}{}^{1}0 \\ -2\,9,5\,1\,6 \\ \hline \mathbf{9,9\,0\,4} \end{array}$$

18.
$$\begin{array}{r} \$60.50 \\ -N \\ \hline \$43.20 \end{array}$$

$$\begin{array}{r} {}^{5} \\ \$\cancel{6}{}^{1}0.\,5\,0 \\ -\$4\,3.\,2\,0 \\ \hline \$1\,7.\,3\,0 \end{array}$$

$$N = \mathbf{\$17.30}$$

19.
$$\begin{array}{r} {}^{3} \\ 604 \\ \times9 \\ \hline \mathbf{5436} \end{array}$$

20.
$$\begin{array}{r} {}^{4} \\ 87 \\ \times60 \\ \hline \mathbf{5220} \end{array}$$

21.
$$\begin{array}{r} {}^{3}{}^{2} \\ \$6.75 \\ \times4 \\ \hline \mathbf{\$27.00} \end{array}$$

22.
$$\begin{array}{r} \mathbf{206} \\ 3\overline{)618} \\ \underline{6} \\ 01 \\ \underline{0} \\ 18 \\ \underline{18} \\ 0 \end{array}$$

23.
$$\begin{array}{r} \mathbf{\$4.30} \\ 5\overline{)\$21.50} \\ \underline{20} \\ 1\,5 \\ \underline{1\,5} \\ 00 \\ \underline{0} \\ 0 \end{array}$$

24.
$$\begin{array}{r} N \\ +1467 \\ \hline 2459 \end{array}$$

$$\begin{array}{r} {}^{1}{}^{1}3 \\ 2\,\cancel{4}{}^{1}5\,9 \\ -1\,4\,6\,7 \\ \hline 9\,9\,2 \end{array}$$

$$N = \mathbf{992}$$

25.
$$\begin{array}{r} \mathbf{150} \\ 4\overline{)600} \\ \underline{4} \\ 20 \\ \underline{20} \\ 00 \\ \underline{0} \\ 0 \end{array}$$

26.
$$\begin{array}{r} \mathbf{90\ R\ 3} \\ 6\overline{)543} \\ \underline{54} \\ 03 \\ \underline{0} \\ 3 \end{array}$$

27.
$$\begin{array}{r} \mathbf{59} \\ 8\overline{)472} \\ \underline{40} \\ 72 \\ \underline{72} \\ 0 \end{array}$$

28.
$$9 \times 9 = 81$$
$$9 \times 2 = 18$$

$$\begin{array}{r} 81 \\ +18 \\ \hline 99 \end{array}$$

$$9 \times W = 99$$
$$9 \times 11 = 99$$
$$W = \mathbf{11}$$

29. **600**

30.

INVESTIGATION 8

1. Pay Schedule

Hours Worked	Total Pay
1	$10
2	$20
3	$30
4	$40
5	$50
6	$60
7	$70
8	$80

2.

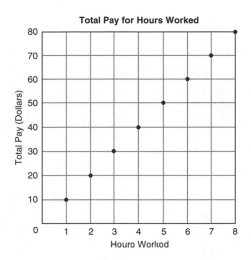

Total Pay for Hours Worked

3. 20 Questions

Number of Correct Answers	Score
1	5%
2	10%
3	15%
4	20%
5	25%
6	30%
7	35%
8	40%
9	45%
10	50%
11	55%
12	60%
13	65%
14	70%
15	75%
16	80%
17	85%
18	90%
19	95%
20	100%

4. **90%**

5. **15 answers**

6. **(6, 8)**

7. **(8, 3)**

Activity

1.

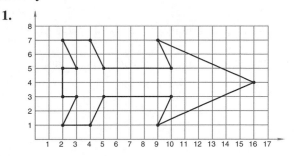

2. See student work.

LESSON 81, WARM-UP

a. **84**

b. **91**

c. **72**

d. **15,000**

e. **$1.85**

Problem Solving

$AB = 1\frac{1}{2}$ inches; $BC = 1$ inch; Segment AB is $\frac{1}{2}$ inch longer than segment BC.

LESSON 81, LESSON PRACTICE

a. $45° +$ about $15° =$ **about 60°**

b. $90° -$ about $5° =$ **about 85°**

c. $135° -$ about $15° =$ **about 120°**

d. **About 45°**

e. **90°**

LESSON 81, MIXED PRACTICE

1. Pattern: Some
             + Some more
               Total

   Problem: ¹
              27 times
            + 33 times
              60 times

2. Pattern:
   Number of groups × number in each group
                  = total
   Problem: 3 lb × $0.68 per lb = N dollars

             ²
            $0.68
          ×   3
          $2.04

3.   N
        + 55
        ─────
         113

          ⁰ ¹0
         X̷ X̷'3
        − 5 5
        ─────
          5 8

4.   **28 volunteers**
   3)84̄
     6
     ──
     24
     24
     ──
      0

5. 2250 is closer to 2000 than 3000, so it rounds to
   2000.

6. Five out of seven letters are not A's.
   $\frac{5}{7}$

7. 1 ton = 2000 lb

       2000 lb
     ×   7
     ─────────
     14,000 lb

8. **15 mm**

9. (a) **400 g**

   (b) **10 L**

10. **February 26, 2019**

11.   ³ ¹
      4̷ 2̷,'7 0 0
    − 3 4, 9 0 0
    ─────────────
      7 8 0 0

12. Perimeter: 10 units + 5 units + 10 units
                    + 5 units = **30 units**

    Area: 10 units × 5 units = **50 square units**

13. **West**

14.   ⁴
        507
      ×   6
      ──────
        3042

        6743
      − 3042
      ──────
      3701

15.   ⁶⁹ ⁹
       $7̷0̷.0̷'0
     − $63.1 7
     ─────────
       $6.8 3

16. 3 × 7 × 0 = **0**

17.   ³
         $8.15
       ×   6
       ───────
       $48.90

18. 67¢ × 10 = **$6.70**

19.   ¹ ¹
       4.5
       0.52
     + 1.39
     ───────
       6.41

20.   **$6.08**
     2)$12.16̄
       12
       ──
       0 1
        0
       ──
       16
       16
       ──
        0

21.   **720 R 1**
     6)4321̄
       42
       ──
       12
       12
       ──
       01
        0
       ──
        1

22.

$$
\begin{array}{r}
\mathbf{600} \\
8\overline{)4800} \\
\underline{48} \\
00 \\
\underline{00} \\
00 \\
\underline{00} \\
0
\end{array}
$$

23. $\sqrt{9} = 3$

$$
\begin{array}{r}
\mathbf{321} \\
3\overline{)963} \\
\underline{9} \\
06 \\
\underline{6} \\
03 \\
\underline{3} \\
0
\end{array}
$$

24. $5^3 \div 5$
$5 \times 5 \times 5 = 125$
$125 \div 5 = \mathbf{25}$

25.

$$
\begin{array}{r}
\mathbf{\$0.73} \\
9\overline{)\$6.57} \\
\underline{6\,3} \\
27 \\
\underline{27} \\
0
\end{array}
$$

26. $200 = 4 \times B$

$$
\begin{array}{r}
50 \\
4\overline{)200} \\
\underline{20} \\
00 \\
\underline{0} \\
0
\end{array}
$$

$B = \mathbf{50}$

27. $D \times 7 = 105$

$$
\begin{array}{r}
15 \\
7\overline{)105} \\
\underline{7} \\
35 \\
\underline{35} \\
0
\end{array}
$$

$D = \mathbf{15}$

28.

$$
\begin{array}{r}
\overset{1}{4}73 \\
+\ 286 \\
\hline
759
\end{array}
$$

$$
\begin{array}{r}
759 \\
+\ \ N \\
\hline
943
\end{array}
$$

$$
\begin{array}{r}
\overset{8}{9}\overset{13}{4}3 \\
-\ 7\,5\,9 \\
\hline
1\,8\,4
\end{array}
$$

$N = \mathbf{184}$

29.

$$
\begin{array}{r}
\overset{2}{1}\overset{1}{} \\
12 \\
3 \\
14 \\
5 \\
+\ 26 \\
\hline
61
\end{array}
$$

30.

$$
\begin{array}{r}
\overset{2}{} 2 \\
33 \\
4 \\
25 \\
6 \\
+\ 27 \\
\hline
97
\end{array}
$$

LESSON 82, WARM-UP

a. 27

b. 25

c. 26

d. 92

e. 3650

f. $9.50

Patterns

15   21   28

SOLUTIONS

LESSON 82, LESSON PRACTICE

a. See student work.

b. No

LESSON 82, MIXED PRACTICE

1. Pattern:
$$\begin{array}{r} \text{Larger} \\ -\ \text{Smaller} \\ \hline \text{Difference} \end{array}$$

Problem:
$$\begin{array}{r} \overset{2}{\cancel{3}}{}^{1}5 \text{ members} \\ -\ 2\ 8 \text{ books} \\ \hline \textbf{7 books} \end{array}$$

2. Pattern:
Number of groups $\times$ number in each group
$$=\ \text{total}$$
Problem: 7 children $\times$ 11 times per child
$$=\ \textbf{77 times}$$

3.
$$1 \text{ yd } = 3 \text{ ft}$$
$$60 \text{ ft} \div 3 = \textbf{20 yd}$$

4. Pattern:
$$\begin{array}{r} \text{Some} \\ +\ \text{Some more} \\ \hline \text{Total} \end{array}$$

Problem:
$$\begin{array}{r} \overset{2}{4}40 \text{ yd} \\ {}_{2}880 \text{ yd} \\ 1320 \text{ yd} \\ +\ 1760 \text{ yd} \\ \hline \textbf{4400 yd} \end{array}$$

5. Two out of every three members did not vote no.
$$\mathbf{\frac{2}{3}}$$

6. 6821 rounds to 7000 and 4963 rounds to 5000
$$\textbf{7000 + 5000 = 12,000}$$

7. Saturday and Sunday start with the letter S.
Two out of seven days start with S.
$$\mathbf{\frac{2}{7}}$$

8.
$$1 \text{ kg } = 1000 \text{ g}$$
$$500 \text{ g} + 500 \text{ g} = 1000 \text{ g}$$
About 500 g

9. **About 25 lb**

10. $35 \text{ lb} - 25 \text{ lb} = 10 \text{ lb}$
About 10 lb

11.

Juan's Growth	
Age	Weight
At birth	6 pounds
1 year	20 pounds
2 years	25 pounds
3 years	30 pounds
4 years	35 pounds
5 years	40 pounds

12. $100\% - 65\% = \mathbf{35\%}$

13.
$$\begin{array}{r} \overset{1\ 1}{\$60.75} \\ +\ \$95.75 \\ \hline \mathbf{\$156.50} \end{array}$$

14.
$$\begin{array}{r} \$1\overset{5}{6}.\overset{9}{\cancel{0}}{}^{1}0 \\ -\ \$1\ 5.\ 4\ 3 \\ \hline \mathbf{\$0.\ 5\ 7} \end{array}$$

15.
$$\begin{array}{r} 3.15 \\ -\ 3.12 \\ \hline \mathbf{0.03} \end{array}$$

16.
$$\begin{array}{r} 320 \\ \times\ 30 \\ \hline \mathbf{9600} \end{array}$$

17.
$$\begin{array}{r} \overset{4\ 3}{465} \\ \times\ 7 \\ \hline \mathbf{3255} \end{array}$$

18.
$$\begin{array}{r} \overset{5\ 4}{\$0.98} \\ \times\ 6 \\ \hline \mathbf{\$5.88} \end{array}$$

19.
$$\begin{array}{r} \mathbf{70\ R\ 5} \\ 6\overline{)425} \\ \underline{42} \\ 05 \\ \underline{0} \\ 5 \end{array}$$

20.
$$\begin{array}{r} \mathbf{\$0.75} \\ 8\overline{)\$6.00} \\ \underline{5\ 6} \\ 40 \\ \underline{40} \\ 0 \end{array}$$

21. $5)\overline{625}$

$$\frac{125}{5)\overline{625}}$$
$\frac{5}{12}$
$\frac{10}{25}$
$\frac{25}{0}$

22. $3)\overline{150}$

$\frac{50}{3)\overline{150}}$
$\frac{15}{00}$
$\frac{0}{0}$

$R = \mathbf{50}$

23.
$10 \times 10 = 100$
$100 + T = 150$
$150 - 100 = 50$
$T = \mathbf{50}$

24. $29 + N = 37$

$\overset{2}{\cancel{3}}{}^{1}7$
$- \ 2\ 9$
$\overline{\ \ \ 8}$

$N = \mathbf{8}$

25. **9 1-inch squares**

26.
1.6 cm
1.2 cm
+ 2.0 cm
4.8 cm

27. **T**

28.

29. **D.**

30. A diameter is equal to two radii.

1.2 cm
+ 1.2 cm
2.4 cm

a. 27

b. 17

c. 43

d. 33

e. 49

f. 39

_____

LESSON 83, LESSON PRACTICE

a.
$\overset{1}{}$
$\$2.24$
$\times \quad\quad 3$
$\overline{\$6.72}$

$\overset{1}{}$
$\$6.72$
$+ \ \$0.34$
$\overline{\mathbf{\$7.06}}$

b.
$\overset{1\ \ 1}{}$
$\$6.95$
$+ \ \$0.49$
$\overline{\$7.44}$

$\$\overset{9}{\cancel{1}}0.\ \overset{9}{\cancel{0}}{}^{1}0$
$- \ \ \$7.\ 4\ 4$
$\overline{\mathbf{\$2.\ 5\ 6}}$

_____

LESSON 83, MIXED PRACTICE

1. Pattern:
Number of groups $\times$ number in each group
$=$ total
Problem: 30 bags $\times$ 320 coins in each bag
$= N$ coins

320
$\times \quad\ 30$
$\overline{\mathbf{9600\ coins}}$

2. Counting forward 3 hours from 11:10 a.m. makes it **2:10 p.m.**

3. Pattern: Some
$$\underline{-\;\text{Some went away}}$$
What is left

Problem: $\overset{1}{\cancel{2}}\,\overset{\overset{1}{0}}{\cancel{1}}2$ pages
$$\underline{-\;1\;3\;5}\text{ pages}$$
7 7 pages

4. $\overset{\textbf{14 points}}{3\overline{)42}\text{ points}}$
$$\underline{3}$$
$$12$$
$$\underline{12}$$
$$0$$

5. 4286 is closer to 4000 than 5000, so it rounds to **4000**

6. $\overset{1\;\;1}{\$16.98}$
$$\underline{+\;\;\$1.02}$$
$$\$18.00$$

$\$20.00\;-\;\$18.00\;=\;\textbf{\$2.00}$

7. Seven out of 34 letters are I's.
$$\frac{\textbf{7}}{\textbf{34}}$$

8. 6 games $+$ 1 game $=$ **7 games**

9. 24 points $-$ 20 points $=$ **4 points**

10. 12 games $-$ 2 games $=$ **10 games**

11. $4\times5=20$
$20\times3=60$
$60\;\bigcirc\!\!\!=\;60$

12. $M-137=257$

$$\overset{1}{257}$$
$$\underline{+\;137}$$
$$394$$

$M=\textbf{394}$

13. $N+137=257$

$$257$$
$$\underline{-\;137}$$
$$120$$

$N=\textbf{120}$

14. $\overset{2\;\;1}{1.45}$
$$2.4$$
$$0.56$$
$$\underline{+\;7.6}$$
$$\textbf{12.01}$$

15. 3.12
$$\underline{+\;0.5}$$
$$3.62$$

$$5.75$$
$$\underline{-\;3.62}$$
$$\textbf{2.13}$$

16. $\overset{1\;4}{638}$
$$\underline{\times\;\;\;\;50}$$
$$\textbf{31,900}$$

17. $\overset{6\,1}{472}$
$$\underline{\times\;\;\;\;9}$$
$$\textbf{4248}$$

18. $\overset{5}{\$6.09}$
$$\underline{\times\;\;\;\;\;6}$$
$$\textbf{\$36.54}$$

19. $\overset{\textbf{307}}{3\overline{)921}}$
$$\underline{9}$$
$$02$$
$$\underline{0}$$
$$21$$
$$\underline{21}$$
$$0$$

20. $\overset{\textbf{135 R 3}}{5\overline{)678}}$
$$\underline{5}$$
$$17$$
$$\underline{15}$$
$$28$$
$$\underline{25}$$
$$3$$

21. $\overset{\textbf{600}}{4\overline{)2400}}$
$$\underline{24}$$
$$00$$
$$\underline{00}$$
$$00$$
$$\underline{00}$$
$$0$$

22.
$$\begin{array}{r} \$2.52 \\ 5)\overline{\$12.60} \\ \underline{10} \\ 2\,6 \\ \underline{2\,5} \\ 10 \\ \underline{10} \\ 0 \end{array}$$

23.
$$\begin{array}{r} \$2.39 \\ 6)\overline{\$14.34} \\ \underline{12} \\ 2\,3 \\ \underline{1\,8} \\ 54 \\ \underline{54} \\ 0 \end{array}$$

24.
$$\begin{array}{r} \$5.75 \\ 8)\overline{\$46.00} \\ \underline{40} \\ 6\,0 \\ \underline{5\,6} \\ 40 \\ \underline{40} \\ 0 \end{array}$$

25. $9 \times 9 = 81$
$81 = 9 \times 9$
$N = \mathbf{9}$

26. $10 \times 10 = 100$
$5 \times 100 = 500$
$W = \mathbf{100}$

27. One fourth equals **25%**.

28. $2\,\text{cm} + 1\,\text{cm} + 2\,\text{cm} + 1\,\text{cm} = 6\,\text{cm}$
$1\,\text{cm} = 10\,\text{mm}$
$6\,\text{cm} = \mathbf{60\ mm}$

29. 4 cm

  [ ] 2 cm

$4\,\text{cm} + 2\,\text{cm} + 4\,\text{cm} + 2\,\text{cm} = \mathbf{12\ cm}$

30. One spin is 360°.
Three spins are

$$\begin{array}{r} \overset{1}{360°} \\ \times 3 \\ \hline \mathbf{1080°} \end{array}$$

LESSON 84, WARM-UP

a. 51

b. 72

c. 63

d. 16,000

e. $7.25

f. 33

Problem Solving

The darkly shaded portion is less than $\frac{1}{2}$ (50%) of the circle, and it is greater than $\frac{1}{4}$ (25%) of the circle. The darkly shaded portion makes up **about 40%** of the circle.

LESSON 84, LESSON PRACTICE

a. **0.425**

b. **3.875**

c. **0.035**

d. **2.007**

e. $\dfrac{214}{1000}$
Two hundred fourteen thousandths

f. $4\dfrac{321}{1000}$
Four and three hundred twenty-one thousandths

g. $\dfrac{25}{1000}$
Twenty-five thousandths

h. $5\dfrac{12}{1000}$
Five and twelve thousandths

i. $\dfrac{3}{1000}$

Three thousandths

j. $9\dfrac{999}{1000}$

Nine and nine hundred ninety-nine thousandths

LESSON 84, MIXED PRACTICE

1. January = 31 days
    February = 28 days
    March = 31 days

    $$\begin{array}{r} \overset{1}{3}1 \text{ days} \\ 28 \text{ days} \\ + \ 31 \text{ days} \\ \hline \textbf{90 days} \end{array}$$

2. 1 pair of pants + 2 shirts = 3 pieces of clothing
    for each child

    Pattern:
    Number of groups × number in each group
    = total

    Problem:
    12 children × 3 pieces of clothing for each child
    = **36 pieces of clothing**

3. Pattern: Larger
    $\underline{-\ \text{Smaller}}$
    Difference

    Problem: 18 chin-ups
    $\underline{-\ N\ \text{chin-ups}}$
    7 chin-ups

    18 chin-ups − 7 chin-ups = **11 chin-ups**

4. Pattern:
    Number of groups × number in each group
    = total
    Problem: 8 gallons × N miles per gallon
    = 200 miles

    $$\begin{array}{r} \textbf{25 miles} \\ 8\overline{)200} \\ \underline{16} \\ 40 \\ \underline{40} \\ 0 \end{array}$$

5. $$\begin{array}{r} \overset{1}{\$}8.95 \\ +\ \$0.54 \\ \hline \$9.49 \end{array}$$

    $$\begin{array}{r} \$2\overset{1}{0}.\overset{9}{0}\overset{9}{0}^{1}0 \\ -\ \ \$9.4\ 9 \\ \hline \$10.5\ 1 \end{array}$$

6. |||| ||||

7. 1 cm = 10 mm
    An octagon has eight sides.
    8 × 10 mm = **80 mm**

8. $\dfrac{1}{3}$ were cat's-eyes. $\left\{\begin{array}{|c|}\hline 6 \text{ marbles} \\\hline\end{array}\right.$ 18 marbles

    $\dfrac{2}{3}$ were not cat's-eyes. $\left\{\begin{array}{|c|}\hline 6 \text{ marbles} \\\hline 6 \text{ marbles} \\\hline\end{array}\right.$

    6 marbles

9. Pattern:
    Number of groups × number in each group
    = total
    Problem: 6 days × 46 peaches per day
    = N peaches

    $$\begin{array}{r} \overset{3}{4}6 \\ \times\ \ \ 6 \\ \hline \textbf{276 peaches} \end{array}$$

10. Pattern:
    Number of groups × number in each group
    = total
    Problem: 7 days × N peaches per day
    = 3640 peaches

    $$\begin{array}{r} \textbf{520 peaches} \\ 7\overline{)3640} \\ \underline{35} \\ 14 \\ \underline{14} \\ 00 \\ \underline{0} \\ 0 \end{array}$$

11. Jack did 129 out of 1000 push-ups with one arm.
    $$\dfrac{129}{1000}$$

12. **0.129**
    One hundred twenty-nine thousandths

13.
$$\overset{\scriptscriptstyle 1}{2.3}$$
$$+\ 1.75$$
$$\overline{4.05}$$

$$4.56$$
$$-\ 4.05$$
$$\mathbf{0.51}$$

14. $\sqrt{36} = 6$
$7 \times 8 = 56$
$6 + N = 56$
$56 - 6 = 50$
$\qquad N = \mathbf{50}$

15. $3 \times 3 = 9$
$3 \times 6 = 18$

$$\overset{\scriptscriptstyle 7}{18}$$
$$\times \quad 9$$
$$\overline{\mathbf{162}}$$

16. $\sqrt{9} = 3$

$$\overset{\scriptscriptstyle 1}{462}$$
$$\times \quad 3$$
$$\overline{\mathbf{1386}}$$

17. $7 \times 7 = 49$
$\sqrt{49} = 7$

$$49$$
$$-\ 7$$
$$\overline{\mathbf{42}}$$

18.
$$\overset{\scriptscriptstyle 3}{36}$$
$$\times \quad 50$$
$$\overline{\mathbf{1800}}$$

19.
$$\overset{\scriptscriptstyle 5\ 4}{\$4.76}$$
$$\times \qquad 7$$
$$\overline{\mathbf{\$33.32}}$$

20.
$$30$$
$$+\ N$$
$$\overline{47}$$

$$47$$
$$-\ 30$$
$$\overline{17}$$
$$N = \mathbf{17}$$

21.
$$\begin{array}{r} \mathbf{131} \\ 4\overline{)524} \\ \underline{4} \\ 12 \\ \underline{12} \\ 04 \\ \underline{4} \\ 0 \end{array}$$

22. **700**

23.
$$\begin{array}{r} \mathbf{\$5.26} \\ 5\overline{)\$26.30} \\ \underline{25} \\ 1\ 3 \\ \underline{1\ 0} \\ 30 \\ \underline{30} \\ 0 \end{array}$$

24.
$$\begin{array}{r} \mathbf{\$1.85} \\ 2\overline{)\$3.70} \\ \underline{2} \\ 1\ 7 \\ \underline{1\ 6} \\ 10 \\ \underline{10} \\ 0 \end{array}$$

25.
$$\begin{array}{r} \mathbf{262} \\ 3\overline{)786} \\ \underline{6} \\ 18 \\ \underline{18} \\ 06 \\ \underline{6} \\ 0 \end{array}$$

26.
$$\begin{array}{r} \mathbf{700\ R\ 2} \\ 7\overline{)4902} \\ \underline{49} \\ 00 \\ \underline{0} \\ 02 \\ \underline{0} \\ 2 \end{array}$$

27. $\dfrac{\mathbf{321}}{\mathbf{1000}}$

28. Perimeter: 3 yd + 3 yd + 3 yd
$\qquad\qquad\qquad$ + 3 yd = **12 yd**

$\quad$ Area: $\quad$ 3 yd $\times$ 3 yd = **9 sq. yd**

29. Possibilities include:
- **Rotation and translation**
- **Rotation and reflection**

30. Angle *C*

LESSON 85, WARM-UP

a. 26

b. 47

c. 39

d. 750

e. $9.53

f. 28

Patterns

$$\overset{+3}{\frown}\ \overset{+5}{\frown}\ \overset{+7}{\frown}\ \overset{+9}{\frown}\ \overset{+11}{\frown}$$
1,   4,   9,   16,   **25,**   **36,** …

$$\overset{+2}{\frown}\ \overset{+3}{\frown}\ \overset{+4}{\frown}\ \overset{+5}{\frown}\ \overset{+6}{\frown}\ \overset{+7}{\frown}\ \overset{+8}{\frown}$$
1,   3,   6,   10,   15,   21,   **28,**   **36,** …

LESSON 85, LESSON PRACTICE

a. $365 \times 10 = \textbf{3650}$

b. $52 \times 100 = \textbf{5200}$

c. $7 \times 1000 = \textbf{7000}$

d. $3.60 \times 10 = \textbf{\$36.00}$

e. $420 \times 100 = \textbf{42,000}$

f. $2.50 \times 1000 = \textbf{\$2500.00}$

LESSON 85, MIXED PRACTICE

1. $4 \times 2 = 8$
Maria

2. 10 books − 6 books = **4 books**

---

3. Pattern:
Number of groups $\times$ number in each group
$\qquad = $ total
Problem: 9 books $\times$ 160 pages in each book
$\qquad = N$ pages

$$\begin{array}{r} \overset{5}{1}60 \\ \times\quad 9 \\ \hline \textbf{1440 pages} \end{array}$$

4. A pentagon has five sides.
100 sides ÷ 5 sides = **20 pentagons**

5. 3 miles $\times$ 2 miles = 6 square miles
50% means one half
Half of 6 square miles is **3 square miles**

6. 24¢ $\times$ 10 = 240¢ or $2.40
$2.40 + $0.14 = **$2.54**

7. B. 2 liters

8. One possibility:

Right triangle

9.

	48 gems
$\frac{1}{4}$ were rubies.	12 gems
	12 gems
$\frac{3}{4}$ were not rubies.	12 gems
	12 gems

12 gems

10. Three fourths of the gems were not rubies. One fourth = 25%, so three fourths = 3 $\times$ 25%, which is **75%**.

11. 81 out of 1000 fans were pleased with the outcome.
$$\frac{\textbf{81}}{\textbf{1000}}$$

12. 0.081; eighty-one thousandths

13.
$$\begin{array}{r} 3.68 \\ + 10.2 \\ \hline 13.88 \end{array}$$

$$\begin{array}{r} \overset{459}{4\cancel{6}.0^{1}1} \\ - 1\,3.8\,8 \\ \hline \textbf{3 2.1 3} \end{array}$$

166

14. $728 + C = 1205$

$$
\begin{array}{r}
{}^{11}\;{}^{9}\\
\cancel{12}\,\cancel{0}\,{}^{1}5\\
-\;\;\;7\,2\,8\\
\hline
4\,7\,7
\end{array}
$$

$C = \textbf{477}$

15. $36 \times 10 = \textbf{360}$

16. $100 \times 42 = \textbf{4200}$

17. $\$2.75 \times 1000 = \textbf{\$2750.00}$

18.
$$
\begin{array}{r}
{}^{2}\;\;\;\\
\$3.17\\
\times\quad\;\;4\\
\hline
\$12.68
\end{array}
$$

19.
$$
\begin{array}{r}
{}^{3}\;\;\;\\
206\\
\times\quad5\\
\hline
1030
\end{array}
$$

20.
$$
\begin{array}{r}
{}^{2}\;\;\;\\
37\\
\times\;40\\
\hline
1480
\end{array}
$$

21.
$$
\begin{array}{r}
164\\
3\overline{)492}\\
\underline{3}\quad\;\;\\
19\;\;\\
\underline{18}\;\;\\
12\\
\underline{12}\\
0
\end{array}
$$

22.
$$
\begin{array}{r}
172\\
5\overline{)860}\\
\underline{5}\quad\;\;\\
36\;\;\\
\underline{35}\;\;\\
10\\
\underline{10}\\
0
\end{array}
$$

23.
$$
\begin{array}{r}
\$1.55\\
6\overline{)\$9.30}\\
\underline{6}\quad\;\;\\
3\,3\;\;\\
\underline{3\,0}\;\;\\
30\\
\underline{30}\\
0
\end{array}
$$

24. $2 \times 2 \times 2 = 8$

$$
\begin{array}{r}
21\\
8\overline{)168}\\
\underline{16}\;\;\\
08\\
\underline{8}\\
0
\end{array}
$$

25.
$$
\begin{array}{r}
\$2.50\\
8\overline{)\$20.00}\\
\underline{16}\quad\;\;\;\\
4\,0\;\;\\
\underline{4\,0}\;\;\\
00\\
\underline{0}\\
0
\end{array}
$$

26. $\sqrt{16} = 4$

$1600 \div 4 = \textbf{400}$

27. Perimeter: 10 ft $+ 6$ ft $+ 10$ ft
$+ 6$ ft $= \textbf{32 ft}$

Area: 10 ft $\times 6$ ft $= \textbf{60 sq. ft}$

28. H

29. Angle D

30. C. ⬠

LESSON 86, WARM-UP

a. 68

b. 64

c. 91

d. 3000

e. \$0.62

f. 93

Problem Solving

$AB = 1\frac{1}{4}$ inches; $BC = 1\frac{1}{2}$ inches;

Segment BC is $\frac{1}{4}$ inch longer than segment AB.

LESSON 86, LESSON PRACTICE

a. $70 \times 80 = $ **5600**

b. $40 \times 50 = $ **2000**

c. $40 \times \$6.00 = $ **\$240.00**

d. $30 \times 800 = $ **24,000**

LESSON 86, MIXED PRACTICE

1. Counting back 20 minutes from 8:10 a.m. makes it **7:50 a.m.**

2. Pattern:
Some
$-$ Some went away
What is left

Problem:
$$\begin{array}{r} 125 \text{ lb} \\ - \quad N \text{ lb} \\ \hline 118 \text{ lb} \end{array}$$

$$\begin{array}{r} 1\ \overset{1}{\cancel{2}}\,5 \text{ lb} \\ -\ 1\ 1\ 8 \text{ lb} \\ \hline 7 \text{ lb} \end{array}$$

3.
$$\begin{array}{r} \overset{2\ 3}{\$2.89} \\ \$0.89 \\ \$0.79 \\ + \ \$0.28 \\ \hline \$4.85 \end{array}$$

$$\begin{array}{r} \$\overset{49}{\cancel{5}}.\cancel{0}^{1}0 \\ -\ \$4.8\ 5 \\ \hline \$0.1\ 5 \text{ or } \textbf{15¢} \end{array}$$

4. **Sunday**

5. 卌 卌 卌 ||

6. 3782 is closer to 4000 than 3000, so it rounds to **4000.**

7.
$$1 \text{ ton} = 2000 \text{ lb}$$
$$2000 \text{ lb} \times 2 \text{ tons} = \textbf{4000 lb}$$

8.

$\frac{1}{5}$ pintos	45 horses
	9 horses
$\frac{4}{5}$ not pintos	9 horses
	9 horses
	9 horses
	9 horses

9 horses

9. $100\% \div 5 = $ **20%**

10. 23,650 is between 20,000 and 30,000. Because 23,560 is less than 25,000, it is closer to 20,000 than to 30,000.
Point B

11. (a) $\dfrac{1}{10}$

(b) $\dfrac{1}{100}$

(c) $\dfrac{1}{1000}$

12.
$$\begin{array}{r} \overset{1\,1\ \ 1}{\$36.47} \\ + \quad \$9.68 \\ \hline \$46.15 \end{array}$$

13.
$$\begin{array}{r} \$\overset{29\ \ 9}{\cancel{3}\cancel{0}}.\cancel{0}^{1}0 \\ -\ \$13.4\ 5 \\ \hline \$16.5\ 5 \end{array}$$

14.
$$\begin{array}{r} 6 \\ \overset{3}{8} \\ 17 \\ 23 \\ 110 \\ 25 \\ + \ 104 \\ \hline 293 \end{array}$$

15.
$$\begin{array}{r} \overset{5\,4}{476} \\ \times \quad 7 \\ \hline 3332 \end{array}$$

16.
$$\begin{array}{r} \overset{2}{804} \\ \times \quad 5 \\ \hline 4020 \end{array}$$

17.
$$\begin{array}{r} 7.43 \\ -\ 2.1 \\ \hline 5.33 \end{array}$$

$$\begin{array}{r} {}^{0}\!\!\not{1}^{1}2.\,6\,5 \\ -\ \ \ 5.\,3\,3 \\ \hline \mathbf{7.\,3\,2} \end{array}$$

18.
$$5 \times 5 = 25$$
$$10 \times 10 = 100$$
$$25 + 25 + N = 100$$
$$50 + N = 100$$
$$100 - 50 = 50$$
$$N = \mathbf{50}$$

19. (a) **Two and one tenth**

(b) **Two and one tenth**

20. $100 \times 23¢ = \mathbf{\$23.00}$

21. $60 \times 30 = \mathbf{1800}$

22. $70 \times \$2.00 = \mathbf{\$140.00}$

23.
$$\begin{array}{r} \mathbf{\$2.09} \\ 3\overline{)\$6.27} \\ \underline{6} \\ 0\ 2 \\ \underline{0} \\ 27 \\ \underline{27} \\ 0 \end{array}$$

24.
$$\begin{array}{r} \mathbf{117\ R\ 1} \\ 7\overline{)820} \\ \underline{7} \\ 12 \\ \underline{7} \\ 50 \\ \underline{49} \\ 1 \end{array}$$

25.
$$\begin{array}{r} \mathbf{55\ R\ 3} \\ 6\overline{)333} \\ \underline{30} \\ 33 \\ \underline{30} \\ 3 \end{array}$$

26. $\sqrt{25} = 5$

$$\begin{array}{r} \mathbf{125} \\ 5\overline{)625} \\ \underline{5} \\ 12 \\ \underline{10} \\ 25 \\ \underline{25} \\ 0 \end{array}$$

27. $2 \times 2 \times 2 = 8$
$$4000 \div 8 = \mathbf{500}$$

28.
$$\begin{array}{r} \mathbf{685} \\ 2\overline{)1370} \\ \underline{12} \\ 17 \\ \underline{16} \\ 10 \\ \underline{10} \\ 0 \end{array}$$

29. Perimeter: $10\,m + 10\,m + 10\,m + 10\,m = \mathbf{40\,m}$
Area: $10\,m \times 10\,m = \mathbf{100\ sq.\ m}$

30. The 8-sided polygon is an **octagon.** The 4-sided polygon is a **quadrilateral (or square).**

LESSON 87, WARM-UP

a. $2.50

b. 8000

c. 93

d. 2500

e. $9.63

f. 47

Problem Solving

$$\$12.95 + \$1.10 = \$14.05$$
$$\$15.00 - \$14.05 = \$0.95$$
$$\begin{array}{r} 3Q = 75¢ \\ +\ 2D = 20¢ \\ \hline 95¢\ or\ \$0.95 \end{array}$$

Sandra received 3 quarters and 2 dimes in change.

LESSON 87, LESSON PRACTICE

a.
$$\begin{array}{r} 32 \\ \times\ 23 \\ \hline 96 \\ 640 \\ \hline \mathbf{736} \end{array}$$

b.
$$\begin{array}{r} 23 \\ \times\ 32 \\ \hline 46 \\ 690 \\ \hline \mathbf{736} \end{array}$$

c.
$$\begin{array}{r} 43 \\ \times\ 12 \\ \hline 86 \\ 430 \\ \hline \mathbf{516} \end{array}$$

d.
$$\begin{array}{r} 34 \\ \times\ 21 \\ \hline 34 \\ 680 \\ \hline \mathbf{714} \end{array}$$

e.
$$\begin{array}{r} 32 \\ \times\ 32 \\ \hline 64 \\ 960 \\ \hline \mathbf{1024} \end{array}$$

f.
$$\begin{array}{r} 22 \\ \times\ 14 \\ \hline 88 \\ 220 \\ \hline \mathbf{308} \end{array}$$

g.
$$\begin{array}{r} 13 \\ \times\ 32 \\ \hline 26 \\ 390 \\ \hline \mathbf{416} \end{array}$$

h.
$$\begin{array}{r} 33 \\ \times\ 33 \\ \hline 99 \\ 990 \\ \hline \mathbf{1089} \end{array}$$

LESSON 87, MIXED PRACTICE

1. 2 mi + 4 mi + 4 mi + 2 mi = **12 mi**

2. Step 1: Counting forward 30 minutes from 3:30 p.m. makes it 4:00 p.m.
Step 2: Counting forward 1 hour from 4:00 p.m. makes it **5:00 p.m.**

3. $F \times 2 = 8$ fish
4 fish $\times\ 2 = 8$ fish

4.
$$\begin{array}{r} \overset{1}{\ }\$12.97 \\ +\ \ \$0.91 \\ \hline \$13.88 \end{array}$$

$$\begin{array}{r} \$2\overset{19}{\cancel{0}}.\overset{9}{\cancel{0}}{}^{1}0 \\ -\ \$13.8\ 8 \\ \hline \mathbf{\$6.1\ 2} \end{array}$$

5. 4876 rounds to 5000 and 3149 rounds to 3000
5000 + 3000 = **8000**

6. 14

7. A pentagon has 5 sides.
20 cm × 5 = **100 cm**

8. $3\frac{1}{2}$ **in.**

9.
$\frac{1}{2}$ were on the field. $\left\{\begin{array}{|c|}\hline 18\ \text{players} \\ \hline 9\ \text{players} \\ \hline\end{array}\right.$
$\frac{1}{2}$ were off the field. $\left\{\begin{array}{|c|}\hline 9\ \text{players} \\ \hline\end{array}\right.$

9 players

10. There are 100 pennies in a dollar, so a penny is $\frac{1}{100}$ of a dollar.

11. $\frac{1}{10} = \mathbf{10\%}$

12. 283 miles out of 1000 miles were through the desert.
$$\frac{\mathbf{283}}{\mathbf{1000}}$$

13. 0.283
Two hundred eighty-three thousandths

14.
$$\begin{array}{r} 31 \\ \times\ 21 \\ \hline 31 \\ 620 \\ \hline \mathbf{651} \end{array}$$

15.
$$\begin{array}{r} 32 \\ \times\ 31 \\ \hline 32 \\ 960 \\ \hline \mathbf{992} \end{array}$$

16.
$$\begin{array}{r} 13 \\ \times\ 32 \\ \hline 26 \\ 390 \\ \hline \mathbf{416} \end{array}$$

17.
$$\begin{array}{r} 11 \\ \times\ 11 \\ \hline 11 \\ 110 \\ \hline \mathbf{121} \end{array}$$

18.
$$\begin{array}{r} 12 \\ \times\ 14 \\ \hline 48 \\ 120 \\ \hline \mathbf{168} \end{array}$$

19. $30 \times 800 = \mathbf{24,000}$

20.
$$\begin{array}{r} \mathbf{142\ R\ 6} \\ 7\overline{)1000} \\ \underline{7} \\ 30 \\ \underline{28} \\ 20 \\ \underline{14} \\ 6 \end{array}$$

21.
$$\begin{array}{r} \mathbf{159} \\ 3\overline{)477} \\ \underline{3} \\ 17 \\ \underline{15} \\ 27 \\ \underline{27} \\ 0 \end{array}$$

22.
$$\begin{array}{r} \mathbf{507} \\ 5\overline{)2535} \\ \underline{25} \\ 03 \\ \underline{0} \\ 35 \\ \underline{35} \\ 0 \end{array}$$

23.
$$\begin{array}{r} \mathbf{\$7.20} \\ 9\overline{)\$64.80} \\ \underline{63} \\ 1\ 8 \\ \underline{1\ 8} \\ 00 \\ \underline{0} \\ 0 \end{array}$$

24.
$$\begin{array}{r} \mathbf{179} \\ 4\overline{)716} \\ \underline{4} \\ 31 \\ \underline{28} \\ 36 \\ \underline{36} \\ 0 \end{array}$$

25.
$$\begin{array}{r} \mathbf{44} \\ 8\overline{)352} \\ \underline{32} \\ 32 \\ \underline{32} \\ 0 \end{array}$$

26. Perimeter: 20 in. + 10 in. + 20 in.

+ 10 in. = **60 in.**

Area: 20 in. × 10 in. = **200 sq. in.**

27.

2 cm   2 cm

2 cm

28. 2 cm + 2 cm + 2 cm = 6 cm
6 cm = **60 mm**

29. Side *CD* (or side *DC*)

30. Angle *D*

LESSON 88, WARM-UP

a. 250

b. 2500

c. 25,000

d. $3.25

e. 93

f. 18

Problem Solving

The darkly shaded portion is greater than $\frac{3}{4}$ (75%) of the circle. It makes up **about 80%** of the circle.

_____

LESSON 88, LESSON PRACTICE

a.
$$5\overline{)32} \quad 6\text{ R }2$$
$$\underline{30}$$
$$2$$

6 cars can be filled with 5 employees each with 2 employees left over.

b. Another car is needed to carry the 2 remaining employees, so **7 cars** are needed.

c.
$$4\overline{)31} \quad 7\text{ R }3$$
$$\underline{28}$$
$$3$$

He made **7 stacks** of 4 quarters each.

d. There are **3 quarters** left over.

e. 7 stacks + 1 short stack = **8 stacks**

_____

LESSON 88, MIXED PRACTICE

1. (a)
$$6\overline{)100} \quad 16\text{ R }4$$
$$\underline{6}$$
$$40$$
$$\underline{36}$$
$$4$$

16 packages can be filled with 6 table-tennis balls each.

(b) There are **4 table-tennis balls** left over.

2. Pattern:
$$\begin{array}{r} \text{Larger} \\ - \text{ Smaller} \\ \hline \text{Difference} \end{array}$$

Problem:
$$\begin{array}{r} \overset{2}{\cancel{3}}\,\overset{1\!1}{\cancel{2}}1 \\ - 1\,2\,3 \\ \hline \mathbf{1\,9\,8} \end{array}$$

3.
$$\begin{array}{r} \overset{3}{\$0.59} \\ \times \quad 4 \\ \hline \$2.36 \end{array}$$

$$\begin{array}{r} \overset{1}{\$2.36} \\ + \ \$0.16 \\ \hline \mathbf{\$2.52} \end{array}$$

4. 12 in. = 1 ft
   24 in. = **2 ft**

5. (a) **40 mm**

   (b) 40 mm = **4 cm**

6. Step 1: Counting forward 20 minutes from 7:03 a.m. makes it 7:23 a.m.
   Step 2: Counting forward 5 hours from 7:23 a.m. makes it **12:23 p.m.**

7. **7000 + 500 + 20 + 8**
   Seven thousand, five hundred twenty-eight

8.
	25 members
$\frac{1}{5}$ missed the note.	5 members
$\frac{4}{5}$ did not miss the note.	5 members
	5 members
	5 members
	5 members

5 band members

9. $\frac{1}{5}$ = **20%**

10.
$$\begin{array}{r} \overset{1\ \ 1}{\$_1 6.35} \\ \$14.25 \\ \$0.97 \\ + \ \$5.00 \\ \hline \mathbf{\$26.57} \end{array}$$

11.

$$\overset{1}{1}.4$$
$$+\ 2.75$$
$$\overline{4.15}$$

$$4.\overset{5}{\cancel{6}}{}^1 0$$
$$-\ 4.1\ 5$$
$$\overline{\mathbf{0.\ 4\ 5}}$$

12.

$$\$0.46$$
$$+\ \$1.30$$
$$\overline{\$1.76}$$

$$\$\overset{9}{\cancel{1}}0.\overset{9}{\cancel{0}}{}^1 0$$
$$-\ \ \$1.7\ 6$$
$$\overline{\mathbf{\$8.\ 2\ 4}}$$

13. $28 \times 1000 = \mathbf{28{,}000}$

14.

$$13$$
$$\times\ 13$$
$$\overline{39}$$
$$130$$
$$\overline{\mathbf{169}}$$

15.

$$12$$
$$\times\ 11$$
$$\overline{12}$$
$$120$$
$$\overline{\mathbf{132}}$$

16.

$$\overset{6\ \ 6}{\$8.67}$$
$$\times\ \ \ \ \ 9$$
$$\overline{\mathbf{\$78.03}}$$

17.

$$31$$
$$\times\ 31$$
$$\overline{31}$$
$$930$$
$$\overline{\mathbf{961}}$$

18.

$$12$$
$$\times\ 31$$
$$\overline{12}$$
$$360$$
$$\overline{\mathbf{372}}$$

19.

$$\begin{array}{r} \mathbf{506} \\ 7\overline{)3542} \\ \underline{35}\ \ \ \ \\ 04 \\ \underline{\ 0}\ \ \\ 42 \\ \underline{42} \\ 0 \end{array}$$

20.

$$\begin{array}{r} \mathbf{\$5.50} \\ 6\overline{)\$33.00} \\ \underline{30}\ \ \ \ \ \\ 3\ 0 \\ \underline{3\ 0}\ \\ 00 \\ \underline{\ 0} \\ 0 \end{array}$$

21.

$$\begin{array}{r} \mathbf{620\ R\ 5} \\ 8\overline{)4965} \\ \underline{48}\ \ \ \ \\ 16 \\ \underline{16}\ \ \\ 05 \\ \underline{\ 0}\ \\ 5 \end{array}$$

22.

$$\begin{array}{r} \mathbf{96\ R\ 2} \\ 5\overline{)482} \\ \underline{45}\ \ \\ 32 \\ \underline{30} \\ 2 \end{array}$$

23. $2700 \div 9 = \mathbf{300}$

24.

$$\sqrt{9} = 3$$
$$2700 \div 3 = \mathbf{900}$$

25.

$$7 \times 7 = 49$$
$$14 + N = 49$$

$$49$$
$$-\ 14$$
$$\overline{35}$$

$$N = \mathbf{35}$$

26.

$$6 \times 6 = 36$$
$$3 \times N = 36$$
$$3 \times 12 = 36$$
$$N = \mathbf{12}$$

27. **One possibility:**

28. 40 ft $\times$ 30 ft $=$ 1200 square feet, which is **1200 floor tiles**

29. **Side *BC* (or side *CB*)**

30. 18 answers $-$ 15 answers $=$ **3 answers**

LESSON 89, WARM-UP

a. 5

b. 20

c. 24

d. 32

e. 43

f. $6.53

g. 112

h. 48

Patterns

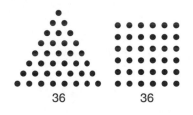

    36   36

LESSON 89, LESSON PRACTICE

a.

b.

c.

LESSON 89, MIXED PRACTICE

1. (a)
$$\begin{array}{r} 8\ R\ 1 \\ 4\overline{)33} \\ \underline{32} \\ 1 \end{array}$$

Because there was one extra player, **3 teams** out of the 4 teams had exactly 8 players.

(b) **1 team** had 9 (8 + 1) players.

2. Pattern:
$$\begin{array}{r} \text{Some} \\ +\ \text{Some more} \\ \hline \text{Total} \end{array}$$

Problem:
$$\begin{array}{r} \overset{1\ \ 1}{\$0.37} \\ \$0.37 \\ \$0.20 \\ \$0.20 \\ +\ \$0.15 \\ \hline \mathbf{\$1.29} \end{array}$$

3. 2 weeks = 14 days
Pattern:
Number of groups × number in each group
= total
Problem: 14 days × 20 pages per day
= N pages

$$\begin{array}{r} 14 \\ \times\ \ 20 \\ \hline \mathbf{280\ pages} \end{array}$$

4.   3 ft = 1 yard
27 ft ÷ 3 ft = **9 yd**

5. 20 mm + 20 mm + 30 mm = 70 mm
                10 mm = 1 cm
                70 mm = **7 cm**

6. 18

7. A. **2 milliliters**

8. $87 = 87 \times N$
$87 = 87 \times 1$
  $N = \mathbf{1}$

9.
$\frac{1}{3}$ finished early. $\left\{\vphantom{\begin{array}{c}a\end{array}}\right.$

24 volunteers
8 volunteers

$\frac{2}{3}$ did not finish early. $\left\{\vphantom{\begin{array}{c}a\\a\end{array}}\right.$

8 volunteers
8 volunteers

8 volunteers

10. **25%**

11.
$$\begin{array}{r} \overset{1\ 1\ 1\ \ 1}{\$478.63} \\ +\ \ \ \$32.47 \\ \hline \mathbf{\$511.10} \end{array}$$

12.
$$\begin{array}{r} 1\ \overset{2}{\cancel{3}}\ \overset{'6}{\cancel{7}},\overset{}{\cancel{1}}\ \overset{3}{\cancel{4}}{}^{1}0 \\ -\ 1\ 2\ 9,5\ 3\ 6 \\ \hline \mathbf{7\ 6\ 0\ 4} \end{array}$$

13.
$$\begin{array}{r} {}^{59}\;{}^{9} \\ \$\cancel{6}0.\,\cancel{0}{}^{1}0 \\ -\ \$24.\,3\,8 \\ \hline \$35.\,6\,2 \end{array}$$

14. $70 \times 90 = \mathbf{6300}$

15.
$$\begin{array}{r} 11 \\ \times\ 13 \\ \hline 33 \\ 110 \\ \hline \mathbf{143} \end{array}$$

16.
$$\begin{array}{r} 12 \\ \times\ 12 \\ \hline 24 \\ 120 \\ \hline \mathbf{144} \end{array}$$

17.
$$\begin{array}{r} {}^{6}\;{}^{4} \\ \$4.76 \\ \times\qquad 8 \\ \hline \$38.08 \end{array}$$

18.
$$\begin{array}{r} 21 \\ \times\ 13 \\ \hline 63 \\ 210 \\ \hline \mathbf{273} \end{array}$$

19.
$$\begin{array}{r} 21 \\ \times\ 21 \\ \hline 21 \\ 420 \\ \hline \mathbf{441} \end{array}$$

20.
$$\begin{array}{r} \mathbf{750} \\ 4)\overline{3000} \\ 28 \\ \hline 20 \\ 20 \\ \hline 00 \\ 0 \\ \hline 0 \end{array}$$

21.
$$\begin{array}{r} \mathbf{127} \\ 5)\overline{635} \\ 5 \\ \hline 13 \\ 10 \\ \hline 35 \\ 35 \\ \hline 0 \end{array}$$

22.
$$\begin{array}{r} \mathbf{60\ R\ 6} \\ 7)\overline{426} \\ 42 \\ \hline 06 \\ 0 \\ \hline 6 \end{array}$$

23.
$$\begin{array}{r} \mathbf{451\ R\ 6} \\ 8)\overline{3614} \\ 32 \\ \hline 41 \\ 40 \\ \hline 14 \\ 8 \\ \hline 6 \end{array}$$

24.
$$\begin{array}{r} \mathbf{456} \\ 6)\overline{2736} \\ 24 \\ \hline 33 \\ 30 \\ \hline 36 \\ 36 \\ \hline 0 \end{array}$$

25.
$$\begin{array}{r} \mathbf{\$2.50} \\ 4)\overline{\$10.00} \\ 8 \\ \hline 2\,0 \\ 2\,0 \\ \hline 00 \\ 0 \\ \hline 0 \end{array}$$

26.

27.

5 cm

4 cm

28. Perimeter: $5\,\text{cm} + 4\,\text{cm} + 5\,\text{cm}$

$+\ 4\,\text{cm} = \mathbf{18\ cm}$

Area: $5\,\text{cm} \times 4\,\text{cm} = \mathbf{20\ sq.\ cm}$

29. **Side *AD* (or side *DA*)**

30. Multiples of 4:
$1 \times 4 = 4$
$2 \times 4 = 8$
$3 \times 4 = \underline{12}$
$4 \times 4 = 16$

Multiples of 6:
$1 \times 6 = 6$
$2 \times 6 = \underline{12}$
$3 \times 6 = 18$

12

---

c.
$$\begin{array}{r} \overset{2}{\overset{2}{84}} \\ \times\ 67 \\ \hline 588 \\ 5040 \\ \hline \mathbf{5628} \end{array}$$

d.
$$\begin{array}{r} \overset{2}{\overset{4}{65}} \\ \times\ 48 \\ \hline 520 \\ 2600 \\ \hline \mathbf{3120} \end{array}$$

LESSON 90, WARM-UP

a. **10**

b. **12**

c. **25**

d. **23**

e. **60**

f. **$1.33**

g. **91**

h. **17**

Problem Solving

1 ton is 2000 pounds.
$\frac{1}{2}$ of 2000 is 1000.
$1000 \div 100 = 10$
The truck can carry **10 sacks** of cement.

LESSON 90, LESSON PRACTICE

a.
$$\begin{array}{r} \overset{1}{\overset{4}{38}} \\ \times\ 26 \\ \hline 228 \\ 760 \\ \hline \mathbf{988} \end{array}$$

b.
$$\begin{array}{r} \overset{7}{\overset{1}{49}} \\ \times\ 82 \\ \hline 98 \\ 3920 \\ \hline \mathbf{4018} \end{array}$$

LESSON 90, MIXED PRACTICE

1. **4 baskets**

2. 7 baskets $\times$ 2 points per basket = **14 points**

3. Game 3: 6 baskets $\times$ 2 points per basket
$= 12$ points

14 points $-$ 12 points = **2 points**

4.
$$\begin{array}{r} \mathbf{\$0.48} \textbf{ per pound} \\ 3\overline{)\$1.44} \\ \underline{1\ 2} \\ 24 \\ \underline{24} \\ 0 \end{array}$$

5. **About 360°**

6.
$$\begin{array}{r} \overset{1\ 1\ \ 1}{\$47.99} \\ +\ \ \$2.88 \\ \hline \$50.87 \end{array}$$

$$\begin{array}{r} \overset{59\ \ \ 9}{\$6\cancel{0}.\cancel{0}^{1}0} \\ -\ \$50.8\ 7 \\ \hline \mathbf{\$9.1\ 3} \end{array}$$

7. 1 ft = 12 in.
A square has 4 sides of equal length.
12 in. $\div$ 4 = **3 in.**

8. 1 kilogram = 1000 grams
If a dollar bill weighs about a gram,
one kilogram equals **about 1000 dollar bills.**

9.

$\frac{1}{4}$ had red noses. $\begin{cases} \end{cases}$

$\frac{3}{4}$ did not have red noses. $\begin{cases} \end{cases}$

64 clowns

16 clowns
16 clowns
16 clowns
16 clowns

16 clowns

10. One fourth = 25%,
so three fourths = 3 × 25%, which is **75%**

11. Counting forward 7 hours from 9:30 a.m. makes it **4:30 p.m.**

12. Pattern:
Number of groups × number in each group
= total
Problem:
42 containers × 8 big fish in each container
= N big fish

$$\begin{array}{r} \overset{1}{42} \\ \times \ \ 8 \\ \hline \textbf{336 big fish} \end{array}$$

13. Pattern:
Number of groups × number in each group
= total
Problem:
N horses × 4 horseshoes for each horse
= 88 horseshoes
88 ÷ 4 = **22 horses**

14. Possibilities include:
- **Reflection and translation**

15. D. Isosceles

16.

$$\begin{array}{r} 0.5 \\ + \ 0.12 \\ \hline 0.62 \end{array}$$

$$\begin{array}{r} 0.625 \\ - \ \ 0.62 \\ \hline \textbf{0.005} \end{array}$$

17. 47 × 1000 = **4700**

18.

$$\begin{array}{r} \overset{1\ 3}{328} \\ \times \ \ \ 4 \\ \hline \textbf{1312} \end{array}$$

19.

$$\begin{array}{r} 43 \\ \times \ 32 \\ \hline 86 \\ 1290 \\ \hline \textbf{1376} \end{array}$$

20.

$$\begin{array}{r} \overset{1}{\overset{2}{25}} \\ \times \ 35 \\ \hline 125 \\ 750 \\ \hline \textbf{875} \end{array}$$

21.

$$\begin{array}{r} \textbf{863 R 2} \\ 5\overline{)4317} \\ \underline{40} \\ 31 \\ \underline{30} \\ 17 \\ \underline{15} \\ 2 \end{array}$$

22.

$$\begin{array}{r} \textbf{\$5.00} \\ 8\overline{)\$40.00} \end{array}$$

23.

$$\begin{array}{r} \textbf{660 R 3} \\ 6\overline{)3963} \\ \underline{36} \\ 36 \\ \underline{36} \\ 03 \\ \underline{0} \\ 3 \end{array}$$

24.

$$\begin{array}{r} \textbf{142} \\ 3\overline{)426} \\ \underline{3} \\ 12 \\ \underline{12} \\ 06 \\ \underline{6} \\ 0 \end{array}$$

25.

$$\begin{array}{r} \textbf{631} \\ 4\overline{)2524} \\ \underline{24} \\ 12 \\ \underline{12} \\ 04 \\ \underline{4} \\ 0 \end{array}$$

26. 60 × 700 = **42,000**

27. ⊘ ⊘ ⊘

28.
```
      4
     2 3
     27
     35
   +  8
   ─────
     77
```

$$77 + N = 112$$

```
   0  0
   1̸ 1̸ 1 2
 −   7 7
 ─────────
     3 5
```

$$N = \mathbf{35}$$

29. **1.8 cm**

30. **1.8 + 1.7 = 3.5**

───────────────

INVESTIGATION 9

1. **4 quarters;** ⊕ = ◯

2. ◓ = ◓

3. **2 pieces;** ◹ = ◸

4. **4 pieces;** ◓ = ◓

5. $\frac{1}{4} + \frac{1}{4} + \frac{1}{4} + \frac{1}{4} = 1$

6. $\frac{1}{2} + \frac{1}{4} + \frac{1}{4} = 1$

7. $\frac{1}{2} + \frac{1}{4} + \frac{1}{8} + \frac{1}{8} = 1$

8. $\frac{1}{2} = \frac{2}{4};$ ◗ = ◔

9. $\frac{1}{2} > \frac{3}{8};$ ◗ > ◕

10. $\frac{1}{4} < \frac{3}{8};$ ◸ < ◕

11. $\frac{3}{4} > \frac{4}{8};$ ◕ > ◕

12. $\frac{2}{8} = \frac{1}{4}$

13. $\frac{4}{8} = \frac{1}{2}$

14. $\frac{6}{8} = \frac{3}{4}$

15. ◹ + ◖ = ◕ ; $\frac{3}{4}$

16. ◹ + ◖ = ◕ ; $\frac{5}{8}$

17. ◕ + ◖ = ✳ ; $\frac{7}{8}$

18. ◕ − ◖ = ◸ ; $\frac{1}{4}$

19. ◕ − ◿ = ◕ ; $\frac{3}{8}$

20. ◐ − ◓ = ◡ ; $\frac{1}{2}$

21. $\frac{1}{2} = \mathbf{50\%}$

22. $\frac{1}{4} = \mathbf{25\%}$

23. $\frac{3}{4} = \mathbf{75\%}$

24. $\frac{1}{2} = \mathbf{0.5}$

25. $\frac{1}{4} = \mathbf{0.25}$

26. $\frac{1}{8} = \mathbf{0.125}$

27. **0.125 < 0.25**

28. ◓ ; **0.25 + 0.25 = 0.5 (or 0.50)**

29. **0.50 = 0.5**

30. ◓ ;
0.125 + 0.125 + 0.125 + 0.125
= 0.5 (or 0.500)

31. **0.500 = 0.5**

32. ◕ ; **0.25 + 0.25 + 0.25 = 0.75;**

◖ ; **0.5 + 0.25 = 0.75**

33. $\quad 1 - \frac{1}{4} = \frac{3}{4}$
$\quad\quad\mathbf{1 - 0.25 = 0.75}$

34. $\quad \frac{1}{2} - \frac{1}{8} = \frac{3}{8}$
$\quad\quad\mathbf{0.5 - 0.125 = 0.375}$

35. $\quad \frac{3}{4} - \frac{1}{2} = \frac{1}{4}$
$\quad\quad\mathbf{0.75 - 0.5 = 0.25}$

LESSON 91, WARM-UP

a. **60**

b. **120**

c. **2400**

d. **$2.50**

e. **295**

f. **219**

g. **XII**

h. **7**

Problem Solving
2 gallons = 4 half gallons
4 half gallons = 8 quarts
8 quarts = 16 pints
16 pints = **32 cups**

LESSON 91, LESSON PRACTICE

a. The hundredths place is the second place to the right of the decimal point.
7

b. **3** is in the same place as 2.

c. **4** is in the third place to the right of the decimal point, which is the **thousandths** place.

d. **A. 12.34 and B. 12.340** are equal because both numbers have the same digits in the same places.

e. 3.25
32.50
3.25 $\lessdot$ 32.50

f. 3.250
3.25
3.250 $\textcircled{=}$ 3.25

g.
$$\begin{array}{r} 1\,2.\,3\,\overset{3}{\cancel{4}}{}^{1}0 \\ -\quad 1.\,2\,3\,4 \\ \hline 1\,1.\,1\,0\,6 \end{array}$$

h.
$$\begin{array}{r} 1.\,\overset{1}{\cancel{2}}{}^{1}0 \\ -\quad 0.\,1\,2 \\ \hline 1.\,0\,8 \end{array}$$

LESSON 91, MIXED PRACTICE

1.
$$\begin{array}{r} {}^{1}{}^{1} \\ \$0.75 \\ \$0.40 \\ \$0.10 \\ +\ \$0.07 \\ \hline \mathbf{\$1.32} \end{array}$$

2. (a)
$$\begin{array}{r} 9\text{ R }1 \\ 4\overline{)37} \\ 36 \\ \hline 1 \end{array}$$
3 stacks had exactly 9 books.

(b) There is 1 book left over, so **1 stack** had 10 books.

3.
$$\begin{array}{r} {}^{0}{}^{9} \\ \$\cancel{1}.\,\cancel{0}{}^{1}0 \\ -\ \$0.\,5\,2 \\ \hline \$0.\,4\,8 \end{array}$$

$$\begin{array}{r} \$0.48 \\ -\ \$0.03 \\ \hline \$0.45 \text{ or } \mathbf{45¢} \end{array}$$

4.
$$\begin{array}{r} {}^{1} \\ 12\text{ words} \\ \times\quad 5 \\ \hline \mathbf{60\text{ words}} \end{array}$$

5. 5456 rounds to 5000 and 2872 rounds to 3000
5000 + 3000 = 8000

6. ⵑⵑⵑⵑⵑ ⵑⵑⵑⵑⵑ

7. (a) 7 out of 10 parts are shaded

$$\frac{7}{10}$$

(b) Seven tenths = **0.7**

8.

$\frac{1}{6}$ are broken.

$\frac{5}{6}$ are not broken.

48 crayons

8 crayons
8 crayons
8 crayons
8 crayons
8 crayons
8 crayons

8 crayons

9.

$$\begin{array}{r} 32 \text{ mm} \\ + \ 26 \text{ mm} \\ \hline 58 \text{ mm} \end{array}$$

$$\begin{array}{r} \overset{8}{\cancel{9}}{}^{1}1 \text{ mm} \\ - \ 5 \ 8 \text{ mm} \\ \hline \mathbf{3 \ 3 \ mm} \end{array}$$

10. The hundredths place is the second place to the right of the decimal point.

2

11. 1 quart = 2 pints

1 pint = about 1 pound

1 quart = **about 2 pounds**

12.

$$\begin{array}{r} \overset{3}{\cancel{4}}.\overset{12}{\cancel{3}}\ \overset{11}{\cancel{2}}0 \\ - \ 0. \ 4 \ 3 \ 2 \\ \hline \mathbf{3. \ 8 \ 8 \ 8} \end{array}$$

13.

$$5 \times 5 = 25$$
$$\sqrt{25} = 5$$
$$25 + 5 + N = 30$$
$$30 + N = 30$$
$$N = \mathbf{0}$$

14.

$$\begin{array}{r} \overset{6}{}\$6.08 \\ \times 8 \\ \hline \mathbf{\$48.64} \end{array}$$

15.

$$\begin{array}{r} \overset{1}{\underset{2}{}} \\ 47 \\ \times \ 24 \\ \hline 188 \\ 940 \\ \hline \mathbf{1128} \end{array}$$

16.

$$\begin{array}{r} \overset{3}{\underset{1}{}} \\ 36 \\ \times \ 62 \\ \hline 72 \\ 2160 \\ \hline \mathbf{2232} \end{array}$$

17.

$$\begin{array}{r} 53 \\ \times 30 \\ \hline \mathbf{1590} \end{array}$$

18.

$$\begin{array}{r} \overset{2}{}63 \\ \times \ 37 \\ \hline 441 \\ 1890 \\ \hline \mathbf{2331} \end{array}$$

19. Multiply mentally: **3200**

20.

$$\begin{array}{r} 864 \\ 4\overline{)3456} \\ 32 \\ \hline 25 \\ 24 \\ \hline 16 \\ 16 \\ \hline 0 \end{array}$$

21.

$$\begin{array}{r} 864 \\ 8\overline{)6912} \\ 64 \\ \hline 51 \\ 48 \\ \hline 32 \\ 32 \\ \hline 0 \end{array}$$

22.

$$\begin{array}{r} \mathbf{\$7.20} \\ 7\overline{)\$50.40} \\ 49 \\ \hline 1 \ 4 \\ 1 \ 4 \\ \hline 00 \\ 0 \\ \hline 0 \end{array}$$

23.

24.

4 cm

25. One possibility:

4 cm

4 cm × 4 cm = 16 sq. cm
Half of 16 sq. cm is **8 sq. cm**

26. Twenty-one thousandths is 21 out of 1000 parts

$\frac{21}{1000}$; **0.021**

27. **D.**

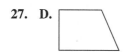

28. A quadrilateral has four sides.
A.

29. One full turn of the clock is 360°.
One half turn is **180°**.

30. 4.2
4.200
4.2 ⊜ 4.200

LESSON 92, WARM-UP

a. 90

b. 220

c. 1300

d. $16.54

e. 19

f. 403

g. XVI

h. 11

Patterns

1, 2, 4, 8, **16, 32, 64,** ...

Saxon Math 5/4—Homeschool

LESSON 92, LESSON PRACTICE

a. There are two pairs of parallel lines. There are four right angles. All sides are equal.
Parallelogram, rhombus, rectangle, square

b. There are two pairs of parallel lines.
Parallelogram

c. There are two pairs of parallel lines. There are four right angles.
Parallelogram, rectangle

d. There is one pair of parallel lines.
Trapezoid

LESSON 92, MIXED PRACTICE

1. 12 tuna sandwiches + 10 bologna sandwiches
+ 8 roast beef sandwiches
= **30 sandwiches**

2. 14 friends + 1 person (Mary) = 15 people
15 people × **2 sandwiches** per person
= 30 sandwiches

3. 12 tuna sandwiches
× 2 halves in each sandwich
24 halves

4. Pattern:
Number of groups × number in each group
= total
Problem: 5 pounds × N dollars per pound
= $2.95

$$\begin{array}{r} \textbf{\$0.59 per pound} \\ 5\overline{)\$2.95} \\ \underline{2\ 5} \\ 45 \\ \underline{45} \\ 0 \end{array}$$

5. A hexagon has six sides
4 in. × 6 = 24 in.
12 in. = 1 ft
24 in. = **2 ft**

6. Pattern: Larger
   − Smaller
   Difference

Problem: $\overset{8}{\cancel{9}},{}^{1}4\,0\,0,0\,0\,0$
   $-\ 2,7\,0\,0,0\,0\,0$
   $\overline{6,7\,0\,0,0\,0\,0}$

7. (a) 7 out of 100 parts are shaded
   $$\frac{7}{100}$$

   (b) Seven hundredths = **0.07**

8. **Seven thousand, five hundred seventy-two and one eighth**

9.

$\frac{1}{5}$ lost wheels. $\Bigl\{$

$\frac{4}{5}$ did not lose wheels. $\Biggl\{$

80 chariots
16 chariots
16 chariots
16 chariots
16 chariots
16 chariots

16 chariots

10. One fifth of the chariots lost wheels.
   One fifth = **20%**

11. Count forward 2 hours from 2:00 to 4:00.
   Count forward 15 minutes from 4:20 to 4:35.
   2 hours 15 minutes

12. Pattern:
   Number of groups × number in each group
        = total
   Problem: 7 hours × N miles per hour
        = 301 miles
   43 miles per hour

   $\begin{array}{r} 43 \\ 7\overline{)301} \\ \underline{28} \\ 21 \\ \underline{21} \\ 0 \end{array}$

13. $\overset{2\ 2}{\ }$
   $\begin{array}{r} \$1.99 \\ \times\quad 3 \\ \hline \$5.97 \end{array}$

   $\overset{1\ 1}{\ }$
   $\begin{array}{r} \$5.97 \\ +\ \$0.33 \\ \hline \$6.30 \end{array}$

   $\begin{array}{r} \overset{19}{\$2\cancel{0}}.{}^{1}0\,0 \\ -\ \$6.\,3\,0 \\ \hline \$13.\,7\,0 \end{array}$

14. $\overset{1\,1\ \ 1}{\ }$
   $\begin{array}{r} \$25.00 \\ \$2.75 \\ \$15.44 \\ +\ \$0.27 \\ \hline \$43.46 \end{array}$

15. $\overset{5\ \ {}^{1}1}{\ }$
   $\begin{array}{r} \cancel{6}.\,\cancel{2}{}^{1}0 \\ -\ 0.\,2\,6 \\ \hline 5.\,9\,4 \end{array}$

16. $\overset{0\ 9\ 9\ \ 9}{\ }$
   $\begin{array}{r} \$\cancel{1}\,\cancel{0}\,\cancel{0}.\,\cancel{0}{}^{1}0 \\ -\ \$8\,9.\,8\,5 \\ \hline \$1\,0.\,1\,5 \end{array}$

17. Multiply mentally: **54,000**

18. $\begin{array}{r} 42 \\ \times\quad 30 \\ \hline 1260 \end{array}$

19. $\begin{array}{r} 21 \\ \times\quad 17 \\ \hline 147 \\ 210 \\ \hline 357 \end{array}$

20. $\overset{\ \ 4}{\overset{\ \ 2}{36}}$
   $\begin{array}{r} \times\quad 74 \\ \hline 144 \\ 2520 \\ \hline 2664 \end{array}$

21. $\overset{\ \ 1}{\overset{\ \ 4}{48}}$
   $\begin{array}{r} \times\quad 25 \\ \hline 240 \\ 960 \\ \hline 1200 \end{array}$

22. $\overset{4\ 5}{\$4.79}$
   $\begin{array}{r} \times\qquad 6 \\ \hline \$28.74 \end{array}$

23. $\begin{array}{r} 302 \\ 9\overline{)2718} \\ \underline{27} \\ 01 \\ \underline{\ 0} \\ 18 \\ \underline{18} \\ 0 \end{array}$

24.
$$
\begin{array}{r}
963 \\
5\overline{)4815} \\
45 \\
\hline
31 \\
30 \\
\hline
15 \\
15 \\
\hline
0
\end{array}
$$

25.
$$
\begin{array}{r}
804\ R\ 5 \\
6\overline{)4829} \\
48 \\
\hline
02 \\
0 \\
\hline
29 \\
24 \\
\hline
5
\end{array}
$$

26.
$$
\begin{array}{r}
\$6.25 \\
8\overline{)\$50.00} \\
48 \\
\hline
2\ 0 \\
1\ 6 \\
\hline
40 \\
40 \\
\hline
0
\end{array}
$$

27. Divide mentally: **300**

28.
$$
\begin{array}{r}
0.5 \\
+\ 0.375 \\
\hline
0.875
\end{array}
$$
$0.875 - 0.875 = \mathbf{0}$

29. One pair of parallel lines.
Trapezoid

30. **One possibility:**

LESSON 93, WARM-UP

a. **80**

b. **120**

c. **140**

d. **320**

e. **IX**

f. **4**

Problem Solving

The darkly shaded portion is less than $\frac{1}{8}$ ($12\frac{1}{2}\%$). It makes up **about 10%** of the circle.

If about 10% of the circle is darkly shaded, then about $100\% - 10\%$, or **about 90%**, is lightly shaded.

LESSON 93, LESSON PRACTICE

a. $60 \times 20 = \mathbf{1200}$

$$
\begin{array}{r}
\overset{1}{\underset{2}{}}\ \\
58 \\
\times\ 23 \\
\hline
174 \\
1160 \\
\hline
\mathbf{1334}
\end{array}
$$

b. $50 \times 50 = \mathbf{2500}$

$$
\begin{array}{r}
\overset{4}{} \\
49 \\
\times\ 51 \\
\hline
49 \\
2450 \\
\hline
\mathbf{2499}
\end{array}
$$

c. $60 \times 40 = \mathbf{2400}$

$$
\begin{array}{r}
61 \\
\times\ 38 \\
\hline
488 \\
1830 \\
\hline
\mathbf{2318}
\end{array}
$$

d. $1800 \div 9 = \mathbf{200}$

$$
\begin{array}{r}
205 \\
9\overline{)1845} \\
18 \\
\hline
04 \\
0 \\
\hline
45 \\
45 \\
\hline
0
\end{array}
$$

LESSON 93, MIXED PRACTICE

1. (a) $3\overline{)91}$ with quotient 30 R 1

$$\begin{array}{r} 30 \text{ R } 1 \\ 3\overline{)91} \\ \underline{9} \\ 01 \\ \underline{0} \\ 1 \end{array}$$

2 galleries received exactly 30 paintings.

(b) There is 1 painting left over, so **1 gallery** received 31 paintings.

2. Pattern:

Number of groups $\times$ number in each group

$ = $ total

Problem: 20 letters $\times$ 6 cents per letter

$ = N$ cents

$20 \times 6\text{¢} = 120\text{¢}$ or **$1.20**

3. Pattern: $\quad$ Larger

$\underline{- \text{ Smaller}}$

 Difference

Problem: $\quad N$

$\underline{- 210}$

790

$$\begin{array}{r} \overset{1}{7}90 \\ + 210 \\ \hline 1000 \end{array}$$

4. Pattern: $\quad$ Larger

$\underline{- \text{ Smaller}}$

 Difference

Problem: $\quad 1799$

$\underline{- 1732}$

67

About 67 years

5. A $1 bill is the same size as a $5 bill, so a $5 bill would also weigh **about 1 gram.**

6. **19**

7. (a) 9 out of 10 parts are shaded

$$\frac{9}{10}$$

(b) Nine tenths $= $ **0.9**

8. $50 \times 60 = $ **3000**

9. $\frac{1}{2}$ on the board $\left\{\begin{array}{l} \text{32 chess pieces} \\ \boxed{\text{16 chess pieces}} \end{array}\right.$

$\frac{1}{2}$ not on board $\left\{\boxed{\text{16 chess pieces}}\right.$

16 chess pieces

10. Counting forward 7 hours from 10:30 a.m. makes it **5:30 p.m.**

11. Pattern:

Number of groups $\times$ number in each group

$ = $ total

Problem: 20 hours $\times$ 42 miles per hour

$ = N$ miles

$$\begin{array}{r} 42 \\ \times 20 \\ \hline \textbf{840 mi} \end{array}$$

12.

$$\begin{array}{r} \overset{11}{\$29.99} \\ + \$1.80 \\ \hline \$31.79 \end{array}$$

$$\begin{array}{r} \$\overset{3}{\cancel{4}}\,\overset{9}{\cancel{0}}.\,\overset{9}{\cancel{0}}{}^{1}0 \\ - \$3\,1.\,7\,9 \\ \hline \textbf{\$8. 2 1} \end{array}$$

13. A 90-degree turn counterclockwise from the west places Connor toward the **south.**

14. $N + 20 = 24$

$4 + 20 = 24$

$N = $ **4**

15.

$$\begin{array}{r} 3.6 \\ 0.2 \\ + 0.125 \\ \hline 3.925 \end{array}$$

$$\begin{array}{r} \overset{3}{\cancel{4}}.\,\overset{1}{\cancel{0}}\,\overset{1}{\cancel{2}}{}^{1}0 \\ - 3.\,9\,2\,5 \\ \hline \textbf{0. 1 9 5} \end{array}$$

16.

$$\begin{array}{r} \$1\,\overset{7}{\cancel{8}}.\,\overset{9}{\cancel{0}}{}^{1}0 \\ - \$1\,5.\,6\,3 \\ \hline \textbf{\$2. 3 7} \end{array}$$

17.
$$
\begin{array}{r}
\overset{1\ 1\ \ 1}{\$15.27} \\
+\ \ \$85.75 \\
\hline
\$101.02
\end{array}
$$

18. $2 \times 2 \times 2 = 8$
$\sqrt{25} = 5$
$8 \times 5 = \textbf{40}$

19. Multiply mentally: **2700**

20.
$$
\begin{array}{r}
\overset{4}{\$7.50} \\
\times\ \ \ \ \ 8 \\
\hline
\$60.00
\end{array}
$$

21.
$$
\begin{array}{r}
\overset{5}{\overset{1}{49}} \\
\times\ \ 62 \\
\hline
98 \\
2940 \\
\hline
3038
\end{array}
$$

22.
$$
\begin{array}{r}
\overset{1}{54} \\
\times\ \ 23 \\
\hline
162 \\
1080 \\
\hline
1242
\end{array}
$$

23.
$$
\begin{array}{r}
\overset{1}{74} \\
\times\ \ \ 40 \\
\hline
2960
\end{array}
$$

24.
$$
\begin{array}{r}
\$1.59 \\
4\overline{)\$6.36} \\
\underline{4} \\
2\ 3 \\
\underline{2\ 0} \\
36 \\
\underline{36} \\
0
\end{array}
$$

25.
$$
\begin{array}{r}
160 \\
5\overline{)800} \\
\underline{5} \\
30 \\
\underline{30} \\
00 \\
\underline{0} \\
0
\end{array}
$$

26.
$$
\begin{array}{r}
591\ R\ 7 \\
8\overline{)4735} \\
\underline{40} \\
73 \\
\underline{72} \\
15 \\
\underline{8} \\
7
\end{array}
$$

27. Divide mentally: **600**

28. $1500 \div 5 = \textbf{300}$

$$
\begin{array}{r}
304 \\
5\overline{)1520} \\
\underline{15} \\
02 \\
\underline{0} \\
20 \\
\underline{20} \\
0
\end{array}
$$

29.

30. Perimeter: $50\text{ ft} + 20\text{ ft} + 50\text{ ft}$
$+\ 20\text{ ft} = \textbf{140 ft}$
Area: $50\text{ ft} \times 20\text{ ft} = \textbf{1000 sq. ft}$

LESSON 94, WARM-UP

a. **230**

b. **310**

c. **1200**

d. **$26.82**

e. **38**

f. **114**

g. **XXX**

h. **14**

Patterns

$\ldots, 64, 32, 16, 8, \underline{\textbf{4}}, \underline{\textbf{2}}, \underline{\textbf{1}}, \underline{\frac{1}{2}}, \underline{\frac{1}{4}}, \ldots$

LESSON 94, LESSON PRACTICE

a.
$$\begin{array}{r} \$5.00 \\ - \ \$2.00 \\ \hline \$3.00 \end{array}$$

4 pounds of peaches costs $2.00

$2.00 ÷ 4 = **$0.50**

b. A square has four equal sides.

Each side = 12 in. ÷ 4 = 3 in.

Area: 3 in. × 3 in. = **9 sq. in.**

c. John:

13 years + 10 years = 23 years old

Jenny:

23 years − 2 years = **21 years old**

LESSON 94, MIXED PRACTICE

1.
$$\begin{array}{r} \overset{1}{}\$1.06 \\ + \ \$2.39 \\ \hline \$3.45 \end{array}$$

$$\begin{array}{r} \$\overset{4}{\cancel{5}}.\,\overset{9}{\cancel{0}}{}^{1}0 \\ - \ \$3.\,4\ 5 \\ \hline \$1.\,5\ 5 \end{array}$$

2.
$$\begin{array}{r} 27 \text{ billy goats} \\ 3\overline{)81} \\ \underline{6} \\ 21 \\ \underline{21} \\ 0 \end{array}$$

$$\begin{array}{r} \overset{7}{\cancel{8}}{}^{1}1 \\ - \ 2\ 7 \\ \hline 5\ 4 \text{ bears} \end{array}$$

3. Pattern:

Number of groups × number in each group
= total

Problem: 8 rows × 15 trees in each row
= N trees

$$\begin{array}{r} \overset{4}{1}5 \\ \times \ \ 8 \\ \hline 120 \text{ trees} \end{array}$$

4. Pattern:

Number of groups × number in each group
= total

Problem: 4 pounds × N rubles per pound
= 156 rubles

$$\begin{array}{r} \mathbf{39 \text{ rubles}} \\ 4\overline{)156} \\ \underline{12} \\ 36 \\ \underline{36} \\ 0 \end{array}$$

5. Count up by fifties from 400: 400, 450, 500, 550

550 g

6. 𝍸𝍸𝍸 |

7. (a) 11 out of 100 parts are shaded

$$\frac{11}{100}$$

(b) Eleven hundredths = **0.11**

8. 30 × 50 = **1500**

$$\begin{array}{r} \overset{1}{3}2 \\ \times \ 48 \\ \hline 256 \\ 1280 \\ \hline \mathbf{1536} \end{array}$$

9.

$\frac{1}{3}$ were Bactrian.

$\frac{2}{3}$ were not Bactrian.

24 camels
8 camels
8 camels
8 camels

8 camels

10. A quarter is **25%.**

11. (a) All squares have four right angles, so every square is a rectangle.
True

(b) Rectangles may not have four equal sides, so every rectangle is not a square.
False

12. 471 out of 1000 contestants were girls.

$$\frac{471}{1000}$$

13. **0.471; four hundred seventy-one thousandths**

14. The tenths place is the first place to the right of the decimal point.
8

15. Pattern:
Number of groups $\times$ number in each group
$\qquad = $ total
Problem: 8 hours $\times$ N miles per hour
$\qquad = $ 496 miles

$$\begin{array}{r} \textbf{62 miles per hour} \\ 8\overline{)496} \\ \underline{48} \\ 16 \\ \underline{16} \\ 0 \end{array}$$

16.
$$\begin{array}{r} {}^{1} \\ 1.74 \\ +\ 0.9 \\ \hline 2.64 \end{array} \qquad \begin{array}{r} {}^{7}{}^{1}2 \\ \cancel{8}.\cancel{3}{}^{1}0 \\ -\ 2.6\ 4 \\ \hline \textbf{5. 6 6} \end{array}$$

17. Multiply mentally: **63,000**

18. Multiply mentally: **$40.00**

19.
$$\begin{array}{r} {}^{3} \\ 37 \\ 81 \\ {}_{2}45 \\ 139 \\ 7 \\ 15 \\ +\ \ 60 \\ \hline 384 \end{array}$$

20.
$$\begin{array}{r} {}^{1} \\ 52 \\ \times\ 15 \\ \hline 260 \\ 520 \\ \hline \textbf{780} \end{array}$$

21.
$$\begin{array}{r} {}^{1} \\ {}^{4} \\ 36 \\ \times\ 27 \\ \hline 252 \\ 720 \\ \hline \textbf{972} \end{array}$$

22.
$$\begin{array}{r} 357 \\ 2\overline{)714} \\ \underline{6} \\ 11 \\ \underline{10} \\ 14 \\ \underline{14} \\ 0 \end{array}$$

23.
$$\begin{array}{r} \textbf{131 R 3} \\ 6\overline{)789} \\ \underline{6} \\ 18 \\ \underline{18} \\ 09 \\ \underline{6} \\ 3 \end{array}$$

24. $3 \times N = 624$
$$\begin{array}{r} 208 \\ 3\overline{)624} \\ \underline{6} \\ 02 \\ \underline{0} \\ 24 \\ \underline{24} \\ 0 \end{array}$$
$N = \textbf{208}$

25.
$5 \times 5 = 25$
$5 + W = 25$
$5 + 20 = 25$
$\qquad W = \textbf{20}$

26. One possibility:

27. 5 yards $\times$ 4 yards = **20 square yards**

28. The diameter equals two radii.
15 mm + 15 mm = 30 mm
$\qquad$ 10 mm = 1 cm
$\qquad$ 30 mm = **3 cm**

29. X has two lines of symmetry.

30. The angle formed by the letter V measures about 45°.
A. 45°

LESSON 95, WARM-UP

a. 600

b. 1200

c. 2400

d. **$5.11**

e. **124**

f. **600**

g. **XV**

h. **25**

Problem Solving

$7 \times 15¢ \times 2 = 2.10

LESSON 95, LESSON PRACTICE

a.

24 checkers

$\frac{1}{4}$ off the board	6 checkers
	6 checkers
$\frac{3}{4}$ on the board	6 checkers
	6 checkers

3×6 checkers $=$ **18 checkers**

b.

30 soldiers

$\frac{2}{5}$ guarded.	6 soldiers
	6 soldiers
	6 soldiers
$\frac{3}{5}$ did not guard.	6 soldiers
	6 soldiers

2×6 soldiers $=$ **12 soldiers**

c.

40 pennies

$\frac{3}{8}$ were shiny.	5 pennies
	5 pennies
	5 pennies
$\frac{5}{8}$ were not shiny.	5 pennies
	5 pennies
	5 pennies
	5 pennies
	5 pennies

3×5 pennies $=$ **15 pennies**

LESSON 95, MIXED PRACTICE

1. **Thanh** had the second highest number of votes.

2. 2×6 votes $= 12$ votes. **Marisol** had 12 votes.

3. 7 votes $+$ 6 votes $+$ 8 votes
       $+$ 12 votes $=$ **33 votes**

4.

20 balloons

$\frac{2}{5}$ were yellow.	4 balloons
	4 balloons
	4 balloons
$\frac{3}{5}$ were not yellow.	4 balloons
	4 balloons

2×4 balloons $=$ **8 balloons**

5. Brad: 11 years $+$ 2 years $=$ 13 years old
    Tim: 13 years $-$ 5 years $=$ **8 years old**

6. (a) 3 out of 10 parts are shaded

    $$\frac{3}{10}$$

    (b) Three tenths $=$ **0.3**

7. $3 \times 10\% =$ **30%**

8. $90 \times 60 =$ **5400**

$$
\begin{array}{r}
\overset{4}{} \\
\overset{7}{88} \\
\times\ 59 \\
\hline
792 \\
4400 \\
\hline
\mathbf{5192}
\end{array}
$$

9. May 2 is a **Tuesday** in 2045.

10.
$$
\begin{array}{r}
\overset{1}{17}\ \text{mm} \\
+\ 36\ \text{mm} \\
\hline
53\ \text{mm}
\end{array}
$$

$$
\begin{array}{r}
89\ \text{mm} \\
-\ 53\ \text{mm} \\
\hline
\mathbf{36\ mm}
\end{array}
$$

11.
$$
\begin{array}{r}
\overset{1\ 1}{\$32.63} \\
\$42.00 \\
+\ \ \$7.56 \\
\hline
\mathbf{\$82.19}
\end{array}
$$

12.
$$
\begin{array}{r}
\$74.50 \\
+\ \ \$5.00 \\
\hline
\$79.50
\end{array}
$$

$$
\begin{array}{r}
\$8\overset{7}{\cancel{6}}.\overset{1}{4}5 \\
-\ \$79.50 \\
\hline
\mathbf{\$6.95}
\end{array}
$$

13.
$$\overset{1}{8}3$$
$$\times\ 40$$
$$\overline{\mathbf{3320}}$$

14. Multiply mentally: **53,000**

15. $9 \times 9 = 81$

$\sqrt{81} = 9$

$$\overset{7}{\cancel{8}}{}^{1}1$$
$$-\quad 9$$
$$\overline{\mathbf{7\ 2}}$$

16.
$$\overset{1}{3}2$$
$$\times\ 16$$
$$\overline{192}$$
$$320$$
$$\overline{\mathbf{512}}$$

17.
$$\overset{2}{\underset{1}{}}67$$
$$\times\ 32$$
$$\overline{134}$$
$$2010$$
$$\overline{\mathbf{2144}}$$

18.
$$\overset{3\ 2}{\$8.95}$$
$$\times\qquad 4$$
$$\overline{\mathbf{\$35.80}}$$

19.
$$\begin{array}{r} \mathbf{208\ R\ 1} \\ 3\overline{)625} \\ \underline{6} \\ 02 \\ \underline{0} \\ 25 \\ \underline{24} \\ 1 \end{array}$$

20.
$$\begin{array}{r} \mathbf{178\ R\ 2} \\ 4\overline{)714} \\ \underline{4} \\ 31 \\ \underline{28} \\ 34 \\ \underline{32} \\ 2 \end{array}$$

21.
$$\begin{array}{r} \mathbf{230\ R\ 5} \\ 6\overline{)1385} \\ \underline{12} \\ 18 \\ \underline{18} \\ 05 \\ \underline{0} \\ 5 \end{array}$$

22.
$$\begin{array}{r} \mathbf{180} \\ 5\overline{)900} \\ \underline{5} \\ 40 \\ \underline{40} \\ 00 \\ \underline{0} \\ 0 \end{array}$$

23.
$$\begin{array}{r} \mathbf{416\ R\ 4} \\ 9\overline{)3748} \\ \underline{36} \\ 14 \\ \underline{9} \\ 58 \\ \underline{54} \\ 4 \end{array}$$

24.
$$\begin{array}{r} \mathbf{\$3.57} \\ 8\overline{)\$28.56} \\ \underline{24} \\ 4\ 5 \\ \underline{4\ 0} \\ 56 \\ \underline{56} \\ 0 \end{array}$$

25.

26. Perimeter: 50 mi + 40 mi + 50 mi
+ 40 mi = **180 mi**

Area: 50 mi $\times$ 40 mi = **2000 sq. mi**

27.

Trapezoid

There are one pair of parallel sides.

28.

29. 0.05
0.050
0.05 $\bigoplus$ 0.050

30. 2400 ÷ 6 = **400**

$$\begin{array}{r} 402 \\ 6\overline{)2412} \\ \underline{24} \\ 01 \\ \underline{0} \\ 12 \\ \underline{12} \\ 0 \end{array}$$

LESSON 96, WARM-UP

a. **800**

b. **2200**

c. **1300**

d. **3400**

e. **$37.47**

f. **67**

g. **201**

h. **XIV**

i. **33**

Patterns

Each term is double the preceding term.
1, 2, 4, 8, 16, **32**, **64**, **128**, **256**, **512**, …

LESSON 96, LESSON PRACTICE

a.
$$\begin{array}{r} \overset{1}{24}\text{ children} \\ 26\text{ children} \\ + 28\text{ children} \\ \hline 78\text{ children} \end{array}$$

$$\begin{array}{r} \textbf{26 children} \\ 3\overline{)78} \\ \underline{6} \\ 18 \\ \underline{18} \\ 0 \end{array}$$

b.
$$\begin{array}{r} 17\text{ books} \\ + 11\text{ books} \\ \hline 28\text{ books} \end{array}$$

$$\begin{array}{r} \textbf{14 books} \\ 2\overline{)28} \\ \underline{2} \\ 08 \\ \underline{8} \\ 0 \end{array}$$

c.
$$\begin{array}{r} \overset{1}{85} \\ \overset{1}{85} \\ + 100 \\ \hline 270 \end{array}$$

270 ÷ 3 = **90**

LESSON 96, MIXED PRACTICE

1. Francesca: 2 × 6 years = 12 years old
Freddie: 12 years + 2 years = **14 years old**

2. In a leap year, February has 29 days.
January = 31 days
March = 31 days

$$\begin{array}{r} \overset{1}{31}\text{ days} \\ 29\text{ days} \\ + 31\text{ days} \\ \hline \textbf{91 days} \end{array}$$

3.
$$\begin{array}{r} \overset{2}{\$0.37} \\ \times \quad 3 \\ \hline \$1.11 \end{array}$$

$$\begin{array}{r} \$1.52 \\ - \$1.11 \\ \hline \textbf{\$0.41} \end{array}$$

4. (a)
$$\begin{array}{r} 5\text{ R }2 \\ 6\overline{)32} \\ \underline{30} \\ 2 \end{array}$$

4 rows had exactly 5 chairs.

(b) There are 2 chairs left over, so **2 rows** had
6 chairs.

5.

	21 riders
$\frac{2}{3}$ rode bareback. {	7 riders
	7 riders
$\frac{1}{3}$ did not ride bareback. {	7 riders

2 × 7 riders = **14 riders**

6. (a) 5 out of 100 parts are shaded, so
   five hundredths are shaded.
   0.05

   (b) 95 out of 100 parts are not shaded, so
   ninety-five hundredths are not shaded.
   0.95

7. $5 \times 1\% =$ **5%**

8. 3874 is closer to 4000 than to 3000, so it rounds
to **4000.**

9. 1 L = 1000 mL
   half of a liter = 500 mL
   About 500 mL
   If half was poured into the pitcher, then half was
   still in the container.
   One half = **50%**

10. Count forward 9 hours from 11:00 to 8:00.
   Count forward 10 minutes from 30 minutes to
   40 minutes
   9 hours 10 minutes

11.
$$
\begin{array}{r}
\overset{2}{7}9°F \\
82°F \\
84°F \\
81°F \\
+\ 74°F \\
\hline
400°F
\end{array}
$$

   $400°F \div 5 =$ **80°F**

12. Pattern:
   Number of groups × number in each group
                     = total
   Problem: 8 hours × N miles in each hour
                     = 368 miles

   46 miles each hour
$$
\begin{array}{r}
8)\overline{368} \\
32 \\
\hline
48 \\
48 \\
\hline
0
\end{array}
$$

13.
$$
\begin{array}{r}
\overset{1\,1\,1\ 1}{496,325} \\
+\ \ \ \ 3,680 \\
\hline
\mathbf{500,005}
\end{array}
$$

14.
$$
\begin{array}{r}
\$3\, \overset{5}{\cancel{6}}.\, \overset{9}{\cancel{0}}{}^{1}0 \\
-\ \$3\,0.\,7\,8 \\
\hline
\mathbf{\$5.\,2\,2}
\end{array}
$$

15.
$$
\begin{array}{r}
\overset{1\ 2}{\$12.45} \\
\$1.30 \\
\$2.00 \\
\$0.25 \\
\$0.04 \\
\$0.32 \\
+\ \$1.29 \\
\hline
\mathbf{\$17.65}
\end{array}
$$

16.
$$
\begin{array}{r}
\overset{1}{\overset{2}{26}} \\
\times\ 24 \\
\hline
104 \\
520 \\
\hline
\mathbf{624}
\end{array}
$$

17.
$$
\begin{array}{r}
\overset{1}{\overset{2}{25}} \\
\times\ 25 \\
\hline
125 \\
500 \\
\hline
\mathbf{625}
\end{array}
$$

18.
$$
\begin{array}{r}
\mathbf{\$2.05} \\
8)\overline{\$16.40} \\
16 \\
\hline
0\ 4 \\
0 \\
\hline
40 \\
40 \\
\hline
0
\end{array}
$$

19. Multiply mentally: **18,000**

20.
$$
\begin{array}{r}
\overset{3\ 4}{\$8.56} \\
\times\ \ \ \ 7 \\
\hline
\mathbf{\$59.92}
\end{array}
$$

21.
$$
\begin{array}{r}
\mathbf{120\ R\ 5} \\
7)\overline{845} \\
7 \\
\hline
14 \\
14 \\
\hline
05 \\
0 \\
\hline
5
\end{array}
$$

22.
$$
\begin{array}{r}
\mathbf{111\ R\ 1} \\
9)\overline{1000} \\
9 \\
\hline
10 \\
9 \\
\hline
10 \\
9 \\
\hline
1
\end{array}
$$

23.
$$6\overline{)432}$$ with quotient **72**
$$\frac{42}{12}$$
$$\frac{12}{0}$$

24.

25.
$$\begin{array}{r} \overset{1}{12} \text{ ft} \\ \times \quad 8 \text{ ft} \\ \hline \mathbf{96 \text{ sq. ft}} \end{array}$$

26. **80, 85, 85, 85, 90, 100, 100**

27. **85** is the middle score in the list.

28. **85** occurs most frequently—three times.

29. $900 \div 3 = \mathbf{300}$

$$3\overline{)912} \text{ with quotient } \mathbf{304}$$
$$\frac{9}{01}$$
$$\frac{0}{12}$$
$$\frac{12}{0}$$

30. Pattern:
Number of groups × number in each group
= total
Problem: *N* glasses × 8 ounces in each glass
= 64 ounces
$64 \div 8 = \mathbf{8 \text{ glasses}}$

Lesson 97, Warm-Up

a. **6**

b. **8**

c. **30**

d. **100**

e. **50**

f. **$0.58**

g. **255**

h. **18**

i. **XXV**

j. **19**

Problem Solving

364 days after Tuesday will be Tuesday. So 365 days after Tuesday will be **Wednesday.**

Lesson 97, Lesson Practice

a.
$$\begin{array}{r} \overset{1}{50} \\ 80 \\ 90 \\ 85 \\ 90 \\ 95 \\ {}_5 90 \\ + \; 100 \\ \hline 680 \end{array} \qquad \begin{array}{r} 85 \\ 8\overline{)680} \\ \frac{64}{40} \\ \frac{40}{0} \end{array}$$

Mean: 85

50, 80, 85, 90, 90, 90, 95, 100
$90 + 90 = 180$
$180 \div 2 = 90$
Median: 90

90 is the score that occurs most often
Mode: 90

$100 - 50 = 50$
Range: 50

b.
$$\begin{array}{r} \overset{1}{31} \\ 28 \\ 31 \\ 30 \\ + \; 25 \\ \hline 145 \end{array} \qquad \begin{array}{r} 29 \\ 5\overline{)145} \\ \frac{10}{45} \\ \frac{45}{0} \end{array}$$

Mean: 29

25, 28, 30, 31, 31
Median: 30

31 is the number that occurs most often
Mode: 31

$31 - 25 = 6$
Range: 6

c. $80 + 90 = 170$

$$\begin{array}{r} 85 \\ 2\overline{)170} \\ \underline{16} \\ 10 \\ \underline{10} \\ 0 \end{array}$$

The median is 85. There is an even number of scores, so there is no one middle score. The average of the two middle scores, 80 and 90, is 85.

LESSON 97, MIXED PRACTICE

1. The title of the graph is "Activities of 100 Children at the Park," so there were **100 children** altogether at the park.

2.
$$\begin{array}{r} \overset{0}{\cancel{1}}\,\overset{9}{\cancel{0}}{}^{1}0 \text{ children} \\ - \quad 1\,9 \text{ children} \\ \hline \mathbf{8\,1 \text{ children}} \end{array}$$

3. $12 \text{ children} + 15 \text{ children} = \textbf{27 children}$

4. $20 \text{ children} - 19 \text{ children} = 1 \text{ child}$
1 more child

5.

1000 coins

$\frac{3}{4}$ were doubloons.	250 coins
	250 coins
$\frac{1}{4}$ were not doubloons.	250 coins
	250 coins

$$\begin{array}{r} \overset{1}{2}50 \\ \times \quad 3 \\ \hline \textbf{750 doubloons} \end{array}$$

6. One fourth is 25%, so three fourths is $3 \times 25\%$, which is **75%**.

7. (a) Use one decimal place.
3.5

(b) Use three decimal places.
14.021

(c) Use two decimal places.
9.04

8. $40 \times 400 = \textbf{16,000}$

$$\begin{array}{r} \overset{1}{}\overset{5}{} \\ 406 \\ \times \quad 39 \\ \hline 3\,654 \\ 12,180 \\ \hline \mathbf{15,834} \end{array}$$

9. (a) $30 \text{ mm} + 10 \text{ mm} + 30 \text{ mm} + 10 \text{ mm} = \textbf{80 mm}$

(b) $10 \text{ mm} = 1 \text{ cm}$
$80 \text{ mm} = \textbf{8 cm}$

10. (a) $30 \text{ mm} \times 10 \text{ mm} = \textbf{300 sq. mm}$

(b) $10 \text{ mm} = 1 \text{ cm}$
$30 \text{ mm} = 3 \text{ cm}$
$3 \text{ cm} \times 1 \text{ cm} = \textbf{3 sq. cm}$

11. Step 1: Counting forward 30 minutes from 7:00 a.m. makes it 7:30 a.m.
Step 2: Counting forward 7 hours from 7:30 a.m. makes it **2:30 p.m.**

12. January has 31 days, February has 28 days, and March has 31 days.

$$\begin{array}{r} \overset{1}{3}1 \text{ days} \\ 28 \text{ days} \\ + \quad 31 \text{ days} \\ \hline 90 \text{ days} \end{array}$$

$90 \text{ days} \div 3 = \textbf{30 days}$

13.
$$\begin{array}{r} \overset{2}{2}5 \\ \times \quad 40 \\ \hline \mathbf{1000} \end{array}$$

14.
$$\begin{array}{r} \overset{5}{\$0.9}8 \\ \times \quad 7 \\ \hline \mathbf{\$6.86} \end{array}$$

15. $\sqrt{36} = 6$
$\sqrt{4} = 2$
$6 \times 2 = \textbf{12}$

16. $3 \times 3 = 9$
$9 \times 3 = 27$
$27 \div 3 = \textbf{9}$

17.
$$\begin{array}{r} \overset{1}{\underset{2}{}}36 \\ \times\ 34 \\ \hline 144 \\ 1080 \\ \hline \mathbf{1224} \end{array}$$

18.
$$\begin{array}{r} \overset{1}{\underset{2}{}}35 \\ \times\ 35 \\ \hline 175 \\ 1050 \\ \hline \mathbf{1225} \end{array}$$

19.
$$\begin{array}{r} 32 \\ +\ X \\ \hline 42 \end{array}$$

$$\begin{array}{r} 42 \\ -\ 32 \\ \hline 10 \end{array}$$

$$X\ =\ \mathbf{10}$$

20.
$$\begin{array}{r} \mathbf{\$8.75} \\ 8\overline{)\$70.00} \\ \underline{64} \\ 6\,0 \\ \underline{5\,6} \\ 40 \\ \underline{40} \\ 0 \end{array}$$

21.
$$\begin{array}{r} \mathbf{205\ R\ 4} \\ 6\overline{)1234} \\ \underline{12} \\ 03 \\ \underline{0} \\ 34 \\ \underline{30} \\ 4 \end{array}$$

22.
$$\begin{array}{r} \mathbf{114\ R\ 2} \\ 7\overline{)800} \\ \underline{7} \\ 10 \\ \underline{7} \\ 30 \\ \underline{28} \\ 2 \end{array}$$

23.
$$\begin{array}{r} \mathbf{162\ R\ 1} \\ 3\overline{)487} \\ \underline{3} \\ 18 \\ \underline{18} \\ 07 \\ \underline{6} \\ 1 \end{array}$$

24.
$$\begin{array}{r} \overset{1\ 2}{}\$2.74 \\ \$0.27 \\ \$6.00 \\ +\ \$0.49 \\ \hline \mathbf{\$9.50} \end{array}$$

25.
$$\begin{array}{r} \overset{1}{}3.7 \\ +\ 2.36 \\ \hline 6.06 \end{array}$$

$$\begin{array}{r} 9.487 \\ -\ 6.060 \\ \hline \mathbf{3.427} \end{array}$$

26.

27. 7 is the number that occurs most often

Mode: 7

28. 6, 7, 7, 7, 8, 8, 9, 10, 10

Median: 8

29. $10\ -\ 6\ =\ 4$

Range: 4

30.
$$\begin{array}{r} 8 \\ 7 \\ 7 \\ 8 \\ {}_{5}6 \\ 10 \\ 9 \\ 10 \\ +\ 7 \\ \hline 72 \end{array}$$

$$72\ \div\ 9\ =\ 8$$

Mean: 8

LESSON 98, WARM-UP

a. **42**

b. **64**

c. **110**

d. **1200**

e. **XXVIII**

f. **27**

LESSON 98, LESSON PRACTICE

a. **Sphere**

b. **Rectangular solid**

c. **Cone**

d. **Cylinder**

e. **Pyramid**

f. 4 edges around the bottom
+ 4 edges running from top to bottom
= **8 edges**

g. **18 small cubes**

LESSON 98, MIXED PRACTICE

1. James: 10 years + 4 years = **14 years old**

2. Jill: 14 years + 2 years = **16 years old**

3. 16 years + 2 years = **18 years old**

4.
$$\begin{array}{r} \overset{1}{\$2}.\overset{1}{7}6 \\ -\ \$0.84 \\ \hline \$1.92 \end{array}$$

$$\begin{array}{r} \$0.32 \text{ or } \mathbf{32}\cancel{c} \\ 6\overline{)\$1.92} \\ \underline{1\ 8} \\ 12 \\ \underline{12} \\ 0 \end{array}$$

5. (a) Use three decimal places.
5.031

(b) Use one decimal place.
16.7

(c) Use two decimal places.
5.07

6.
$\frac{3}{5}$ were scored in the first half.
$\frac{2}{5}$ were not scored in the first half.

40 points
8 points
8 points
8 points
8 points
8 points

3 × 8 points = **24 points**

7. 3 × 20% = **60%**

8. Two decimal places show hundredths.
Seven and sixty-eight hundredths

9. One decimal place shows tenths.
Seventy-six and eight tenths

10. 80 × 90 = **7200**

11. (a) One and three tenths are shaded.
$1\dfrac{3}{10}$

(b) One and three tenths = **1.3**

12.
$$\begin{array}{r} \overset{1}{2}4 \text{ people} \\ +\ 16 \text{ people} \\ \hline 40 \text{ people} \end{array}$$

40 people ÷ 2 = **20 people**

13. Step 1: Counting forward 20 minutes from 9:40 p.m. makes it 10:00 p.m.
Step 2: Counting forward 5 hours from 10:00 p.m. makes it **3:00 a.m.**

14. (a) 27 pizzas ÷ 3 hours = **9 pizzas** in 1 hour

(b) 5 hours × 9 pizzas per hour = **45 pizzas**

15.
$$\begin{array}{r} \overset{1}{3}.65 \\ 4.2 \\ +\ 0.625 \\ \hline \mathbf{8.475} \end{array}$$

16.
$$\begin{array}{r} \$\overset{0}{1}\overset{1}{3}.\overset{1}{7}0 \\ -\ \$6.85 \\ \hline \$6.85 \end{array}$$

17. Multiply mentally: **2600**

18.
$$\overset{6}{\$0.87}$$
$$\times \quad 9$$
$$\overline{\$7.83}$$

19.
$$\overset{2}{14}$$
$$\times 16$$
$$\overline{84}$$
$$\underline{140}$$
$$\mathbf{224}$$

20.
$$\overset{2}{15}$$
$$\times 15$$
$$\overline{75}$$
$$\underline{150}$$
$$\mathbf{225}$$

21.
$$\begin{array}{r} \mathbf{76} \\ 6\overline{)456} \\ \underline{42} \\ 36 \\ \underline{36} \\ 0 \end{array}$$

22.
$$\overset{4}{47}$$
$$\times \quad 60$$
$$\overline{\mathbf{2820}}$$

23.
$$\begin{array}{r} \mathbf{708} \\ 6\overline{)4248} \\ \underline{42} \\ 04 \\ \underline{0} \\ 48 \\ \underline{48} \\ 0 \end{array}$$

24. $163 \div 1 = \mathbf{163}$

25.
$$\begin{array}{r} \mathbf{\$9.80} \\ 5\overline{)\$49.00} \\ \underline{45} \\ 4\,0 \\ \underline{4\,0} \\ 00 \\ \underline{0} \\ 0 \end{array}$$

26. $24 + P = 44$
$44 - 24 = 20$
$P = \mathbf{20}$

27. $15\text{ ft} \times 10\text{ ft} = 150\text{ sq. ft}$
$150\text{ sq. ft} = \mathbf{150\text{ floor tiles}}$

28. 1, 1, 2, 3, 5, 8, 13
Median: 3
1 is the number that occurs most often
Mode: 1

29. **Sphere**

30. (a) **Cube**

(b) A cube has **6 faces.**

LESSON 99, WARM-UP

a. **54**

b. **900**

c. **2400**

d. **$3.25**

e. **424**

f. **28**

g. **XXXI**

h. **24**

Problem Solving

1 minute $\times$ 2 = 2 minutes to run a quarter mile

2 minutes $\times$ 2 = 4 minutes to walk a quarter mile

4 minutes $\times$ 2 = **8 minutes** to swim a quarter mile

LESSON 99, LESSON PRACTICE

| Corn 20¢ each ear | Corn $0.20 each ear |

LESSON 99, MIXED PRACTICE

1. (a) $6\overline{)53}$ with quotient 8 R 5
$$\frac{48}{5}$$

She could make **8 groups** with exactly 6 suffragettes.

(b) **5 suffragettes** were left over.

(c) 8 groups + 1 group = **9 groups**

2. Pattern:
$$\begin{array}{r} \text{Larger} \\ - \text{ Smaller} \\ \hline \text{Difference} \end{array}$$

Problem:
$$\begin{array}{r} 1\,8\,\overset{5}{\cancel{6}}{}^{1}5 \\ -\,1\,8\,0\,9 \\ \hline 5\,6 \end{array}$$

About 56 years

3.
$$\begin{array}{r} \overset{1}{}\overset{1}{} \\ \$1.25 \\ \$0.75 \\ +\ \$0.75 \\ \hline \mathbf{\$2.75} \end{array}$$

4.
$\frac{2}{3}$ scored in second half
$\frac{1}{3}$ not scored in second half

45 points
15 points
15 points
15 points

2×15 points = **30 points**

5.
ICE CREAM BARS 75¢ each
ICE CREAM BARS $0.75 each

6.
$$\begin{array}{r} \$30.00 \\ \$4.00 \\ \$0.50 \\ +\ \$0.02 \\ \hline \mathbf{\$34.52} \end{array}$$

7. One decimal place indicates tenths.
Six thousand, four hundred twelve and five tenths

8. 5139 rounds to 5000 and 6902 rounds to 7000
5000 + 7000 = 12,000

9. 1 gallon = 4 quarts
4 quarts − 1 quart = **3 quarts**

10. 1 quart is 25% of a gallon
$3 \times 25\% = \mathbf{75\%}$

11. $40 \times 40 = \mathbf{1600}$
$$\begin{array}{r} \overset{3}{39} \\ \times\ 41 \\ \hline 39 \\ 1560 \\ \hline \mathbf{1599} \end{array}$$

12. A quarter turn is **90°**.

13.
48 tourists
$$5\overline{)240}$$
$$\begin{array}{r} 20 \\ \hline 40 \\ 40 \\ \hline 0 \end{array}$$

14.
$$\begin{array}{r} \overset{1}{}\overset{1}{}\overset{1}{} \\ \$68.57 \\ +\ \$36.49 \\ \hline \mathbf{\$105.06} \end{array}$$

15.
$$\begin{array}{r} \$\overset{0}{\cancel{1}}\overset{9}{\cancel{0}}\overset{9}{\cancel{0}}.\overset{9}{\cancel{0}}{}^{1}0 \\ -\ \ \$5.\,4\,3 \\ \hline \mathbf{\$9\,4.\,5\,7} \end{array}$$

16.
$$\begin{array}{r} \overset{3}{15} \\ 24 \\ 36 \\ 75 \\ 21 \\ 8 \\ \overset{2}{}36 \\ +\ 420 \\ \hline \mathbf{635} \end{array}$$

17.
$$\begin{array}{r} 12 \\ \times\ 12 \\ \hline 24 \\ 120 \\ \hline \mathbf{144} \end{array}$$

18.
$$\begin{array}{r} \overset{5}{\$5.08} \\ \times\ \ \ 7 \\ \hline \mathbf{\$35.56} \end{array}$$

19. $50 \times 50 = \mathbf{2500}$

20. $12 \times 12 = 144$
12

21.
$$\begin{array}{r} 9.61 \\ -\ 2.4 \\ \hline 7.21 \end{array}$$

$$\begin{array}{r} ^0^1 \\ \cancel{1}\ \cancel{2}.^1 0\ 8 \\ -\quad 7.\ 2\ 1 \\ \hline \mathbf{4.\ 8\ 7} \end{array}$$

22.
$$\begin{array}{r} ^4 \\ 49 \\ \times\ 51 \\ \hline 49 \\ 2450 \\ \hline \mathbf{2499} \end{array}$$

23.
$$\begin{array}{r} ^1 \\ 33 \\ \times\ 25 \\ \hline 165 \\ 660 \\ \hline \mathbf{825} \end{array}$$

24.
$$\begin{array}{r} \mathbf{106} \\ 8\overline{)848} \\ \underline{8} \\ 04 \\ \underline{0} \\ 48 \\ \underline{48} \\ 0 \end{array}$$

25. Divide mentally: **700**

26.

27.
3 in.
| | **1 in.**

28. Perimeter: 3 in. + 1 in. + 3 in.
+ 1 in. = **8 in.**
Area: 3 in. $\times$ 1 in. = **3 sq. in.**

29. **8 small cubes**

30. 4 vertices around the bottom + 1 vertex on top
= **5 vertices**

LESSON 100, WARM-UP

a. 56

b. 2000

c. 3200

d. $1.84

e. $7.63

f. 27

g. XIX

h. 29

Problem Solving

$$\frac{3}{30} = \frac{1}{10}$$

$$\frac{1}{10} = \mathbf{10\%}$$

$$100\% - 10\% = \mathbf{90\%}$$

LESSON 100, ACTIVITY

1. front + back + top + bottom
+ left + right = **6 panels**

2. **front, back, top, bottom, left, right**

3. **3 panels** (front, top or bottom, and left or right)

4. **12 edges**

5. **8 corners**

LESSON 100, LESSON PRACTICE

There is no Lesson Practice for Lesson 100.

LESSON 100, MIXED PRACTICE

1.
$$1 \text{ yd} = 3 \text{ ft}$$
$$150 \text{ ft} \div 3 = \textbf{50 yd}$$

2.
$$\begin{array}{r} \overset{1}{\$0.64} \\ +\ \$0.38 \\ \hline \$1.02 \end{array}$$

$$\begin{array}{r} \$\overset{5}{\cancel{6}}.\overset{9}{\cancel{0}}{}^{1}0 \\ -\ \$1.\ 0\ 2 \\ \hline \mathbf{\$4.\ 9\ 8} \end{array}$$

3. Rebecca: 12 years $-$ 2 years $=$ 10 years old
Dina: 10 years $\div$ 2 $=$ **5 years old**

4. (a) Three decimal places indicate thousandths.
$$3\frac{295}{1000}$$

(b) One decimal place indicates tenths.
$$32\frac{9}{10}$$

(c) Two decimal places indicate hundredths.
$$3\frac{9}{100}$$

5.

84 contestants

$\frac{3}{4}$ guessed incorrectly. $\left\{\begin{array}{|c|}\hline \text{21 contestants} \\ \hline \text{21 contestants} \\ \hline\end{array}\right.$

$\frac{1}{4}$ did not guess incorrectly. $\left\{\begin{array}{|c|}\hline \text{21 contestants} \\ \hline \text{21 contestants} \\ \hline\end{array}\right.$

3×21 contestants $=$ **63 contestants**

6. One fourth $=$ 25%
$3 \times 25\% = \mathbf{75\%}$

7. (a) The diameter is **1 in.**

(b) The radius is half the diameter.
$$\frac{1}{2} \textbf{ in.}$$

8. Two decimal places indicate hundredths.
Eight and seventy-five hundredths

9. $50 \times 60 = \mathbf{3000}$

$$\begin{array}{r} \overset{4}{\underset{}{}}\overset{1}{47} \\ \times\ 62 \\ \hline 94 \\ 2820 \\ \hline \mathbf{2914} \end{array}$$

10. $1 + 3 + 5 + 7 + 9 = 25$
$$25 \div 5 = 5$$
Mean: 5
1, 3, 5, 7, 9
Median: 5

11. **Cylinder**

12. A parallelogram has exactly two pairs of parallel sides.
B. $\boxed{}$

13. $\$6.00 - \$0.50 = \$5.50$

$$\begin{array}{r} \$1\overset{5}{\cancel{6}}.{}^{1}2\ 5 \\ -\ \ \$5.\ 5\ 0 \\ \hline \mathbf{\$1\ 0.\ 7\ 5} \end{array}$$

14. $5 \times 7 = 35$

$$\begin{array}{r} \overset{4}{35} \\ \times\ \ 9 \\ \hline \mathbf{315} \end{array}$$

15.
$$\begin{array}{r} \overset{4}{}\overset{1}{\$7.83} \\ \times\ \ \ \ 6 \\ \hline \mathbf{\$46.98} \end{array}$$

16. Multiply mentally: **54,000**

17.
$$\begin{array}{r} \overset{2}{\underset{}{}}\overset{2}{45} \\ \times\ 45 \\ \hline 225 \\ 1800 \\ \hline \mathbf{2025} \end{array}$$

18.
$$\begin{array}{r} 32 \\ \times\ 40 \\ \hline \mathbf{1280} \end{array}$$

19.
$$\begin{array}{r} \overset{2}{\underset{}{}}\overset{2}{46} \\ \times\ 44 \\ \hline 184 \\ 1840 \\ \hline \mathbf{2024} \end{array}$$

SOLUTIONS

20.
$$
\begin{array}{r}
604\ \text{R}\ 1 \\
6\overline{)3625} \\
\underline{36} \\
02 \\
\underline{0} \\
25 \\
\underline{24} \\
1
\end{array}
$$

21. Divide mentally: **600**

22.
$$
\begin{array}{r}
141 \\
7\overline{)987} \\
\underline{7} \\
28 \\
\underline{28} \\
07 \\
\underline{7} \\
0
\end{array}
$$

23.
$$
\begin{aligned}
10 \times 10 &= 100 \\
100 \times 10 &= 1000 \\
\sqrt{25} &= 5 \\
1000 \div 5 &= \textbf{200}
\end{aligned}
$$

24.
$$
\begin{array}{r}
\$1.72 \\
8\overline{)\$13.76} \\
\underline{8} \\
5\ 7 \\
\underline{5\ 6} \\
16 \\
\underline{16} \\
0
\end{array}
$$

25.
$$
\begin{array}{r}
58\ \text{R}\ 2 \\
4\overline{)234} \\
\underline{20} \\
34 \\
\underline{32} \\
2
\end{array}
$$

26.

27. A square has four equal sides.
$$
\begin{aligned}
40\ \text{cm} \div 4 &= 10\ \text{cm} \\
10\ \text{cm} \times 10\ \text{cm} &= \textbf{100 sq. cm}
\end{aligned}
$$

28. Similar figures have the same shape, but not necessarily the same size.

Triangle may be larger or smaller than the triangle shown.

29. (a) 0.25
0.250
0.25 $=$ 0.250

(b) 0.25¢ = $0.0025
$0.25 $>$ $0.0025

30. A.

INVESTIGATION 10

1. A, B, or C

2. The arrow is most likely to stop on A because the A sectors cover more of the spinner than the B sectors or the C sector.

3. C, B, A

4. B

5. A. outcome A

6. A. $\dfrac{1}{6}$

7. One of the six sectors is labeled C. We expect the arrow to land in sector C about $\frac{1}{6}$ of the time, so in 60 spins this would be **B. about 10 times.**

8. 0; The letter D is not on the spinner, so D is an impossible outcome.

9. 1; Every sector of the spinner is labeled with one of the first three letters of the alphabet. So that outcome is certain to happen.

10. $\dfrac{1}{6}$

11. 1, 2, 3, 4, 5, and 6

12. None of the numbers is most likely to end up on top, since each number is equally likely to end up on top after one roll.

13. In the numbers 1 through 6, half of the numbers are greater than 3, so we expect a number greater than three **B. about half the time.**

14. The number 7 is not a possible outcome, so the probability of rolling a 7 is **0.**

15. The number 1 is one of six outcomes, so the probability of rolling a 1 is $\frac{1}{6}$.

16. The likelihood of rolling a 6 is less than $\frac{1}{2}$, so it is **C. unlikely.**

17. 100% − 40% = 60%
Since 60% (the chance of no rain) is greater than 40% (the chance of rain), it is more likely **not to rain.**

18. 100% − 80% = **20%**

19. **The frequencies of the outcomes are likely to be roughly equal.**

20. **Seven is likely to be the most frequent outcome. There are more dot combinations that total 7 than any other outcome.**

21. **Two and 12 will likely be the least frequent outcomes. There are fewer dot combinations that total 2 and 12 than any other outcome.**

22. **The more combinations that add up to the outcome's number, the more likely that outcome is to happen. The lowest and highest numbers in the set of outcomes are least likely to happen, while those in the middle are most likely.**

LESSON 101, WARM-UP

a. 525

b. 175

c. $3.00

d. 3300

e. $8.05

f. 18

g. XXXVI

h. <

Problem Solving
A common year ends on the same day of the week as it starts. A leap year ends one day of the week later than it starts.

LESSON 101, LESSON PRACTICE

a.
$$\begin{array}{r} 5\,8\,\overset{8}{\cancel{9}}{}^{1}5\text{ m} \\ -\ 3\,7\,7\,6\text{ m} \\ \hline 2\,1\,1\,9\text{ m} \end{array}$$

b.
$$\begin{array}{r} 2\,\overset{8}{\cancel{9}},\overset{9}{\cancel{0}}{}^{1}3\,5\text{ ft} \\ -\ 1\,4,6\,9\,1\text{ ft} \\ \hline 1\,4,3\,4\,4\text{ ft} \end{array}$$

c. Count forward 9 hours and 30 minutes from 9:00 p.m. to 6:30 a.m.

$9\frac{1}{2}$ **hours (or 9 hours 30 minutes)**

LESSON 101, MIXED PRACTICE

1. 1 foot = 12 inches
2 feet = 24 inches
Denver and San Francisco average less than 24 inches of rain per year.

2.
$$\begin{array}{r} \overset{5}{\cancel{6}}{}^{1}2\text{ inches} \\ -\ 4\,8\text{ inches} \\ \hline 1\,4\text{ inches} \end{array}$$

3.

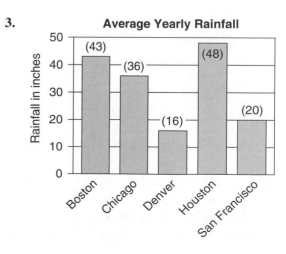

Average Yearly Rainfall

4.

288 marchers

 $\frac{5}{6}$ were out of step.

| 48 marchers |
| 48 marchers |
| 48 marchers |
| 48 marchers |
| 48 marchers |
| 48 marchers |

$\frac{1}{6}$ were not out of step.

$$\begin{array}{r} \overset{4}{48} \text{ marchers} \\ \times \quad 5 \\ \hline \textbf{240 marchers} \end{array}$$

5.

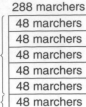

GAS
99¢
per gallon

GAS
$0.99
per gallon

6. The radius is half the diameter.
1 cm = **10 mm**

7. 60% > 50%, so it is more likely **that it will rain.**

8. 90 × 20 = **1800**

$$\begin{array}{r} \overset{1}{\underset{}{\overset{1}{88}}} \\ \times \quad 22 \\ \hline 176 \\ 1760 \\ \hline \textbf{1936} \end{array}$$

9. Pattern:
Number of groups × number in each group
= total
Problem: 5 pounds × $0.53 per pound
= N dollars

$$\begin{array}{r} \overset{1}{\$0.53} \\ \times \quad 5 \\ \hline \textbf{\$2.65} \end{array}$$

10. 3708 = **3000 + 700 + 8**
Three thousand, seven hundred eight

11. 1 m = 100 cm
2 m = **200 cm**

12. Pattern:
Number of groups × number in each group
= total
Problem: 4 pounds × N dollars for 1 pound
= $1.20
$1.20 ÷ 4 = **$0.30**
6 pounds × $0.30 per pound = **$1.80**

13. Pattern:
Number of groups × number in each group
= total
Problem: 3 hours × N miles per hour
= 150 miles
150 ÷ 3 = **50 miles per hour**

14.
$$\begin{array}{r} \$4\,\overset{5}{\cancel{6}}.\,\overset{9}{\cancel{0}}{}^{1}0 \\ -\ \$4\,5.\,5\,6 \\ \hline \$0.\,\textbf{4}\,\textbf{4} \end{array}$$

15.
$$\begin{array}{r} \overset{0}{\cancel{1}}\,\overset{9}{\cancel{0}},{}^{1}1\,\overset{5}{\cancel{6}}{}^{1}5 \\ -\ \quad\quad 8\,5\,6 \\ \hline \textbf{9}\,\textbf{3}\,\textbf{0}\,\textbf{9} \end{array}$$

16.
$$\begin{array}{r} \overset{2}{}\overset{3}{} \\ \$0.63 \\ \$_{1}1.49 \\ \$12.24 \\ \$0.38 \\ \$0.06 \\ \$5.00 \\ +\ \$1.20 \\ \hline \textbf{\$21.00} \end{array}$$

17. 70 × 70 = **4900**

18.
$$\begin{array}{r} 71 \\ \times \quad 69 \\ \hline 639 \\ 4260 \\ \hline \textbf{4899} \end{array}$$

19.
$$\begin{array}{r} \textbf{\$7.50} \\ 4\overline{)\$30.00} \\ 28 \\ \hline 2\,0 \\ 2\,0 \\ \hline 00 \\ 0 \\ \hline 0 \end{array}$$

20.
$$\begin{array}{r} \textbf{87 R 2} \\ 3\overline{)263} \\ 24 \\ \hline 23 \\ 21 \\ \hline 2 \end{array}$$

21.
$$\begin{array}{r} \textbf{816} \\ 5\overline{)4080} \\ 40 \\ \hline 08 \\ 5 \\ \hline 30 \\ 30 \\ \hline 0 \end{array}$$

22.

$$\begin{array}{r} 43 \\ 8\overline{)344} \\ \underline{32} \\ 24 \\ \underline{24} \\ 0 \end{array}$$

23.

$$\begin{array}{r} \overset{4}{37} \\ \times\ \ 60 \\ \hline \mathbf{2220} \end{array}$$

24.

$$\begin{array}{r} \overset{2}{\underset{1}{56}} \\ \times\ \ 42 \\ \hline 112 \\ 2240 \\ \hline \mathbf{2352} \end{array}$$

25.

$$\begin{array}{r} \overset{7\ 5}{\$5.97} \\ \times\ \ \ \ \ 8 \\ \hline \mathbf{\$47.76} \end{array}$$

26.

$$\begin{array}{r} 4.468 \\ -\ 2.300 \\ \hline 2.168 \end{array}$$

$$\begin{array}{r} \overset{0\ \ 9\ \ 9\ \ 9}{\cancel{1}\ \cancel{0}.\ \cancel{0}\ \cancel{0}\,^1 0} \\ -\ \ \ \ 2.\,1\ 6\ 8 \\ \hline \mathbf{7.\ 8\ 3\ 2} \end{array}$$

27. $3 + 1 + 4 + 1 + 6 = 15$
$15 \div 5 = 3$
Mean: 3
1, 1, 3, 4, 6
Median: 3
1 is the number that occurs most often
Mode: 1
$6 - 1 = 5$
Range: 5

28.

29.   **4 cm**

30. Perimeter: 4 cm + 4 cm + 4 cm
                    + 4 cm = **16 cm**
     Area: 4 cm × 4 cm = **16 sq. cm**

Saxon Math 5/4—Homeschool

LESSON 102, WARM-UP

a. **400**

b. **250**

c. **$5.50**

d. **72**

e. **$5.63**

f. **417**

g. **XXIV**

h. **<**

Problem Solving
  $10\% = \frac{1}{10}$
  1 foot equals 12 inches, so 1 inch equals $\frac{1}{12}$ of a foot.
  $\frac{1}{12} < \frac{1}{10}$
  Explanations may vary, but Amber's statement is true.

LESSON 102, LESSON PRACTICE

a. 100 cm = 1 m
     167 cm = **1.67 m**

b. The number line is divided into tenths.
     5.5

c. The number line is divided into tenths.
     6.8

d. The number line is divided into tenths.
     7.4

e. The number line is divided into hundredths.
     4.05

f. The number line is divided into hundredths.
     4.44

g. The number line is divided into hundredths.
     4.95

h. $6.8 > 6.5$, so 6.8 rounds to **7**

i. 4.44 is closer to 4 than to 5, so it rounds to **4**

j. 4.44 is closer to 4.4 than to 4.5, so it rounds to **4.4**

LESSON 102, MIXED PRACTICE

1. (a)
$$
\begin{array}{r}
13\text{ R }6 \\
8\overline{)110} \\
\underline{8} \\
30 \\
\underline{24} \\
6
\end{array}
$$

13 boxes can be filled with 8 books.

(b) One box is needed to hold the 6 extra books.
13 boxes + 1 box = **14 boxes**

2. $6 \times 7 = 42$
$42 + 5 = \mathbf{47}$

3. $\$0.25 + \$0.10 + \$0.10 = \$0.45$

$$
\begin{array}{r}
\$\overset{6}{7}.\overset{9}{\cancel{0}}{}^{1}0 \\
- \ \$0.\ 4\ 5 \\
\hline
\$6.\ 5\ 5
\end{array}
$$

$$
\begin{array}{r}
\$6.55 \\
- \ \$0.42 \\
\hline
\mathbf{\$6.13}
\end{array}
$$

4.

600 gymnasts

120 gymnasts
120 gymnasts
120 gymnasts
120 gymnasts
120 gymnasts

$\frac{4}{5}$ did back handsprings.

$\frac{1}{5}$ did not do back handsprings.

$$
\begin{array}{r}
120\ \text{gymnasts} \\
\times 4 \\
\hline
\mathbf{480\ gymnasts}
\end{array}
$$

5. One fifth = **20%**

6.
$$
\begin{array}{r}
\$200.00 \\
\$50.00 \\
\$4.00 \\
\$0.30 \\
+ \ \ \ \$0.01 \\
\hline
\mathbf{\$254.31}
\end{array}
$$

7. (a) Use the top ruler.
35 mm

(b) Use the bottom ruler.
3.5 cm

8. Two decimal places indicate hundredths.
Twelve and sixty-seven hundredths

9. (a) 3834 is closer to 4000 than to 3000, so round up to **4000**.

(b) 38.34 is closer to 38 than to 39, so round down to **38**.

10. 1 meter = 100 cm
The radius is half the diameter.
Half of 100 cm is **50 cm**.

11.
$$
\begin{array}{r}
\overset{1\ 1}{286{,}514} \\
+ \ 137{,}002 \\
\hline
\mathbf{423{,}516}
\end{array}
$$

12. Pattern:
Number of groups $\times$ number in each group
$= $ total
Problem:
7 whirligigs $\times$ N dollars for 1 whirligig
$= \$56$
$56 \div 7 = \mathbf{\$8}$

$$
\begin{array}{r}
\overset{1}{12} \\
\times \ \ \$8 \\
\hline
\mathbf{\$96}
\end{array}
$$

13.
$$
\begin{array}{r}
\overset{1}{36}\text{ children} \\
+ \ 24\text{ children} \\
\hline
60\text{ children}
\end{array}
$$

60 children $\div$ 2 = **30 children**

14. Sector C is one out of four equal-sized sectors.
$$\frac{1}{4}$$

15.
$$
\begin{array}{r}
6.47 \\
+ \ 0.5 \\
\hline
6.97
\end{array}
$$

$$
\begin{array}{r}
\overset{6}{7}.{}^{1}4\ 8\ 6 \\
- \ 6.\ 9\ 7\ 0 \\
\hline
\mathbf{0.\ 5\ 1\ 6}
\end{array}
$$

16. Multiply mentally: **2000**

17.
$$\begin{array}{r} 41 \\ \times\ 49 \\ \hline 369 \\ 1640 \\ \hline \mathbf{2009} \end{array}$$

18. $2 \times 2 \times 2 = 8$
$\sqrt{49} = 7$
$8 \times 5 \times 7$
$8 \times 5 = 40$
$40 \times 7 = \mathbf{280}$

19.
$$\begin{array}{r} \overset{1}{32} \\ \times\ 17 \\ \hline 224 \\ 320 \\ \hline \mathbf{544} \end{array}$$

20.
$$\begin{array}{r} \overset{3}{38} \\ \times\ \ 40 \\ \hline \mathbf{1520} \end{array}$$

21. $42 + K = 47$
$47 - 42 = 5$
$K = \mathbf{5}$

22.
$$\begin{array}{r} \mathbf{452} \\ 8\overline{)3616} \\ 32 \\ \hline 41 \\ 40 \\ \hline 16 \\ 16 \\ \hline 0 \end{array}$$

23.
$$\begin{array}{r} \mathbf{620\ R\ 2} \\ 4\overline{)2482} \\ 24 \\ \hline 08 \\ 8 \\ \hline 02 \\ 0 \\ \hline 2 \end{array}$$

24.
$$\begin{array}{r} \mathbf{502\ R\ 2} \\ 7\overline{)3516} \\ 35 \\ \hline 01 \\ 0 \\ \hline 16 \\ 14 \\ \hline 2 \end{array}$$

25.
$$\begin{array}{r} \mathbf{\$0.73} \\ 6\overline{)\$4.38} \\ 4\,2 \\ \hline 18 \\ 18 \\ \hline 0 \end{array}$$

26.
$$\begin{array}{r} \mathbf{795\ R\ 7} \\ 9\overline{)7162} \\ 63 \\ \hline 86 \\ 81 \\ \hline 52 \\ 45 \\ \hline 7 \end{array}$$

27.
$$\begin{array}{r} \mathbf{707} \\ 2\overline{)1414} \\ 14 \\ \hline 01 \\ 0 \\ \hline 14 \\ 14 \\ \hline 0 \end{array}$$

28.

29. $100\ \text{cm} = 1\ \text{m}$
$211\ \text{cm} = \mathbf{2.11\ m}$

30. 15 yards $\times$ 10 yards = **150 square yards**

LESSON 103, WARM-UP

a. **1000**

b. **275**

c. **$9.25**

d. **2100**

e. **$0.38**

f. **418**

g. **XXIX**

h. **=**

Problem Solving
February 29

LESSON 103, LESSON PRACTICE

a. A fraction equals 1 if its numerator and denominator are the same.

$$\frac{6}{6}$$

b. A fraction equals 1 if its numerator and denominator are the same.

B. $\frac{10}{10}$

c. There are six equal parts.

$$\frac{6}{6}$$

d. There are nine equal parts.

$$\frac{9}{9}$$

e. Since 6 is half of 12, $\frac{6}{12}$ is equal to $\frac{1}{2}$.

f. The fraction $\frac{1}{2}$ is equal to $\frac{10}{20}$. Since $\frac{9}{20} < \frac{10}{20}$, we know that $\frac{9}{20}$ $\bigcirc\!\!\!<$ $\frac{1}{2}$.

g. $\frac{3}{8}$ is less than $\frac{4}{8}$, which equals $\frac{1}{2}$. $5\frac{3}{8}$ is less than $5\frac{1}{2}$, so it rounds down to **5**.

LESSON 103, MIXED PRACTICE

1.
$$6\overline{)84}$$
$$\begin{array}{r} 14 \\ 6\overline{)84} \\ \underline{6} \\ 24 \\ \underline{24} \\ 0 \end{array}$$
84

2.
$$1 \text{ hour} = 60 \text{ minutes}$$
$$3 \times 60 \text{ minutes} = \textbf{180 minutes}$$

3. Mary: $8 - $2 = $6
$$\$8 + \$6 = \textbf{\$14}$$

4. (a) Tenths have one decimal place.
0.3

(b) Hundredths have two decimal places.
4.99

(c) Thousandths have three decimal places.
12.001

5.

40 attendees	
$\frac{5}{8}$ gave money.	5 attendees
	5 attendees
	5 attendees
	5 attendees
	5 attendees
$\frac{3}{8}$ did not give money.	5 attendees
	5 attendees
	5 attendees

5×5 attendees = **25 attendees**

6. (a) 10 mm = 1 cm
30 mm = **3 cm**

(b) The radius is half the diameter.
1.5 cm

7. The radius is half the diameter.
One half = **50%**

8. $50 \times 70 = \textbf{3500}$

$$\begin{array}{r} \overset{5}{\overset{7}{4}9} \\ \times\ 68 \\ \hline 392 \\ 2940 \\ \hline \textbf{3332} \end{array}$$

9.
20 blocks ÷ 4 containers
= **5 blocks** in 1 container
20 containers × 5 blocks in each container
= **100 blocks**

10. 6 + 4 + 6 + 4 = 20 reporters
20 reporters ÷ 4 = **5 reporters**

11. $20 - $6 = $14

$$\begin{array}{r} \$2.80 \\ 5\overline{)\$14.00} \\ \underline{10} \\ 4\,0 \\ \underline{4\,0} \\ 00 \\ \underline{0} \\ 0 \end{array}$$

12. 80, 85, 90, 90, 100, 100, 100, 100
90 + 100 = 190
190 ÷ 2 = 95
Median: 95
100 is the score that occurs most often
Mode: 100
100 − 80 = 20
Range: 20

13.
$$
\begin{array}{r}
\overset{5\ \ 3}{\ \$3.85} \\
\times \quad\quad 7 \\
\hline
\$26.95
\end{array}
$$

14.
$$
\begin{array}{r}
\overset{1}{\overset{7}{\ \ 48}} \\
\times\ \ 29 \\
\hline
432 \\
960 \\
\hline
\mathbf{1392}
\end{array}
$$

15.
$$
\begin{array}{r}
\overset{4}{16} \\
15 \\
23 \\
\overset{1}{}\ 8 \\
217 \\
20 \\
6 \\
+\ 317 \\
\hline
\mathbf{622}
\end{array}
$$

16.
$$
\begin{array}{r}
39 \\
+\ \ N \\
\hline
45
\end{array}
$$

$$
\begin{array}{r}
39 \\
+\ \ 6 \\
\hline
45
\end{array}
$$

$$N = \mathbf{6}$$

17. $60 \times 60 = \mathbf{3600}$

18.
$$
\begin{array}{r}
\overset{5}{59} \\
\times\ \ 61 \\
\hline
59 \\
3540 \\
\hline
\mathbf{3599}
\end{array}
$$

19. Divide mentally: **80**

20.
$$
\begin{array}{r}
\mathbf{970\ R\ 4} \\
6\overline{)5824} \\
54 \\
\hline
42 \\
42 \\
\hline
04 \\
0 \\
\hline
4
\end{array}
$$

21.
$$
\begin{array}{r}
\mathbf{\$4.17} \\
9\overline{)\$37.53} \\
36 \\
\hline
1\ 5 \\
9 \\
\hline
63 \\
63 \\
\hline
0
\end{array}
$$

22.
$$
\begin{array}{r}
\mathbf{600\ R\ 5} \\
7\overline{)4205} \\
42 \\
\hline
00 \\
0 \\
\hline
05 \\
0 \\
\hline
5
\end{array}
$$

23.
$$
\begin{array}{r}
3.\ 2\ \overset{4}{\cancel{8}}{}^{1}0 \\
-\ 0.\ 1\ 2\ 5 \\
\hline
3.\ 1\ 2\ 5
\end{array}
$$

$$
\begin{array}{r}
7.\ \overset{4}{\cancel{8}}\ \overset{9}{\cancel{0}}{}^{1}0 \\
-\ 3.\ 1\ 2\ 5 \\
\hline
\mathbf{4.\ 3\ 7\ 5}
\end{array}
$$

24.

25. A square has four equal sides.
20 in. ÷ 4 = **5 in.**
Area: 5 in. × 5 in. = **25 sq. in.**

26. A fraction equals 1 if its numerator and denominator are the same.
$$\mathbf{\frac{8}{8}}$$

27. The only way that two dot cubes will total 12 is 6 + 6.
Two dot cubes can total 7 in several different ways (1 + 6, 2 + 5, etc. . . .).
Dots totaling 7; There are more ways for the dots on two dot cubes to total 7 than to total 12.

28. 100 cm = 1 m
28 cm = **0.28 m**

29. $\frac{5}{12}$ is less than $\frac{6}{12}$, which equals $\frac{1}{2}$. $12\frac{5}{12}$ is less than $12\frac{1}{2}$, so it rounds down to **12.**

30. (a) **Rectangular solid (or rectangular prism)**

   (b)   4 edges on the bottom
          4 edges on the top
       + 4 edges running from top to bottom
          12 edges

LESSON 104, WARM-UP

a. **799**

b. **226**

c. **$6.75**

d. **600**

e. **$7.70**

f. **48**

g. **XXXIV**

h. **<**

Problem Solving

100 yards = 300 feet
300 feet ÷ 50 feet = 6
There are six 50-foot driveway lengths in
100 yards. Todd counted 8 full turns of the
wheel on his driveway. So the number of times
the wheel would turn in 100 yards is 6 × 8, or
48 times.

LESSON 104, LESSON PRACTICE

a. $2\overline{)7}$ with quotient $3\frac{1}{2}$

   $\frac{6}{1}$

b. **12 ÷ 3 = 4**

c. $3\overline{)8}$ with quotient $2\frac{2}{3}$

   $\frac{6}{2}$

d. **15 ÷ 5 = 3**

LESSON 104, MIXED PRACTICE

1. A square has four equal sides.
   280 ft ÷ 4 = **70 ft**

2. There are 7 days in one full week.

   $\frac{52 \text{ R } 1}{7\overline{)365}}$
   $\frac{35}{15}$
   $\frac{14}{1}$

   There are **52 full weeks** in a common year.

3. 6 friends × 3 cookies for each friend
               = 18 cookies
   18 cookies + 2 cookies = **20 cookies**

4.

60 kangaroos	
$\frac{3}{5}$ were less than 2 ft tall.	12 kangaroos
	12 kangaroos
	12 kangaroos
$\frac{2}{5}$ were greater than 2 ft tall.	12 kangaroos
	12 kangaroos

   3 × 12 kangaroos = **36 kangaroos**

5. (a) Use the top ruler.
       43 mm

   (b) Use the bottom ruler.
       4.3 cm

6. There are eight equal parts.
   $\frac{8}{8}$

7. (a) $7.86 > $7.50, so $7.86 rounds up to **$8**

   (b) 7.86 > 7.5, so 7.86 rounds up to **8**

8. $90 \times 70 =$ **6300**

$$
\begin{array}{r}
\overset{4}{87} \\
\times\ 71 \\
\hline
87 \\
6090 \\
\hline
\textbf{6177}
\end{array}
$$

9. $4\overline{)5}$ with quotient $1\frac{1}{4}$

$$
\begin{array}{r}
1\frac{1}{4} \\
4\overline{)5} \\
4 \\
\hline
1
\end{array}
$$

10. $100\% - 10\% =$ **90%**

11. Pattern:
Number of groups $\times$ number in each group
$\qquad\qquad$ = total
Problem: 73 days $\times$ 30 lb each day = N lb

$$
\begin{array}{r}
73 \\
\times\ 30 \\
\hline
\textbf{2190 lb}
\end{array}
$$

12. Pattern:
Number of groups $\times$ number in each group
$\qquad\qquad$ = total
Problem: 6 days $\times$ N lb each day = 132 lb

$$
\begin{array}{r}
\textbf{22 lb} \\
6\overline{)132} \\
12 \\
\hline
12 \\
12 \\
\hline
0
\end{array}
$$

13.
$$
\begin{array}{r}
\overset{1\ 1}{\$_1 6.52} \\
\$12.00 \\
\$1.74 \\
+\ \$0.26 \\
\hline
\textbf{\$20.52}
\end{array}
$$

14.
$$
\begin{array}{r}
\overset{1\ 1}{3.65} \\
2.7 \\
0.454 \\
+\ 2.0 \\
\hline
\textbf{8.804}
\end{array}
$$

15. $\$63.72 + \$2.00 =$ $\$65.72$

$$
\begin{array}{r}
\overset{7\ \ 9\ \ 9}{\$8\,0.\,\cancel{0}{}^1 0} \\
-\ \$6\,5.\,7\,2 \\
\hline
\textbf{\$1\,4.\,2\,8}
\end{array}
$$

16.
$$
\begin{array}{r}
\overset{2}{\cancel{3}}{}^1 7,\ \overset{5}{\cancel{6}}\ \overset{1}{\cancel{1}}{}^1 4 \\
-\ 2\,9,1\,4\,8 \\
\hline
\textbf{8\ 4\ 6\ 6}
\end{array}
$$

17. $9 \times 26 = 9 \times 26$
$\qquad W =$ **26**

18. $3 \times 3 = 9$
$\qquad 9 \times 3 = 27$

$$
\begin{array}{r}
\overset{2}{27} \\
\times\ \ 3 \\
\hline
\textbf{81}
\end{array}
$$

19. Multiply mentally: **24,000**

20.
$$
\begin{array}{r}
\overset{5}{\$0.79} \\
\times\ \ \ \ 6 \\
\hline
\textbf{\$4.74}
\end{array}
$$

21. Multiply mentally: **2500**

22.
$$
\begin{array}{r}
51 \\
\times\ 49 \\
\hline
459 \\
2040 \\
\hline
\textbf{2499}
\end{array}
$$

23.
$$
\begin{array}{r}
\overset{4}{\overset{2}{47}} \\
\times\ 63 \\
\hline
141 \\
2820 \\
\hline
\textbf{2961}
\end{array}
$$

24.
$$
\begin{array}{r}
\textbf{576} \\
4\overline{)2304} \\
20 \\
\hline
30 \\
28 \\
\hline
24 \\
24 \\
\hline
0
\end{array}
$$

25.
$$
\begin{array}{r}
\textbf{963} \\
5\overline{)4815} \\
45 \\
\hline
31 \\
30 \\
\hline
15 \\
15 \\
\hline
0
\end{array}
$$

26.

$$
\begin{array}{r}
604\ R\ 5 \\
6\overline{)3629} \\
\underline{36} \\
02 \\
\underline{0} \\
29 \\
\underline{24} \\
5
\end{array}
$$

27. $\sqrt{49} = 7$

$$
\begin{array}{r}
205 \\
7\overline{)1435} \\
\underline{14} \\
03 \\
\underline{0} \\
35 \\
\underline{35} \\
0
\end{array}
$$

28. 100 cm = 1 m
25 cm = **0.25 m**

29. $\frac{5}{8}$ is greater than $\frac{4}{8}$, which equals $\frac{1}{2}$. $16\frac{5}{8}$ is greater than $16\frac{1}{2}$, so it rounds up to **17.**

30. $3\frac{2}{3}$ is between 3 and 4. $4\frac{1}{2}$ is between 4 and 5. $3 + 4 = 7$ and $4 + 5 = 9$, so the sum is between **C. 7 and 9.**

LESSON 105, WARM-UP

a. 501

b. 426

c. $9.26

d. 3000

e. $3.75

f. 37

g. XXXIX

h. <

Patterns

The pattern is "count up by five then count down by two."

0, 5, 3, 8, 6, 11, 9, 14, **12, 17, 15, 20, 18,** …

LESSON 105, LESSON PRACTICE

a.
$$
\begin{array}{r}
7\ R\ 3 \\
10\overline{)73} \\
\underline{70} \\
3
\end{array}
$$

b.
$$
\begin{array}{r}
34\ R\ 2 \\
10\overline{)342} \\
\underline{30} \\
42 \\
\underline{40} \\
2
\end{array}
$$

c.
$$
\begin{array}{r}
24\ R\ 3 \\
10\overline{)243} \\
\underline{20} \\
43 \\
\underline{40} \\
3
\end{array}
$$

d.
$$
\begin{array}{r}
72 \\
10\overline{)720} \\
\underline{70} \\
20 \\
\underline{20} \\
0
\end{array}
$$

e.
$$
\begin{array}{r}
56\ R\ 1 \\
10\overline{)561} \\
\underline{50} \\
61 \\
\underline{60} \\
1
\end{array}
$$

f.
$$
\begin{array}{r}
38 \\
10\overline{)380} \\
\underline{30} \\
80 \\
\underline{80} \\
0
\end{array}
$$

g. There is no remainder when the last digit of the whole-number dividend is zero.
C. 560

LESSON 105, MIXED PRACTICE

1. 2 quarters = 50¢

$$
\begin{array}{r}
8\ R\ 2 \\
6\overline{)50} \\
\underline{48} \\
2
\end{array}
$$

8 mints can be bought with 2 quarters

2. 50¢ is one half of a dollar
One half = **50%**

3. Jason: $8
David: $8 + $2 = $10
$8 + $10 = **$18**

4.

```
            32 surfers
          ┌ 4 surfers ┐
3/8 caught │ 4 surfers │
the wave.  └ 4 surfers ┘
          ┌ 4 surfers ┐
          │ 4 surfers │
5/8 did not│ 4 surfers │
catch the  │ 4 surfers │
wave.      └ 4 surfers ┘
```

3 × 4 surfers = **12 surfers**

5. **A number card is more likely to be drawn. In a standard deck, there are more number cards than face cards.**

6. A fraction equals 1 if its numerator and denominator are the same.
$$\frac{10}{10}$$

7. Three decimal places indicate thousandths. **Eighty-six and seven hundred forty-three thousandths**

8. 600 − 500 = **100**

9. (a) $1\frac{4}{5}$
$5\overline{)9}$
$\frac{5}{4}$

(b) 9 ÷ 3 = **3**

(c) $4\frac{1}{2}$
$2\overline{)9}$
$\frac{8}{1}$

10. Pattern:
Number of groups × number in each group
= total
Problem: 10 days × N people each day
= 2400 people

2400 ÷ 10 = 240 people;
About 240 people came each day

$\overset{2}{240}$ people
× 7
1680 people
About 1680 people came in 1 week

11. **4.3 cm**

12. $\overset{1}{\$4.20}$
× 6
$25.20

$\overset{1}{\$0.12}$
× 8
$0.96

$\overset{1}{\$25.20}$
+ $0.96
$26.16

13. (a) The tenths place is the first place to the right of the decimal point.
7

(b) 86.74 < 86.75, so 86.74 is closer to **86.7** than to 86.8.

14. A trapezoid has four sides and exactly one pair of parallel sides.
One possibility:

15. $\overset{1}{2.8}$
+ 0.56
3.36

4,867
− 3.360
1.507

16. 30 × 30 = **900**

17. $\overset{3}{54}$
× 29
486
1080
1566

18. $\overset{23}{10\overline{)230}}$
20
30
30
0

Saxon Math 5/4—Homeschool

211

19.
$$\begin{array}{r} 340 \text{ R } 3 \\ 7\overline{)2383} \\ \underline{21} \\ 28 \\ \underline{28} \\ 03 \\ \underline{0} \\ 3 \end{array}$$

20.
$$\begin{array}{r} 37 \text{ R } 2 \\ 10\overline{)372} \\ \underline{30} \\ 72 \\ \underline{70} \\ 2 \end{array}$$

21.
$$\begin{array}{r} \$0.72 \\ 8\overline{)\$5.76} \\ \underline{5\,6} \\ 16 \\ \underline{16} \\ 0 \end{array}$$

22.
$$\begin{array}{r} \overset{2}{12} \\ 26 \\ 13 \\ {}_1 35 \\ 110 \\ 8 \\ +\ 15 \\ \hline 219 \end{array}$$

23.
$$\begin{array}{r} {}^{111\ 11} \\ 351,426 \\ +\ 449,576 \\ \hline 801,002 \end{array}$$

24.
$$\begin{array}{r} \overset{4\ 9\ 9}{\$5\,0.\,0^1 0} \\ -\ \$4\,9.\,4\,9 \\ \hline \$0.\,5\,1 \end{array}$$

25.
$$\begin{array}{r} \overset{1\ 3\ 7}{\$12.49} \\ \times\qquad 8 \\ \hline \$99.92 \end{array}$$

26.
$$\begin{array}{r} \overset{1}{73} \\ \times\ 62 \\ \hline 146 \\ 4380 \\ \hline 4526 \end{array}$$

27. 300 ft + 200 ft + 300 ft + 200 ft = **1000 ft**

28.
$$\begin{array}{r} 2\ \overset{8}{\cancel{9}}\overset{17}{}9 \text{ mi} \\ -\ 2\,7\,8\,6 \text{ mi} \\ \hline 1\,9\,3 \text{ mi} \end{array}$$

29.
$$\begin{array}{r} {}^{1\,1} \\ {}_1 917 \text{ mi} \\ 1387 \text{ mi} \\ +\ 2054 \text{ mi} \\ \hline 4358 \text{ mi} \end{array}$$

30. The distance from Chicago to Chicago is 0 miles, from Los Angeles to Los Angeles is 0 miles, and from New York City to New York City is 0 miles. **0**

LESSON 106, WARM-UP

a. **12**

b. **8**

c. **6**

d. **72**

e. **$8.26**

f. **16**

g. **110**

h. **70**

Problem Solving

Two cups are in a pint, so a cup weighs 1 lb ÷ 2, or $\frac{1}{2}$ **lb.**

Two pints are in a quart, so a quart weighs 1 lb × 2, or **2 lb.**

Two quarts are in a half gallon, so a half gallon weighs 2 lb × 2, or **4 lb.**

Two half gallons are in a gallon, so a gallon weighs 4 lb × 2, or **8 lb.**

LESSON 106, LESSON PRACTICE

a. $M - 10$
  $12 - 10 = $ **2**

b. $A + B$
  $9 + 15 = $ **24**

c. xy

$6 \times 7 = $ **42**

d. w^2

$5 \times 5 = $ **25**

e. $A = lw$

$A = 8 \times 4$

$A = $ **32**

f. $\dfrac{m}{n}$

$12 \div 3 = $ **4**

g. $\sqrt{t}$

$\sqrt{16} = $ **4**

LESSON 106, MIXED PRACTICE

1. 6 cats each eat $\frac{1}{2}$ can each day.

6 half cans $=$ 3 whole cans.

3 cans

2.
$$\begin{array}{r} \overset{2}{\$0.47} \\ \times \quad 3 \\ \hline \mathbf{\$1.41} \end{array}$$

3.
$$\begin{array}{r} \$\overset{2}{1}.41 \\ \times \quad 7 \\ \hline \mathbf{\$9.87} \end{array}$$

4. Each side of the square room is the same length.

$120 \text{ ft} \div 4 = $ **30 ft**

Area: $30 \text{ ft} \times 30 \text{ ft} = $ **900 sq. ft**

5.

28 students

4 students

Math was the favorite subject of $\frac{5}{7}$.
- 4 students
- 4 students
- 4 students
- 4 students
- 4 students

Math was not the favorite subject of $\frac{2}{7}$.
- 4 students
- 4 students

5×4 students $= $ **20 students**

6.

Admission	Admission
75¢	$0.75
☆ each ☆	☆ each ☆

7. A diameter equals two radii.

$1\frac{1}{2}$ in. $+ 1\frac{1}{2}$ in. $= $ **3 in.**

8. Two decimal places indicate hundredths.

Five hundred twenty-three and forty-three hundredths

9. $60 \times 400 = $ **24,000**

10. (a) $10 \div 10 = $ **1**

(b) $10 \div 5 = $ **2**

(c)
$$\begin{array}{r} 3\frac{1}{3} \\ 3\overline{)10} \\ \underline{9} \\ 1 \end{array}$$

11.
$$\begin{array}{r} \overset{2}{\$6}.\overset{1}{8}5 \\ \$4.50 \\ + \quad \$0.75 \\ \hline \$12.10 \end{array}$$

$$\begin{array}{r} \$\overset{1}{2}\overset{9}{\cancel{0}}.\overset{1}{0}0 \\ - \quad \$1\,2.\,1\,0 \\ \hline \mathbf{\$7.\,9\,0} \end{array}$$

12.
$$\begin{array}{r} \$\overset{1}{7}.40 \\ \$7.40 \\ + \quad \$0.98 \\ \hline \$15.78 \end{array}$$

$$\begin{array}{r} \$\overset{1}{2}\overset{9}{\cancel{0}}.\overset{9}{\cancel{0}}\overset{1}{\cancel{0}} \\ - \quad \$1\,5.\,7\,8 \\ \hline \mathbf{\$4.\,2\,2} \end{array}$$

13. Pattern:

Number of groups $\times$ number in each group

$= $ total

Problem: 5 hours $\times$ N miles per hour

$= 140$ miles

$$\begin{array}{r} \textbf{28 miles per hour} \\ 5\overline{)140} \\ \underline{10} \\ 40 \\ \underline{40} \\ 0 \end{array}$$

14. The fraction $\frac{1}{2}$ is equal to $\frac{50}{100}$. Since $\frac{49}{100} < \frac{50}{100}$, we know that $\frac{49}{100} \bigcirc\!\!\!< \frac{1}{2}$.

15. (a) $12.25 < $12.50, so $12.25 rounds down to **$12.**

    (b) 12.25 < 12.5, so 12.25 rounds down to **12.**

16. (a) The tenths place is the first place to the right of the decimal point.
    4

    (b) 36.47 > 36.45, so 36.47 is closer to **36.5** than to 36.4.

17.
$$\begin{array}{r} \overset{1\,2}{7}\overset{1}{3}.48 \\ 5.63 \\ +\ 17.9 \\ \hline \mathbf{97.01} \end{array}$$

18.
$$\begin{array}{r} \$\overset{5}{6}\,\overset{14}{5}.\,\overset{9}{\cancel{0}}\,{}^1 0 \\ -\ \$2\,9.\,8\,7 \\ \hline \$3\,5.\,1\,3 \end{array}$$

19.
$$\begin{array}{r} \overset{1}{2}\,\overset{13}{4},\overset{13}{3}\,\overset{6}{7}{}^1 5 \\ -\ \ \ 8,4\,1\,6 \\ \hline 1\,5,9\,5\,9 \end{array}$$

20.
$$\begin{array}{r} \overset{6\,7}{\$3.68} \\ \times\ \ \ \ 9 \\ \hline \$33.12 \end{array}$$

21.
$$\begin{array}{r} \overset{8}{89} \\ \times\ 91 \\ \hline 89 \\ 8010 \\ \hline \mathbf{8099} \end{array}$$

22.
$$\begin{array}{r} 254\ \text{R}\ 1 \\ 3\overline{)763} \\ \underline{6} \\ 16 \\ \underline{15} \\ 13 \\ \underline{12} \\ 1 \end{array}$$

23.
$$\begin{array}{r} 43 \\ 10\overline{)430} \\ \underline{40} \\ 30 \\ \underline{30} \\ 0 \end{array}$$

24.
$$\begin{array}{r} \$9.54 \\ 6\overline{)\$57.24} \\ \underline{54} \\ 3\ 2 \\ \underline{3\ 0} \\ 24 \\ \underline{24} \\ 0 \end{array}$$

25.
$$\begin{array}{r} 85 \\ 9\overline{)765} \\ \underline{72} \\ 45 \\ \underline{45} \\ 0 \end{array}$$

26.
$$\begin{array}{r} 56\ \text{R}\ 3 \\ 10\overline{)563} \\ \underline{50} \\ 63 \\ \underline{60} \\ 3 \end{array}$$

27. n^2
    $90 \times 90 = \mathbf{8100}$

28. $\dfrac{m}{\sqrt{m}}$

    $\dfrac{36}{6} = \mathbf{6}$

29. $6\frac{3}{4}$ is between 6 and 7. $5\frac{3}{5}$ is between 5 and 6.
    $6 + 5 = 11$ and $7 + 6 = 13$, so the sum is between **D. 11 and 13.**

30.   1 ton = 2000 pounds
    7 tons $\times$ 2000 pounds = **14,000 pounds**

LESSON 107, WARM-UP

a. **15**

b. **10**

c. **6**

d. **1400**

e. **$2.25**

f. **76**

g. 90

h. 65

Problem Solving

The Sunday in spring is 24 hours $-$ 1 hour, or **23 hours long.**

The Sunday in fall is 24 hours $+$ 1 hour, or **25 hours long.**

LESSON 107, LESSON PRACTICE

a. Add only the numerators.

$\dfrac{2}{3}$

b. Add only the numerators.

$\dfrac{3}{4}$

c. Add only the numerators.

$\dfrac{7}{10}$

d. Subtract only the numerators.

$\dfrac{1}{3}$

e. Subtract only the numerators.

$\dfrac{1}{4}$

f. Subtract only the numerators.

$\dfrac{3}{10}$

g.

$$\begin{array}{r} 2\frac{1}{4} \\ + \ 4\frac{2}{4} \\ \hline \mathbf{6\frac{3}{4}} \end{array}$$

h.

$$\begin{array}{r} 5\frac{3}{8} \\ + \ 1\frac{2}{8} \\ \hline \mathbf{6\frac{5}{8}} \end{array}$$

i.

$$\begin{array}{r} 8 \\ + \ 1\frac{2}{5} \\ \hline \mathbf{9\frac{2}{5}} \end{array}$$

j.

$$\begin{array}{r} 4\frac{3}{5} \\ - \ 1\frac{1}{5} \\ \hline \mathbf{3\frac{2}{5}} \end{array}$$

k.

$$\begin{array}{r} 9\frac{3}{4} \\ - \ 4\frac{2}{4} \\ \hline \mathbf{5\frac{1}{4}} \end{array}$$

l.

$$\begin{array}{r} 12\frac{8}{9} \\ - \ 3\frac{3}{9} \\ \hline \mathbf{9\frac{5}{9}} \end{array}$$

m. Add only the numerators.

$\dfrac{3}{8} + \dfrac{4}{8} = \mathbf{\dfrac{7}{8}}$

n.

$$\begin{array}{r} 3\frac{1}{2} \text{ miles} \\ + \ 3\frac{1}{2} \text{ miles} \\ \hline 6\frac{2}{2} \text{ miles} = \mathbf{7 \text{ miles}} \end{array}$$

LESSON 107, MIXED PRACTICE

1.

$$\begin{array}{r} \overset{3\ 2}{\$2.75} \\ \times \quad 5 \\ \hline \$13.75 \end{array}$$

$$\begin{array}{r} \$\overset{1}{2}\overset{9}{\emptyset}.\,\overset{9}{\emptyset}{}^{1}0 \\ - \ \$1\,3.\,7\,5 \\ \hline \mathbf{\$6.\,2\,5} \end{array}$$

2.

$$\begin{array}{r} 16 \text{ R } 2 \\ 3{\overline{)50¢}} \\ \underline{3} \\ 20 \\ \underline{18} \\ 2 \end{array}$$

There will be **2 cents** left over.

3.
$$\begin{array}{r} {}^{8}\cancel{9}\,{}^{9}\cancel{0}\,{}^{1}4 \\ -\ 4\ 0\ 9 \\ \hline \mathbf{4\ 9\ 5} \end{array}$$

4.

45 stamps

$\frac{2}{9}$ were from Brazil.

5 stamps
5 stamps
5 stamps
5 stamps
5 stamps

$\frac{7}{9}$ were not from Brazil.

5 stamps
5 stamps
5 stamps
5 stamps

2×5 stamps $=$ **10 stamps**

5. (a) Use the top ruler.
   27 mm

   (b) Use the bottom ruler.
   2.7 cm

6. A fraction equals 1 if its numerator and denominator are the same.
   $$\frac{\mathbf{10}}{\mathbf{10}}$$

7. One out of ten slices is one tenth of the pizza. One tenth $=$ **10%.**

8. A number cube has six sides labeled 1–6. The most likely outcome if a number cube is tossed once is **C. a number greater than 1.**

9. $5167 < 5500$, so 5167 rounds down to **5000.**

10.
$$\begin{array}{r} \mathbf{2\frac{1}{4}} \\ 4\overline{)9} \\ \underline{8} \\ 1 \end{array}$$

11. A fraction equals 1 if its numerator and denominator are the same.
    $$\text{C. } \frac{\mathbf{11}}{\mathbf{10}}$$

12.
$$\begin{array}{r} {}^{5}17 \text{ stores} \\ \times\quad 8 \\ \hline \mathbf{136} \text{ stores} \end{array}$$

13. Pattern:

    Number of groups $\times$ number in each group
    $=$ total

    Problem: 9 days $\times$ N miles per day
    $=$ 243 miles

$$\begin{array}{r} \mathbf{27}\ \textbf{miles} \\ 9\overline{)243} \\ \underline{18} \\ 63 \\ \underline{63} \\ 0 \end{array}$$

14.
$$\begin{array}{r} 1\frac{1}{2} \text{ hours} \\ +\ 2\frac{1}{2} \text{ hours} \\ \hline 3\frac{2}{2} \text{ hours} = \mathbf{4\ hours} \end{array}$$

15. $\frac{21}{100}$ is less than $\frac{50}{100}$, which equals $\frac{1}{2}$. $8\frac{21}{100}$ is less than $8\frac{1}{2}$, so it rounds down to **8.**

16.
$$\begin{array}{r} {}^{2}\cancel{3}\,{}^{1\!5}\cancel{6}.{}^{1}3\ 1 \\ -\quad 7.\ 4 \\ \hline \mathbf{2\ 8.\ 9\ 1} \end{array}$$

17. Add only the numerators.
    $$\frac{\mathbf{7}}{\mathbf{8}}$$

18.
$$\begin{array}{r} 18 \\ +\ N \\ \hline 25 \end{array}$$

$$\begin{array}{r} 18 \\ +\ 7 \\ \hline 25 \end{array}$$

$N = \mathbf{7}$

19. Subtract only the numerators.
    $$\frac{\mathbf{7}}{\mathbf{10}}$$

20.
$$\begin{array}{r} 3\frac{2}{5} \\ +\ 1\frac{1}{5} \\ \hline \mathbf{4\frac{3}{5}} \end{array}$$

21.
$$\begin{array}{r} \overset{2}{} \\ \overset{1}{27} \\ \times\ 32 \\ \hline 54 \\ 810 \\ \hline \mathbf{864} \end{array}$$

22.
$$\begin{array}{r} \overset{1}{62} \\ \times\ 15 \\ \hline 310 \\ 620 \\ \hline \mathbf{930} \end{array}$$

23. $7 \times 7 = 49$

$\sqrt{49} = 7$

$$\begin{array}{r} \overset{6}{49} \\ \times\ 7 \\ \hline \mathbf{343} \end{array}$$

24.
$$\begin{array}{r} \mathbf{46} \\ 10\overline{)460} \\ \underline{40} \\ 60 \\ \underline{60} \\ 0 \end{array}$$

25.
$$\begin{array}{r} \mathbf{\$3.04} \\ 9\overline{)\$27.36} \\ \underline{27} \\ 0\ 3 \\ \underline{0} \\ 36 \\ \underline{36} \\ 0 \end{array}$$

26.
$$\begin{array}{r} \mathbf{386} \\ 6\overline{)2316} \\ \underline{18} \\ 51 \\ \underline{48} \\ 36 \\ \underline{36} \\ 0 \end{array}$$

27.
$$\begin{array}{r} \mathbf{220\ R\ 3} \\ 7\overline{)1543} \\ \underline{14} \\ 14 \\ \underline{14} \\ 03 \\ \underline{0} \\ 3 \end{array}$$

28.
$$\begin{array}{r} \mathbf{53\ R\ 2} \\ 10\overline{)532} \\ \underline{50} \\ 32 \\ \underline{30} \\ 2 \end{array}$$

29.
$$\begin{array}{r} \mathbf{32} \\ 8\overline{)256} \\ \underline{24} \\ 16 \\ \underline{16} \\ 0 \end{array}$$

30. 60 feet $\times$ 40 feet = **2400 square feet**

LESSON 108, WARM-UP

a. 18

b. 12

c. 9

d. 3600

e. $12.25

f. 15

g. 150

h. 76

Patterns

1, 1, 2, 3, 5, 8, **13**, **21**, **34**, **55**, …

LESSON 108, LESSON PRACTICE

a. $6(10 + 6) = \mathbf{60 + 36 = 96}$

b. $2(15\,\text{cm} + 10\,\text{cm})$

$30\,\text{cm} + 20\,\text{cm} = \mathbf{50\ cm}$

c. $A = S^2$

$A = 20\,\text{ft} \times 20\,\text{ft}$

$A = \mathbf{400\ sq.\ ft}$

LESSON 108, MIXED PRACTICE

1. $$\overset{4}{\cancel{5}}.\overset{9}{\cancel{\emptyset}}{}^{1}0$$
    $$-\ \$1.\ 9\ 6$$
    $$\overline{\$3.\ 0\ 4}$$

    $$
    \begin{array}{r}
    \mathbf{\$0.38} \\
    8)\overline{\$3.04} \\
    \underline{2\ 4} \\
    64 \\
    \underline{64} \\
    0
    \end{array}
    $$

2. A dozen is 12.
    12 cookies − 2 cookies = 10 cookies
    Half of 10 is 5.
    10 cookies − 5 cookies = **5 cookies**

3. 5 × 4 = 20
    20 − 6 = **14**

4. $\frac{2}{3}$ were out of tune.
    $\frac{1}{3}$ were not out of tune.

12 strings
4 strings
4 strings
4 strings

    2 × 4 guitar strings = **8 guitar strings**

5. One side out of six sides on a dot cube has exactly two dots.
    $$\frac{1}{6}$$

6. A fraction equals 1 if its numerator and denominator are the same.
    $$\frac{5}{5}$$

7. **Three hundred ninety-seven and three fourths**

8. 4178 rounds to 4000 and 6899 rounds to 7000
    4000 + 7000 = **11,000**

9. (a) $3)\overline{7}$ → $2\frac{1}{3}$
    $$\underline{6}$$
    $$1$$

    (b) 8 ÷ 4 = **2**

    (c) $5)\overline{9}$ → $1\frac{4}{5}$
    $$\underline{5}$$
    $$4$$

10. $$
    \begin{array}{r}
    {}_{2}8\ \text{miles} \\
    15\ \text{miles} \\
    11\ \text{miles} \\
    +\ \ 18\ \text{miles} \\
    \hline
    52\ \text{miles}
    \end{array}
    $$

    $$
    \begin{array}{r}
    \mathbf{13}\ \textbf{miles} \\
    4)\overline{52} \\
    \underline{4} \\
    12 \\
    \underline{12} \\
    0
    \end{array}
    $$

11. 3 hours × 3 miles per hour = 9 miles
    2 hours × 4 miles per hour = 8 miles
    9 miles + 8 miles = 17 miles
    25 miles − 17 miles = **8 miles**

12. A pint is half of a quart.
    One half = **50%**

13. $$
    \begin{array}{r}
    \overset{1}{4}1.6 \\
    13.17 \\
    +\ \ \ 9.2 \\
    \hline
    \mathbf{63.97}
    \end{array}
    $$

14. $$
    \begin{array}{r}
    \overset{1}{2}\overset{\ 1\ ^{1}5}{\cancel{6}}.{}^{1}4\ 7 \\
    -\ \ \ \ 8.\ 7\ 0 \\
    \hline
    1\ 7.\ 7\ 7
    \end{array}
    $$

15. $$
    \begin{array}{r}
    6\frac{3}{8} \\
    +\ 4\frac{2}{8} \\
    \hline
    10\frac{5}{8}
    \end{array}
    $$

16. $$
    \begin{array}{r}
    4\frac{7}{10} \\
    -\ 1\frac{6}{10} \\
    \hline
    \mathbf{3\frac{1}{10}}
    \end{array}
    $$

17. One dozen = 12

    $$
    \begin{array}{r}
    54\ \text{dozen} \\
    \times\ \ 12\ \text{eggs per dozen} \\
    \hline
    108 \\
    540 \\
    \hline
    \mathbf{648}\ \textbf{eggs}
    \end{array}
    $$

18. $\frac{1}{5} + \frac{2}{5} = \frac{3}{5}$

19.
$$\begin{array}{r} \overset{4}{\$0.48} \\ \times \quad 5 \\ \hline \$2.40 \end{array}$$

20. $80 \times 80 = \mathbf{6400}$

21. $\sqrt{25} = 5$
$5 \times 5 = \mathbf{25}$

22.
$$\begin{array}{r} \$1.59 \\ 4\overline{)\$6.36} \\ \underline{4} \\ 2\ 3 \\ \underline{2\ 0} \\ 36 \\ \underline{36} \\ 0 \end{array}$$

23.
$$\begin{array}{r} 52 \\ 10\overline{)520} \\ \underline{50} \\ 20 \\ \underline{20} \\ 0 \end{array}$$

24.
$$\begin{array}{r} 35 \\ 5\overline{)175} \\ \underline{15} \\ 25 \\ \underline{25} \\ 0 \end{array}$$

25. Perimeter: $10\,m + 10\,m + 10\,m$
$+ 10\,m = \mathbf{40\ m}$
Area: $10\,m \times 10\,m = \mathbf{100\ sq.\ m}$

26. $5(40 + 8)$
$\mathbf{200 + 40 = 240}$

27. $100\,cm = 1\,m$
$76\,cm = \mathbf{0.76\ m}$

28. $A = bh$
$A = 5\,m \times 4\,m$
$A = \mathbf{20\ sq.\ m}$

29.
$$\begin{array}{r} 3\frac{4}{5} \text{ pies} \\ - 1\frac{3}{5} \text{ pies} \\ \hline \mathbf{2\frac{1}{5}} \text{ pies} \end{array}$$

30. $5\frac{3}{8}$ is between 5 and 6. $7\frac{4}{5}$ is between 7 and 8. $5 + 7 = 12$ and $6 + 8 = 14$, so the sum is between **B. 12 and 14.**

LESSON 109, WARM-UP

a. 20

b. 10

c. 4

d. 2400

e. $8.05

f. 210

g. 155

h. 40

Problem Solving

One common year is 365 days, which is 52 weeks plus one day. Fifty-two weeks after Monday is Monday, so 1 year after Monday is **Tuesday.**

LESSON 109, LESSON PRACTICE

a. 6 out of 8 parts are shaded and 3 out of 4 parts are shaded.
$$\frac{6}{8} = \frac{3}{4}$$

b. 3 out of 9 parts are shaded and 1 out of 3 parts is shaded.
$$\frac{3}{9} = \frac{1}{3}$$

c. One possibility:

d. One possibility:

e. One possibility:

f. $\frac{1}{4} \times \frac{2}{2} = \frac{2}{8}$

$\frac{1}{4} \times \frac{3}{3} = \frac{3}{12}$

$\frac{1}{4} \times \frac{4}{4} = \frac{4}{16}$

$\frac{1}{4} \times \frac{5}{5} = \frac{5}{20}$

g. $\frac{5}{6} \times \frac{2}{2} = \frac{10}{12}$

$\frac{5}{6} \times \frac{3}{3} = \frac{15}{18}$

$\frac{5}{6} \times \frac{4}{4} = \frac{20}{24}$

$\frac{5}{6} \times \frac{5}{5} = \frac{25}{30}$

h. $\frac{2}{5} \times \frac{2}{2} = \frac{4}{10}$

$\frac{2}{5} \times \frac{3}{3} = \frac{6}{15}$

$\frac{2}{5} \times \frac{4}{4} = \frac{8}{20}$

$\frac{2}{5} \times \frac{5}{5} = \frac{10}{25}$

i. $\frac{1}{10} \times \frac{2}{2} = \frac{2}{20}$

$\frac{1}{10} \times \frac{3}{3} = \frac{3}{30}$

$\frac{1}{10} \times \frac{4}{4} = \frac{4}{40}$

$\frac{1}{10} \times \frac{5}{5} = \frac{5}{50}$

LESSON 109, MIXED PRACTICE

1. $24 - 10 = $ **14 more cars**

2.
$$\overset{1}{24} \text{ vehicles}$$
$$10 \text{ vehicles}$$
$$2 \text{ vehicles}$$
$$+ \ 7 \text{ vehicles}$$
$$\overline{\textbf{43 vehicles}}$$

3. $7 + 8 = 15$
$15 - 6 = $ **9**

4.

180 pages

18 pages
18 pages
18 pages
18 pages
18 pages
18 pages
18 pages
18 pages
18 pages
18 pages

$\frac{3}{10}$ were read in one day.

$\frac{7}{10}$ were not read in one day.

$$\overset{2}{18} \text{ pages}$$
$$\times \ 3$$
$$\overline{\textbf{54 pages}}$$

5. (a) The diameter of the dime is **18 mm.**

(b) The radius is half the diameter. Half of 18 mm is **9 mm.**

6. $10 \text{ mm} = 1 \text{ cm}$
$18 \text{ mm} = $ **1.8 cm**

7. A fraction equals 1 if its numerator and denominator are the same.

$\frac{4}{4}$

8. $80 \times 30 = $ **2400**

$$\overset{2}{\underset{}{\overset{1}{78}}}$$
$$\times \ 32$$
$$\overline{156}$$
$$2340$$
$$\overline{\textbf{2496}}$$

9. $2\overline{)5}^{\ 2\frac{1}{2}}$

$\frac{4}{1}$

One possibility:

10.
$$\overset{2}{12} \text{ miles}$$
$$14 \text{ miles}$$
$$16 \text{ miles}$$
$$18 \text{ miles}$$
$$+ \ 20 \text{ miles}$$
$$\overline{\textbf{80 miles}}$$

11.
$$
\begin{array}{r}
\textbf{16 miles} \\
5{\overline{)80}} \\
\underline{5} \\
30 \\
\underline{30} \\
0
\end{array}
$$

12.
$$
\begin{array}{r}
14 \text{ red checkers} \\
- \quad 6 \text{ black checkers} \\
\hline
8 \text{ more red checkers}
\end{array}
$$
14 red; 6 black

13.
$$\frac{2}{3} \times \frac{2}{2} = \frac{\textbf{4}}{\textbf{6}}$$
$$\frac{2}{3} \times \frac{3}{3} = \frac{\textbf{6}}{\textbf{9}}$$
$$\frac{2}{3} \times \frac{10}{10} = \frac{\textbf{20}}{\textbf{30}}$$

14. $5(60 + 3)$
$300 + 15 = 315$

15. ac
$18 \times 22 = \textbf{396}$

16. b^2
$20 \times 20 = \textbf{400}$

17. 60, 80, 80, 80, 90, 90, 95, 100, 100
Median: 90
80 is the score that occurs most often.
Mode: 80
$100 - 60 = 40$
Range: 40

18. A parallelogram always has exactly two pairs of parallel sides.
B. Parallelogram

19.
$$
\begin{array}{r}
\overset{1}{\cancel{2}}\,\overset{\overset{1}{3}}{\cancel{4}}.\overset{1}{3}\,4 \\
- \quad 8.\,5\,0 \\
\hline
1\,5.\,8\,4
\end{array}
$$

20.
$$
\begin{array}{r}
2\,6.\overset{\overset{3}{}}{\cancel{4}}{}^{1}0 \\
- \quad 1\,5.\,1\,8 \\
\hline
1\,1.\,2\,2
\end{array}
$$

21. $3 \times 2 = 6$
$6 \times 4 = 24$
$24 \times 1 = \textbf{24}$

22.
$$
\begin{array}{r}
\overset{1}{2}6 \\
\times \quad 30 \\
\hline
780
\end{array}
$$

23.
$$
\begin{array}{r}
\textbf{\$2.06} \\
8{\overline{)\$16.48}} \\
\underline{16} \\
0\,4 \\
\underline{0} \\
48 \\
\underline{48} \\
0
\end{array}
$$

24.
$$
\begin{array}{r}
624 \\
6{\overline{)3744}} \\
\underline{36} \\
14 \\
\underline{12} \\
24 \\
\underline{24} \\
0
\end{array}
$$

25. Add only the numerators.
$$\frac{\textbf{11}}{\textbf{12}}$$

26. Subtract only the numerators.
$$\frac{\textbf{5}}{\textbf{12}}$$

27. $3 \text{ ft} \times 6 \text{ ft} = \textbf{18 sq. ft}$

28. Double the recipe.
$$
\begin{array}{r}
7\frac{1}{2} \text{ cups of flour} \\
+ \quad 7\frac{1}{2} \text{ cups of flour} \\
\hline
14\frac{2}{2} \text{ cups of flour} = \textbf{15 cups of flour}
\end{array}
$$

29. C.

30. (a) $13 - 6 = \textbf{7 red stripes}$

(b) 6 out of 13 stripes are white.
$$\frac{\textbf{6}}{\textbf{13}}$$

(c) 7 out of 13 stripes are red.
$$\frac{\textbf{7}}{\textbf{13}}$$

LESSON 110, WARM-UP

a. 50

b. 25

c. 10

d. 230

e. $7.35

f. 18

g. 1110

h. 41

Problem Solving

$$
\begin{array}{r}
1\text{HD} = 50¢ \\
1\text{Q} = 25¢ \\
1\text{D} = 10¢ \\
2\text{N} = 10¢ \\
+ \quad 4\text{P} = \underline{\ 4¢} \\
99¢
\end{array}
$$

4 pennies, 2 nickels, 1 dime, 1 quarter, and 1 half-dollar

LESSON 110, LESSON PRACTICE

a.
$$
\begin{array}{r}
2\ \text{R}\ 12 \\
30\overline{)72} \\
\underline{60} \\
12
\end{array}
$$

b.
$$
\begin{array}{r}
4\ \text{R}\ 7 \\
20\overline{)87} \\
\underline{80} \\
7
\end{array}
$$

c.
$$
\begin{array}{r}
2\ \text{R}\ 15 \\
40\overline{)95} \\
\underline{80} \\
15
\end{array}
$$

d.
$$
\begin{array}{r}
6\ \text{R}\ 7 \\
20\overline{)127} \\
\underline{120} \\
7
\end{array}
$$

e.
$$
\begin{array}{r}
3\ \text{R}\ 7 \\
40\overline{)127} \\
\underline{120} \\
7
\end{array}
$$

f.
$$
\begin{array}{r}
7\ \text{R}\ 7 \\
30\overline{)217} \\
\underline{210} \\
7
\end{array}
$$

LESSON 110, MIXED PRACTICE

1.
$$
\begin{array}{r}
26\ \text{R}\ 2 \\
3\overline{)80} \\
\underline{6} \\
20 \\
\underline{18} \\
2
\end{array}
$$

26, 26 + 1 = **27,** 26 + 1 = **27**

2. 3 + 4 = 7
3 × 4 = 12
12 − 7 = **5**

3. Inma: 6 years × 2 = 12 years old
Brother: 12 years + 3 years = **15 years old**

4.

513 fans
57 fans
57 fans
57 fans
57 fans
57 fans
57 fans
57 fans
57 fans
57 fans

$\frac{4}{9}$ cheered.

$\frac{5}{9}$ did not cheer.

$$
\begin{array}{r}
\overset{2}{57}\ \text{fans} \\
\times \quad 4 \\
\hline
\textbf{228 fans}
\end{array}
$$

5.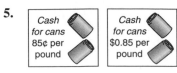

6. 3 out of 6 parts = $\dfrac{3}{6}$

4 out of 8 parts = $\dfrac{4}{8}$

7. If the chance of winning is 1%, then the chance of not winning is 100% − 1%, which is 99%. 99% > 1%, so the chance of **not winning** is more likely.

8. 600 + 400 = **1000**

9.
$$2 \overline{)5} \quad 2\frac{1}{2}$$
$$\frac{4}{1}$$

10. 42 minutes ÷ 7 miles = **6 minutes** for one mile

11.
$$\overset{2\ 2}{\$2.75}$$
$$\$2.75$$
$$\$2.75$$
$$+\ \$0.58$$
$$\overline{\$8.83}$$

$$\overset{0\ 9\ 9}{\$\cancel{1}\ \cancel{0}.\ \cancel{0}{}^{1}0}$$
$$-\ \ \$8.\ 8\ 3$$
$$\overline{\$1.\ 1\ 7}$$

12. If two tickets cost $26, then one ticket costs $26 ÷ 2, which is $13.

$$\$13$$
$$\times\ \ \ 20$$
$$\overline{\$260}$$

13.
$$49\frac{1}{2}\ \text{inches}$$
$$-\ 47\frac{1}{2}\ \text{inches}$$
$$\overline{2\quad\text{inches}}$$

14.
$$\overset{1}{7}.43$$
$$\overset{1}{6}.25$$
$$+\ \ 12.7$$
$$\overline{26.38}$$

15.
$$\overset{0\ \ {}^{1}3}{\cancel{1}\ \cancel{4}.{}^{1}3\ 6}$$
$$-\ \ \ 7.\ 5\ 0$$
$$\overline{6.\ 8\ 6}$$

16. Multiply mentally: **720,000**

17.
$$\overset{5\ 2}{\$0.73}$$
$$\times\ \ \ \ \ \ 8$$
$$\overline{\$5.84}$$

18. The product of any number and 0 is **0.**

19.
$$\overset{2}{15}$$
$$\times\ \ 15$$
$$\overline{75}$$
$$150$$
$$\overline{225}$$

20. 5 × 5 = 25

$$\overset{3}{25}$$
$$\times\ \ \ 60$$
$$\overline{1500}$$

21. $\sqrt{49}$ = 7
    7 × 7 = **49**

22.
$$5\frac{1}{3}$$
$$+\ 3\frac{1}{3}$$
$$\overline{8\frac{2}{3}}$$

23.
$$4\frac{4}{5}$$
$$-\ 3\frac{3}{5}$$
$$\overline{1\frac{1}{5}}$$

24.
$$\begin{array}{r} 124 \\ 10\overline{)1240} \\ \underline{10}\ \ \ \ \\ 24 \\ \underline{20}\ \\ 40 \\ \underline{40} \\ 0 \end{array}$$

25.
$$\begin{array}{r} 4 \\ 60\overline{)240} \\ \underline{240} \\ 0 \end{array}$$

26. A square has four equal sides.
    8 cm ÷ 4 = **2 cm**
    Area: 2 cm × 2 cm = **4 sq. cm**

27. 4 × 2 = 8
    8 × 2 = **16 small cubes**

28. If Nikki catches the 6:50 at 5th and Western, she should arrive at Hill and Lincoln at **7:08 a.m.**

29. There are 18 minutes from 6:50 a.m. to 7:08 a.m.
18 min

30. The next bus arrives at 5th and Western at **7:32 a.m.**

INVESTIGATION 11

1. **8 cubic feet;** multiplicative method:
2 ft × 2 ft × 2 ft = 8 cu. ft

2. **24 cubic centimeters;** multiplicative method:
4 cm × 2 cm × 3 cm = 24 cu. cm

3. **27 cubic inches;** multiplicative method:
3 in. × 3 in. × 3 in. = 27 cu. in.

4. **48 cubic meters;** multiplicative method:
4 m × 3 m × 4 m = 48 cu. m

5. 3 ft × 3 ft × 3 ft = **27 cu. ft**

6. 12 in. × 12 in. × 12 in. = **1728 cu. in.**

7. 100 cm × 100 cm × 100 cm
= **1,000,000 cu. cm**

8. 20 ft × 5 ft × 6 ft = **600 cubic feet**

9.

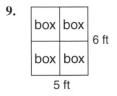

4 boxes

10. Each stack of four boxes takes 2 ft of the 20 ft length of the truck. So ten stacks of four boxes will fit, which is **40 boxes.**

11. **the length, the width, and the height of the room; yards**

12. Answers will vary with size of the room.

13. Answers will vary with size of room. A room that is 5 yd by 3 yd by 3 yd has a volume of 45 cubic yards. Such a room could hold 45 boxes that are one yard on each edge.

14. Answers will vary with size of room. A room that is 15 ft by 10 ft by 8 ft has a volume of 1200 cubic feet. Such a room could hold 1200 boxes that are one foot on each edge.

15. Answers will vary with size of room and arrangement of boxes. Against a wall 15 feet long and 8 feet high, 20 boxes could be stacked.

16. Answers will vary; perhaps 3 or more.

17. Answers will vary; perhaps 60 or more.

LESSON 111, WARM-UP

a. 20

b. 40

c. 60

d. 2300

e. $9.54

f. 37

g. 1090

h. >

Problem Solving
$1.00 − $0.54 = $0.46

$$1Q = 25¢$$
$$2D = 20¢$$
$$+\ 1P = \ \ 1¢$$
$$46¢ \text{ or } \$0.46$$

1 quarter, 2 dimes, and 1 penny

LESSON 111, LESSON PRACTICE

a. 6 full square inches
+ about 3 full square inches
9 square inches

b.   10 full square inches
+ about 11 full square inches
         about 21 square inches
Answers may vary slightly.

c. Guide and monitor student work. **Answers will vary.**

LESSON 111, MIXED PRACTICE

1. Pattern:
Number of groups × number in each group
= total

Problem:
N minutes × 60 seconds per minute
= 300 seconds

$$\begin{array}{r} \textbf{5 minutes} \\ 60\overline{)300} \\ \underline{300} \\ 0 \end{array}$$

2. Ann: 9 marbles × 2 = 18 marbles
Cho: 18 marbles + 5 marbles = **23 marbles**

3. Pattern:
Number of groups × number in each group
= total

Problem:
5 bookshelves × 44 books on each shelf
= *N* books

$$\begin{array}{r} \overset{2}{44} \\ \times\ \ 5 \\ \hline \textbf{220 books} \end{array}$$

4.

30 children	
	3 children
	3 children
	3 children
	3 children
$\frac{9}{10}$ remembered.	3 children
	3 children
	3 children
	3 children
$\frac{1}{10}$ did not remember.	3 children
	3 children

9 × 3 children = **27 children**

5. One tenth = **10%**

6. A fraction equals 1 if its numerator and denominator are the same.
$\frac{3}{3}$

7. 5 out of 10 parts and 1 out of 2 parts
$\frac{5}{10} = \frac{1}{2}$

8. **One possibility:**

9. 6 + 7 + 5 + 2 + 4 + 1 + 3 + 5
+ 3 = 36
36 ÷ 9 = **4**

10. There are 5 hours from 11:00 a.m. to 4:00 p.m. and there are 20 minutes from 4:00 p.m. to 4:20 p.m.
5 hours 20 minutes

11.   1 minute = 60 seconds
60 minutes × 60 seconds per minute
= **3600 seconds**

12. 1 and 5 are factors of both 10 and 15.
The largest factor of both 10 and 15 is **5.**

13. 1 × 8 = 8
2 × 4 = 8
Factors of **8: 1, 2, 4, 8**
1 × 12 = 12
  2 × 6 = 12
  3 × 4 = 12
Factors of **12: 1, 2, 3, 4, 6, 12**
The largest factor of both 8 and 12 is 4.

14. $\overset{1}{\underset{1}{4}.3}$
12.6
+   3.75
———
20.65

15. $3\,\overset{5}{\cancel{6}}\,\overset{13}{\cancel{4}}.\,\overset{10}{\cancel{1}}0$
−   1 6. 4 1
————
3 4 7. 6 9

16. Add only the numerators.
$\frac{7}{8}$

17. Add only the numerators.
$\frac{4}{5}$

18. $1\dfrac{9}{10}$
$-\ 1\dfrac{2}{10}$
$\dfrac{7}{10}$

19. Multiply mentally: **48,000**

20.
$$\begin{array}{r} \overset{\scriptstyle 1}{\overset{\scriptstyle 2}{7}}3 \\ \times\ 48 \\ \hline 584 \\ 2920 \\ \hline \mathbf{3504} \end{array}$$

21.
$$\begin{array}{r} \overset{7\ 7}{\$0.78} \\ \times\quad 9 \\ \hline \mathbf{\$7.02} \end{array}$$

22. $10 \times 10 = 100$
$100 \times 10 = \mathbf{1000}$

23.
$$\begin{array}{r} \mathbf{\$8.75} \\ 4\overline{)\$35.00} \\ \underline{32} \\ 3\,0 \\ \underline{2\,8} \\ 20 \\ \underline{20} \\ 0 \end{array}$$

24.
$$\begin{array}{r} \mathbf{603} \\ 8\overline{)4824} \\ \underline{48} \\ 02 \\ \underline{0} \\ 24 \\ \underline{24} \\ 0 \end{array}$$

25.
$$\begin{array}{r} 9 \\ 60\overline{)540} \\ \underline{540} \\ 0 \end{array}$$

26.
$$\begin{array}{r} \mathbf{46\ R\ 3} \\ 10\overline{)463} \\ \underline{40} \\ 63 \\ \underline{60} \\ 3 \end{array}$$

27.
$$\begin{array}{l} \quad\ 12 \text{ full square inches} \\ +\ \text{about 4 full square inches} \\ \hline \mathbf{16\ square\ inches} \end{array}$$

28. 25% is one fourth, so one out of four parts should be shaded.
One possibility: 4 cm

1 cm

29. 38,274 is between 30,000 and 40,000.
38,274 $>$ 35,000, so it is closer to 40,000 than to 30,000.
D

30. 3 in. $\times$ 2 in. $\times$ 2 in. = **12 cubic inches**

___

LESSON 112, WARM-UP

a. **15**

b. **30**

c. **45**

d. **360**

e. **$4.37**

f. **82**

g. **2150**

h. **>**

Patterns

Each numerator is one half the denominator.

$$\dfrac{1}{2}, \dfrac{2}{4}, \dfrac{3}{6}, \dfrac{4}{8}, \mathbf{\dfrac{5}{10}}, \mathbf{\dfrac{6}{12}}, \mathbf{\dfrac{7}{14}}, \mathbf{\dfrac{8}{16}}, \mathbf{\dfrac{9}{18}}, \cdots$$

___

LESSON 112, LESSON PRACTICE

a. $\dfrac{2}{4} \div \dfrac{2}{2} = \dfrac{2 \div 2}{4 \div 2} = \mathbf{\dfrac{1}{2}}$

b. $\dfrac{2}{6} \div \dfrac{2}{2} = \dfrac{2 \div 2}{6 \div 2} = \mathbf{\dfrac{1}{3}}$

c. $\dfrac{3}{9} \div \dfrac{3}{3} = \dfrac{3 \div 3}{9 \div 3} = \dfrac{1}{3}$

d. The only number that divides 3 and 8 is 1.

$\dfrac{3}{8}$

e. $\dfrac{2}{10} \div \dfrac{2}{2} = \dfrac{2 \div 2}{10 \div 2} = \dfrac{1}{5}$

f. $\dfrac{4}{10} \div \dfrac{2}{2} = \dfrac{4 \div 2}{10 \div 2} = \dfrac{2}{5}$

g. $\dfrac{9}{12} \div \dfrac{3}{3} = \dfrac{9 \div 3}{12 \div 3} = \dfrac{3}{4}$

h. The only number that divides 9 and 10 is 1.

$\dfrac{9}{10}$

LESSON 112, MIXED PRACTICE

1. 10 boards $\div$ 5 feet $=$ 2 boards per foot

50 feet $\times$ 2 boards per foot $=$ **100 boards**

2. $100 \times \$0.90 = \textbf{\$90.00}$

3. Perimeter: $5\,\text{cm} + 3\,\text{cm} + 5\,\text{cm}$
$+\ 3\,\text{cm} = \textbf{16 cm}$

Area: $3\,\text{cm} \times 5\,\text{cm} = \textbf{15 sq. cm}$

4. (a) Use the top ruler.

34 mm

(b) Use the bottom ruler.

3.4 cm

5.

	36 burros
	4 burros
	4 burros
$\frac{5}{9}$ were gray.	4 burros
	4 burros
	4 burros
	4 burros
$\frac{4}{9}$ were not gray.	4 burros
	4 burros
	4 burros

$5 \times 4\ \text{burros} = \textbf{20 burros}$

6. (a) $2)\overline{15} \quad 7\frac{1}{2}$
$\dfrac{14}{1}$

(b) $15 \div 3 = \textbf{5}$

(c) $4)\overline{15} \quad 3\frac{3}{4}$
$\dfrac{12}{3}$

7. 1 out of 2 parts and 6 out of 12 parts

$\dfrac{1}{2} = \dfrac{6}{12}$

8. Each rectangle shows one half.

One half $=$ **50%**

9. (a) $\dfrac{3}{6} \div \dfrac{3}{3} = \dfrac{3 \div 3}{6 \div 3} = \dfrac{1}{2}$

(b) $\dfrac{4}{6} \div \dfrac{2}{2} = \dfrac{4 \div 2}{6 \div 2} = \dfrac{2}{3}$

(c) The only number that divides 5 and 6 is 1.

$\dfrac{5}{6}$

10.
$\overset{1}{23}$ bounces
36 bounces
$\underline{+\ 34\ \text{bounces}}$
93 bounces

$3)\overline{93} \quad \textbf{31 bounces}$
$\dfrac{9}{03}$
$\dfrac{3}{0}$

11. $5 \times \$5 = \$25$
$\$25.00 + \$1.50 = \$26.50$
$\$27.00 - \$26.50 = \textbf{\$0.50}$

12. $3\dfrac{3}{9}$
$+\ 4\dfrac{4}{9}$
$7\dfrac{7}{9}$

13. Add only the numerators.

$\dfrac{6}{7}$

14.
$$\overset{1\,1}{37.2}$$
$$135.7$$
$$10.62$$
$$2.47$$
$$+\ \ 14.0$$
$$\overline{\mathbf{199.99}}$$

15. Subtract only the numerators.

$$\frac{1}{12}$$

16. Subtract only the numerators.

$$\frac{3}{10}$$

17.
$$\overset{\overset{2}{4}}{48}$$
$$\times\ 36$$
$$\overline{288}$$
$$1440$$
$$\overline{\mathbf{1728}}$$

18.
$$\overset{\overset{1}{1}}{72}$$
$$\times\ 58$$
$$\overline{576}$$
$$3600$$
$$\overline{\mathbf{4176}}$$

19.
$$\overset{5}{\$4.08}$$
$$\times\ \ \ \ 7$$
$$\overline{\mathbf{\$28.56}}$$

20.
$$\overset{1\,1}{25.42}$$
$$+\ 24.8$$
$$\overline{\mathbf{50.22}}$$

21.
$$3\,\overset{5}{\cancel{6}}.\,\overset{1}{\cancel{2}}{}^{1}0$$
$$-\ \ \ 4.\ 2\ 7$$
$$\overline{\mathbf{3\ 1.\ 9\ 3}}$$

22.
$$\begin{array}{r}\mathbf{4\ R\ 10}\\ 20\overline{)90}\\ \underline{80}\\ 10\end{array}$$

23. **0**

24.
$$\begin{array}{r}\mathbf{364\ R\ 1}\\ 7\overline{)2549}\\ \underline{21}\\ 44\\ \underline{42}\\ 29\\ \underline{28}\\ 1\end{array}$$

25.
$$\begin{array}{r}\mathbf{\$3.88}\\ 5\overline{)\$19.40}\\ \underline{15}\\ 4\ 4\\ \underline{4\ 0}\\ 40\\ \underline{40}\\ 0\end{array}$$

26. **450,000** is halfway between 400,000 and 500,000.

27. A coin has two sides: heads and tails. The probability that a tossed coin will land heads is 1 out of 2, which is $\frac{1}{2}$.

28. **Rectangular solid or rectangular prism**

29. 6 in. $\times$ 2 in. $\times$ 5 in. = **60 cubic inches**

30.
$$\begin{array}{r}15\ \text{full square inches}\\ +\ \ \text{about 10 full square inches}\\ \hline \mathbf{about\ 25\ sq.\ in.}\end{array}$$
Answers may vary slightly.

LESSON 113, WARM-UP

a. $1\dfrac{1}{2}$

b. $3\dfrac{1}{2}$

c. $5\dfrac{1}{2}$

d. $10\dfrac{1}{2}$

e. $16\dfrac{1}{2}$

f. $<$

g. **49**

Patterns

$4^3 = 4 \times 4 \times 4 = 64$
$5^3 = 5 \times 5 \times 5 = 125$
$6^3 = 6 \times 6 \times 6 = 216$

1, 8, 27, **64**, **125**, **216**, ...

LESSON 113, LESSON PRACTICE

a.
$$
\begin{array}{r}
\scriptstyle 1 \\[-2pt]
\scriptstyle 1\,2 \\[-2pt]
235 \\
\times\ \ \ 24 \\
\hline
940 \\
4700 \\
\hline
\mathbf{5640}
\end{array}
$$

b.
$$
\begin{array}{r}
\scriptstyle 1 \\[-2pt]
430 \\
\times\ \ \ 14 \\
\hline
1720 \\
4300 \\
\hline
\mathbf{6020}
\end{array}
$$

c.
$$
\begin{array}{r}
\scriptstyle 1 \\[-2pt]
\scriptstyle 1\,2 \\[-2pt]
\$1.25 \\
\times\ \ \ \ 24 \\
\hline
5\ 00 \\
25\ 00 \\
\hline
\mathbf{\$30.00}
\end{array}
$$

d.
$$
\begin{array}{r}
\scriptstyle 1 \\[-2pt]
\scriptstyle 1 \\[-2pt]
406 \\
\times\ \ \ 32 \\
\hline
812 \\
12\ 180 \\
\hline
\mathbf{12{,}992}
\end{array}
$$

e.
$$
\begin{array}{r}
\$6.20 \\
\times\ \ \ \ 31 \\
\hline
6\ 20 \\
186\ 00 \\
\hline
\mathbf{\$192.20}
\end{array}
$$

f.
$$
\begin{array}{r}
\scriptstyle 2 \\[-2pt]
\scriptstyle 4\,1 \\[-2pt]
562 \\
\times\ \ \ 47 \\
\hline
3\ 934 \\
22\ 480 \\
\hline
\mathbf{26{,}414}
\end{array}
$$

LESSON 113, MIXED PRACTICE

1.
$$
\begin{array}{r}
\scriptstyle 1\,2 \\[-2pt]
638 \text{ miles} \\
456 \text{ miles} \\
+\ \ 589 \text{ miles} \\
\hline
1683 \text{ miles}
\end{array}
$$

$$
\begin{array}{r}
\scriptstyle 2\ \ 9\ \ 9 \\[-2pt]
\cancel{3}\ \cancel{0}\ \cancel{0}\,{}^1 0 \text{ miles} \\
-\ \ 1\ 6\ 8\ 3 \text{ miles} \\
\hline
\mathbf{1\ 3\ 1\ 7 \text{ miles}}
\end{array}
$$

2. Perimeter: 7 in. + 7 in. + 7 in.
$\qquad\qquad\qquad$ + 7 in. = **28 in.**
Area: 7 in. × 7 in. = **49 sq. in.**

3. 1 m = 100 cm
2 m = 200 cm
A square has four sides of equal length.
200 cm ÷ 4 = **50 cm**

4.

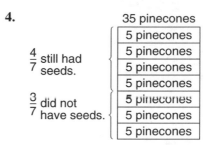

$\frac{4}{7}$ still had seeds. — 5 pinecones, 5 pinecones, 5 pinecones, 5 pinecones

$\frac{3}{7}$ did not have seeds. — 5 pinecones, 5 pinecones, 5 pinecones

35 pinecones

4 × 5 pinecones = **20 pinecones**

5. 6873 > 6500, so it rounds up to **7000.**

6. (a) The only number that divides 4 and 5 is 1.
$$\frac{4}{5}$$

(b) $\dfrac{5}{10} \div \dfrac{5}{5} = \dfrac{5 \div 5}{10 \div 5} = \dfrac{1}{2}$

(c) $\dfrac{4}{10} \div \dfrac{2}{2} = \dfrac{4 \div 2}{10 \div 2} = \dfrac{2}{5}$

7. Three decimal places indicate thousandths.
Three hundred seventy-four and two hundred fifty-one thousandths

8. **One possibility:**

$$\boxed{\ \blacksquare\ |\ |\ } = \boxed{\ \#\#\#\ }$$

9. $\dfrac{1}{4} \times \dfrac{2}{2} = \dfrac{2}{8}$

$\dfrac{1}{4} \times \dfrac{3}{3} = \dfrac{3}{12}$

$\dfrac{1}{4} \times \dfrac{5}{5} = \dfrac{5}{20}$

10. Pattern:
Number of groups $\times$ number in each group
$= $ total
Problem: 5 days $\times$ N lb per day $=$ 750 lb

$$
\begin{array}{r}
\textbf{150 lb} \\
5\overline{)750} \\
\underline{5} \\
25 \\
\underline{25} \\
00 \\
\underline{0} \\
0
\end{array}
$$

11.
$$
\begin{array}{r}
{\scriptstyle 1\,1\,1} \\
8742 \text{ ft} \\
+ \; 5368 \text{ ft} \\
\hline
\textbf{14,110 ft}
\end{array}
$$

12.
$$
\begin{array}{r}
\textbf{806 R 1} \\
6\overline{)4837} \\
\underline{48} \\
03 \\
\underline{0} \\
37 \\
\underline{36} \\
1
\end{array}
$$

13. $\sqrt{16} = 4$

$$
\begin{array}{r}
\textbf{343} \\
4\overline{)1372} \\
\underline{12} \\
17 \\
\underline{16} \\
12 \\
\underline{12} \\
0
\end{array}
$$

14.
$$
\begin{array}{r}
\textbf{24} \\
40\overline{)960} \\
\underline{80} \\
160 \\
\underline{160} \\
0
\end{array}
$$

15.
$$
\begin{array}{r}
\textbf{68} \\
20\overline{)1360} \\
\underline{120} \\
160 \\
\underline{160} \\
0
\end{array}
$$

16.
$$
\begin{array}{r}
{\scriptstyle 2\quad 9} \\
\cancel{3}\,\cancel{0}.{}^{1}0\,7 \\
- \quad 3.\,7 \\
\hline
\textbf{2 6. 3 7}
\end{array}
$$

17.
$$
\begin{array}{r}
{\scriptstyle 5\;\;9} \\
4\,\cancel{6}.\,\cancel{0}{}^{1}0 \\
- \; 1\,2.\,4\,6 \\
\hline
\textbf{3 3. 5 4}
\end{array}
$$

18.
$$
\begin{array}{r}
{\scriptstyle 2\,1\;\,1} \\
37.15 \\
6.84 \\
1.29 \\
29.1 \\
+ \quad 3.6 \\
\hline
\textbf{77.98}
\end{array}
$$

19.
$$
\begin{array}{r}
{\scriptstyle 1} \\
\$3.20 \\
\times \qquad 46 \\
\hline
19\,20 \\
128\,00 \\
\hline
\textbf{\$147.20}
\end{array}
$$

20.
$$
\begin{array}{r}
{\scriptstyle 1} \\
{\scriptstyle 3} \\
307 \\
\times \quad 25 \\
\hline
1535 \\
6140 \\
\hline
\textbf{7675}
\end{array}
$$

21. Add only the numerators.

$\dfrac{\textbf{14}}{\textbf{15}}$

22.
$$
\begin{array}{r}
4\dfrac{4}{5} \\[4pt]
- \; 1\dfrac{3}{5} \\[4pt]
\hline
\textbf{3}\dfrac{\textbf{1}}{\textbf{5}}
\end{array}
$$

23. 8 full sq. cm $+$ about 7 full sq. cm
$= $ **15 sq. cm**

24. The rule is "Count up by 10,000."
40,000; 50,000; 60,000

25. Equilateral triangles have three equal sides.
C. $\triangle$

26. One full turn $= 360°$
Two full turns $= 360° \times 2 = 720°$
C. 720°

27. **K** has no lines of symmetry.

28. **Possibilities include:**
    - **Reflection and translation**
    - **Reflection only, if the reflection is made about a vertical line drawn halfway between the triangles**

29.
$$2\frac{1}{4} \text{ inches}$$
$$2\frac{1}{4} \text{ inches}$$
$$+\ 2\frac{1}{4} \text{ inches}$$
$$\overline{\quad 6\frac{3}{4} \text{ inches}}$$

30. $5 \text{ cm} \times 3 \text{ cm} \times 3 \text{ cm}$
    $= \textbf{45 cubic centimeters}$

LESSON 114, WARM-UP

a. **6**

b. **12**

c. **18**

d. **5000**

e. **$6.64**

f. **18**

g. **1510**

h. **59**

Problem Solving

First think, "8 times what number equals 24?" (3). Then think, "What number minus 24 equals 2?" (26). Bring down the 7. Since the remainder is 0, think, "What number times 3 equals 27?" (9).

$$\begin{array}{r} 89 \\ 3\overline{)267} \\ \underline{24} \\ 27 \\ \underline{27} \\ 0 \end{array}$$

LESSON 114, LESSON PRACTICE

a. $\dfrac{4}{5} + \dfrac{4}{5} = \dfrac{8}{5}$

   $\dfrac{8}{5} = \mathbf{1\dfrac{3}{5}}$

b. $\dfrac{5}{6} - \dfrac{1}{6} = \dfrac{4}{6}$

   $\dfrac{4}{6} = \mathbf{\dfrac{2}{3}}$

c. $3\dfrac{2}{3} + 1\dfrac{2}{3} = 4\dfrac{4}{3}$

   $4\dfrac{4}{3} = \mathbf{5\dfrac{1}{3}}$

d. $5\dfrac{1}{4} + 6\dfrac{3}{4} = 11\dfrac{4}{4}$

   $11\dfrac{4}{4} = \mathbf{12}$

e. $7\dfrac{7}{8} - 1\dfrac{1}{8} = 6\dfrac{6}{8}$

   $6\dfrac{6}{8} = \mathbf{6\dfrac{3}{4}}$

f. $5\dfrac{3}{5} + 1\dfrac{3}{5} = 6\dfrac{6}{5}$

   $6\dfrac{6}{5} = \mathbf{7\dfrac{1}{5}}$

LESSON 114, MIXED PRACTICE

1. $70 \times 6¢ = 420¢$ or $4.20
   $4.20 + $0.25 = $4.45

$$\begin{array}{r} \overset{4}{\$5}.\overset{9}{\cancel{0}}{}^{1}0 \\ -\ \$4.\ 4\ 5 \\ \hline \mathbf{\$0.\ 5\ 5} \end{array}$$

2. (a) $6 \text{ cm} \times 6 \text{ cm} = \textbf{36 sq. cm}$

   (b) $6 \text{ cm} \times 4 = \textbf{24 cm}$

3. David: $9 \times 2 = $18
   Chris: $18 + $6 = **$24**

4.

$$\begin{array}{r} \$_2 9 \\ \$18 \\ + \ \$24 \\ \hline \$51 \end{array}$$

$$\begin{array}{r} \mathbf{\$17} \\ 3)\overline{\$51} \\ \underline{3} \\ 21 \\ \underline{21} \\ 0 \end{array}$$

5. 40 quarters $\times$ 2 = 80 quarters

$$\begin{array}{r} \overset{4}{\$0.25} \\ \times \quad 80 \\ \hline \mathbf{\$20.00} \end{array}$$

6. 30 $\times$ 300 = **9000**

$$\begin{array}{r} \overset{1\,1}{312} \\ \times \quad 29 \\ \hline 2808 \\ 6240 \\ \hline \mathbf{9048} \end{array}$$

7. (a) $\dfrac{2}{12} \div \dfrac{2}{2} = \dfrac{2 \div 2}{12 \div 2} = \mathbf{\dfrac{1}{6}}$

(b) $\dfrac{6}{8} \div \dfrac{2}{2} = \dfrac{6 \div 2}{8 \div 2} = \mathbf{\dfrac{3}{4}}$

(c) $\dfrac{3}{9} \div \dfrac{3}{3} = \dfrac{3 \div 3}{9 \div 3} = \mathbf{\dfrac{1}{3}}$

8. $\dfrac{1}{3} \times \dfrac{2}{2} = \dfrac{2}{6}$

$\dfrac{2}{6} + \dfrac{3}{6} = \mathbf{\dfrac{5}{6}}$

9. Red
Green
Black
Red finished first

10. 7 is not a number on a number cube, so the probability of rolling a 7 is **0.**

11. 50% = one half
Half of $40 is $20.
$40 − $20 = **$20**

12.

$$\begin{array}{r} \overset{2}{_2 4.62} \\ 16.7 \\ + \quad 9.8 \\ \hline \mathbf{31.12} \end{array}$$

13.

$$\begin{array}{r} \overset{5}{\cancel{6}}.\overset{12}{\cancel{3}}0 \\ - \ 2.3\,7 \\ \hline 3.9\,3 \end{array}$$

$$\begin{array}{r} 1\,\overset{3}{\cancel{4}}.\overset{15}{\cancel{6}}2 \\ - \quad 3.9\,3 \\ \hline \mathbf{1\,0.6\,9} \end{array}$$

14. $\dfrac{3}{5} + \dfrac{4}{5} = \dfrac{7}{5}$

$\dfrac{7}{5} = \mathbf{1\dfrac{2}{5}}$

15. $16 + 3\dfrac{3}{4} = \mathbf{19\dfrac{3}{4}}$

16. $1\dfrac{2}{3} + 3\dfrac{1}{3} = 4\dfrac{3}{3}$

$4\dfrac{3}{3} = \mathbf{5}$

17. $\dfrac{2}{5} + \dfrac{3}{5} = \dfrac{5}{5}$

$\dfrac{5}{5} = \mathbf{1}$

18. $7\dfrac{4}{5} + 7\dfrac{1}{5} = 14\dfrac{5}{5}$

$14\dfrac{5}{5} = \mathbf{15}$

19. $6\dfrac{2}{3} + 3\dfrac{2}{3} = 9\dfrac{4}{3}$

$9\dfrac{4}{3} = \mathbf{10\dfrac{1}{3}}$

20.

$$\begin{array}{r} \overset{2}{}\overset{6\,1}{372} \\ \times \quad 39 \\ \hline 3348 \\ 11160 \\ \hline \mathbf{14,508} \end{array}$$

21.

$$\begin{array}{r} \overset{1}{}\overset{2\,1}{142} \\ \times \quad 47 \\ \hline 994 \\ 5680 \\ \hline \mathbf{6674} \end{array}$$

22. $\sqrt{36} = 6$

$$\begin{array}{r} \overset{3}{360} \\ \times \quad 6 \\ \hline \mathbf{2160} \end{array}$$

23.
$$\begin{array}{r} 16 \text{ full sq. cm} \\ +\ \text{about } 12 \text{ full sq. cm} \\ \hline \textbf{about 28 sq. cm} \end{array}$$
Answers may vary.

24.
$$\begin{array}{r} \textbf{604 R 2} \\ 8\overline{)4834} \\ \underline{48} \\ 03 \\ \underline{0} \\ 34 \\ \underline{32} \\ 2 \end{array}$$

25. $2 \times 2 \times 2 = 8$
$$\begin{array}{r} \textbf{355} \\ 8\overline{)2840} \\ \underline{24} \\ 44 \\ \underline{40} \\ 40 \\ \underline{40} \\ 0 \end{array}$$

26.
$$\begin{array}{r} \textbf{32 R 3} \\ 30\overline{)963} \\ \underline{90} \\ 63 \\ \underline{60} \\ 3 \end{array}$$

27. 427,063 is between 400,000 and 500,000.
427,063 < 450,000, so it is closer to 400,000 than to 500,000.
C

28. $1\frac{1}{4}$ inches $+\ 1\frac{1}{4}$ inches $+\ 1\frac{1}{4}$ inches

$+\ 1\frac{1}{4}$ inches $=\ 4\frac{1}{4}$ inches

$4\frac{4}{4}$ inches $=$ **5 inches**

29. **Sphere**

30. $5(20 + 6)$
$100 + 30 = \textbf{130}$

LESSON 115, WARM-UP

a. 9

b. 27

c. 36

d. 1000

e. $0.75

f. 185

g. 1550

h. >

Patterns
Each money amount is one tenth the preceding amount.
…, $1000.00, $100.00, $10.00, **$1.00, $0.10, $0.01,** …

LESSON 115, LESSON PRACTICE

a. $\dfrac{1}{4} \times \dfrac{3}{3} = \dfrac{3}{12}$; **3**

b. $\dfrac{2}{3} \times \dfrac{4}{4} = \dfrac{8}{12}$; **8**

c. $\dfrac{5}{6} \times \dfrac{2}{2} = \dfrac{10}{12}$; **10**

d. $\dfrac{3}{5} \times \dfrac{2}{2} = \dfrac{6}{10}$; **6**

e. $\dfrac{2}{3} \times \dfrac{3}{3} = \dfrac{6}{9}$; **6**

f. $\dfrac{3}{4} \times \dfrac{2}{2} = \dfrac{6}{8}$; **6**

LESSON 115, MIXED PRACTICE

1.
$$\begin{array}{r} \$0.50 \text{ per person} \\ 3\overline{)\$1.50} \\ \underline{1\ 5} \\ 00 \\ \underline{0} \\ 0 \end{array}$$

$$\begin{array}{r} \$0.50 \\ \times\ \ \ \ 12 \\ \hline 1\ 00 \\ 5\ 00 \\ \hline \textbf{\$6.00} \end{array}$$

2. (a) $7 \text{ ft} + 5 \text{ ft} + 7 \text{ ft} + 5 \text{ ft} = \textbf{24 ft}$

   (b) $7 \text{ ft} \times 5 \text{ ft} = \textbf{35 sq. ft}$

3. $9 \times 10 = 90$
   $90 - 8 = \textbf{82}$

4.
$\frac{2}{3}$ were eaten.

3069 bugs
1023 bugs
1023 bugs

$\frac{1}{3}$ were left in the tree.

1023 bugs

   1023 bugs

5. (a) Use the bottom ruler.
   3.7 cm

   (b) Use the top ruler.
   37 mm

6. **Three hundred fifty-six thousand, four hundred twenty**

7. 356,420 is between 300,000 and 400,000. 356,420 $>$ 350,000, so it is closer to 400,000 than to 300,000. 356,420 is closer to 350,000 than to 400,000.
   B

8. (a) $\frac{1}{2} \times \frac{3}{3} = \frac{3}{6}$; **3**

   (b) $\frac{1}{3} \times \frac{2}{2} = \frac{2}{6}$; **2**

   (c) $\frac{2}{3} \times \frac{2}{2} = \frac{4}{6}$; **4**

9. (a) $\frac{2}{6} \div \frac{2}{2} = \frac{2 \div 2}{6 \div 2} = \frac{1}{3}$

   (b) $\frac{6}{9} \div \frac{3}{3} = \frac{6 \div 3}{9 \div 3} = \frac{2}{3}$

   (c) The only number that divides 9 and 16 is 1.
   $\frac{9}{16}$

10. 10 out of 40 workers worked overtime.
    $\frac{10}{40} \div \frac{10}{10} = \frac{10 \div 10}{40 \div 10} = \frac{1}{4}$

11. One fourth = **25%**

12.
$$\begin{array}{r} \overset{1}{\$10} \\ \$11 \\ \$12 \\ \$13 \\ \$14 \\ + \ \$15 \\ \hline \$75 \end{array}$$

13.
$$\begin{array}{r} 2\frac{1}{2} \text{ miles} \\ + \ 2\frac{1}{2} \text{ miles} \\ \hline 4\frac{2}{2} \text{ miles} = \textbf{5 miles} \end{array}$$

14.
$$\begin{array}{r} \overset{3}{\cancel{4}}.\overset{1}{1}3\,7 \\ - \ 3.\,8\,0 \\ \hline 0.\,5\,7 \end{array}$$

$$\begin{array}{r} \overset{8}{\cancel{9}}.\,\overset{{}^{1}2}{\cancel{3}}{}^{1}6 \\ - \ 0.\,5\,7 \\ \hline \textbf{8.\,7\,9} \end{array}$$

15.
$$\begin{array}{r} {}_{1}\overset{1}{8}.61 \\ + \ 12.5 \\ \hline 21.11 \end{array}$$

$$\begin{array}{r} 24.32 \\ - \ 21.11 \\ \hline \textbf{3.21} \end{array}$$

16. $5\frac{5}{8} + 3\frac{3}{8} = 8\frac{8}{8}$
    $8\frac{8}{8} = \textbf{9}$

17. $6\frac{3}{10} + 1\frac{2}{10} = 7\frac{5}{10}$
    $7\frac{5}{10} = \textbf{7}\frac{\textbf{1}}{\textbf{2}}$

18. $8\frac{2}{3} - 5\frac{1}{3} = \textbf{3}\frac{\textbf{1}}{\textbf{3}}$

19. $4\frac{3}{4} - 2\frac{1}{4} = 2\frac{2}{4}$
    $2\frac{2}{4} = \textbf{2}\frac{\textbf{1}}{\textbf{2}}$

20.
$$\begin{array}{r} \overset{1\,3}{125} \\ \times \quad 16 \\ \hline 750 \\ 1250 \\ \hline \textbf{2000} \end{array}$$

21.
$$\begin{array}{r} \overset{1}{\$1.50} \\ \times12 \\ \hline 300 \\ 1500 \\ \hline \$18.00 \end{array}$$

22.
$$\begin{array}{r} 607 \\ 6\overline{)3642} \\ 36 \\ \hline 04 \\ 0 \\ \hline 42 \\ 42 \\ \hline 0 \end{array}$$

23.
$$\begin{array}{r} \$25 \\ 5\overline{)\$125} \\ 10 \\ \hline 25 \\ 25 \\ \hline 0 \end{array}$$

24.
$$\begin{array}{r} 16\text{ R }5 \\ 40\overline{)645} \\ 40 \\ \hline 245 \\ 240 \\ \hline 5 \end{array}$$

25. $3m = 36$

$m = 36 \div 3$

$m = \mathbf{12}$

26. $3n$

$3 \times 16 - \mathbf{48}$

27.
$$\begin{array}{r} \overset{1}{18} \\ 21 \\ +21 \\ \hline 60 \end{array}$$

$60 \div 3 = \mathbf{20\ sailors}$

28.
$$\begin{array}{r} 99.8°F \\ -98.6°F \\ \hline \mathbf{1.2\,°F} \end{array}$$

29. 18 full sq. mi + about 4 full sq. mi = **22 sq. mi**

30. 3 out of 8 numbers are greater than 5.

$\dfrac{\mathbf{3}}{\mathbf{8}}$

LESSON 116, WARM-UP

a. 6

b. 12

c. 18

d. 200

e. $8.61

f. 39

g. 900

h. 69

Problem Solving

8 tapes × 2 columns = 16 tapes

16 tapes + 1 tape between columns

= **17 tapes**

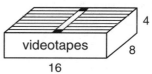

LESSON 116, LESSON PRACTICE

a. $\dfrac{1}{2} \times \dfrac{5}{5} = \dfrac{\mathbf{5}}{\mathbf{10}}$

$\dfrac{1}{5} \times \dfrac{2}{2} = \dfrac{\mathbf{2}}{\mathbf{10}}$

b. $\dfrac{3}{4} \times \dfrac{3}{3} = \dfrac{\mathbf{9}}{\mathbf{12}}$

$\dfrac{5}{6} \times \dfrac{2}{2} = \dfrac{\mathbf{10}}{\mathbf{12}}$

c. Multiples of 2: 2, 4, $\underline{6}$

Multiples of 3: 3, $\underline{6}$

$\dfrac{1}{2} \times \dfrac{3}{3} = \dfrac{\mathbf{3}}{\mathbf{6}}$

$\dfrac{2}{3} \times \dfrac{2}{2} = \dfrac{\mathbf{4}}{\mathbf{6}}$

d. Multiples of 3: 3, 6, 9, <u>12</u>
Multiples of 4: 4, 8, <u>12</u>

$$\frac{1}{3} \times \frac{4}{4} = \frac{4}{12}$$

$$\frac{1}{4} \times \frac{3}{3} = \frac{3}{12}$$

e. Multiples of 2: 2, 4, 6, 8, <u>10</u>
Multiples of 5: 5, <u>10</u>

$$\frac{1}{2} \times \frac{5}{5} = \frac{5}{10}$$

$$\frac{3}{5} \times \frac{2}{2} = \frac{6}{10}$$

f. Multiples of 3: 3, 6, 9, 12, <u>15</u>
Multiples of 5: 5, 10, <u>15</u>

$$\frac{2}{3} \times \frac{5}{5} = \frac{10}{15}$$

$$\frac{2}{5} \times \frac{3}{3} = \frac{6}{15}$$

_____

LESSON 116, MIXED PRACTICE

1.

24 polliwogs	
$\frac{1}{4}$ were let go.	6 polliwogs
	6 polliwogs
$\frac{3}{4}$ were kept.	6 polliwogs
	6 polliwogs

3 × 6 polliwogs = **18 polliwogs**

2. One time around:
2 mi + 1 mi + 2 mi + 1 mi = 6 mi
Two times around: 2 × 6 mi = **12 mi**

3. 42¢ ÷ 2 = 21¢ per orange

$$\begin{array}{r} 21¢ \\ \times\ \ 8 \\ \hline 168¢ \end{array}$$ or **$1.68**

4.

64 cards	
$\frac{3}{4}$ showed AL players.	16 cards
	16 cards
	16 cards
$\frac{1}{4}$ didn't show AL players.	16 cards

$$\begin{array}{r} \overset{1}{1}6 \text{ baseball cards} \\ \times\ \ \ 3 \\ \hline \textbf{48 baseball cards} \end{array}$$

5. One fourth = 25%, so
three fourths = 3 × 25%, or **75%**

6. A fraction is equivalent to one half if the denominator is exactly twice the numerator.

C. $\dfrac{10}{21}$

7. (a) $\dfrac{1}{2} \times \dfrac{6}{6} = \dfrac{6}{12};$ **6**

(b) $\dfrac{1}{3} \times \dfrac{4}{4} = \dfrac{4}{12};$ **4**

(c) $\dfrac{1}{4} \times \dfrac{3}{3} = \dfrac{3}{12};$ **3**

8. (a) $\dfrac{5}{10} \div \dfrac{5}{5} = \dfrac{5 \div 5}{10 \div 5} = \dfrac{1}{2}$

(b) The only number that divides 8 and 15 is 1.
$\dfrac{8}{15}$

(c) $\dfrac{6}{12} \div \dfrac{6}{6} = \dfrac{6 \div 6}{12 \div 6} = \dfrac{1}{2}$

9. 42¢ ÷ 6 = **7¢ for each clip**
64¢ ÷ 8 = **8¢ for each eraser**
7¢ × 10 = 70¢ or $0.70
8¢ × 20 = 160¢ or $1.60

$$\begin{array}{r} \overset{1}{\$}0.70 \\ +\ \ \$1.60 \\ \hline \textbf{\$2.30} \end{array}$$

10. 14, 16, 18, 20, 22, 24, 26, 28, 30, 32
32 volunteers

11. (a) Multiples of 3: 3, 6, 9, <u>12</u>
Multiples of 4: 4, 8, <u>12</u>

$$\frac{1}{4} \times \frac{3}{3} = \frac{3}{12}$$

$$\frac{2}{3} \times \frac{4}{4} = \frac{8}{12}$$

(b) Multiples of 3: 3, 6, 9, <u>12</u>
Multiples of 4: 4, 8, <u>12</u>

$$\frac{1}{3} \times \frac{4}{4} = \frac{4}{12}$$

$$\frac{3}{4} \times \frac{3}{3} = \frac{9}{12}$$

12. Every number on a number cube is less than 7, so it is certain that the number rolled will be less than 7.

1

13.
```
   3.63
+ 36.3
------
 39.93
```

```
  3 |6
  4 7.¹1 4
- 3 9. 9 3
---------
    7. 2 1
```

14.
```
  5 ¹3
  6. 4¹0
- 1. 4 6
--------
  4. 9 4
```

```
   1
  50.1
+  4.94
-------
 55.04
```

15. $\dfrac{3}{4} + \dfrac{3}{4} + \dfrac{3}{4} = \dfrac{9}{4}$

$\dfrac{9}{4} = \mathbf{2\dfrac{1}{4}}$

16. $4\dfrac{1}{6} + 1\dfrac{1}{6} = 5\dfrac{2}{6}$

$5\dfrac{2}{6} = \mathbf{5\dfrac{1}{3}}$

17. $5\dfrac{3}{5} + 1\dfrac{2}{5} = 6\dfrac{5}{5}$

$6\dfrac{5}{5} = \mathbf{7}$

18. Add only the numerators.

$\dfrac{6}{6} = \mathbf{1}$

19. $12\dfrac{3}{4} - 3\dfrac{1}{4} = 9\dfrac{2}{4}$

$9\dfrac{2}{4} = \mathbf{9\dfrac{1}{2}}$

20. $6\dfrac{1}{5} - 1\dfrac{1}{5} = \mathbf{5}$

21.
```
    2
   340
 ×  15
 -----
  1700
  3400
 -----
  5100
```

22.
```
     1
     4
    307
 ×   26
 ------
   1842
   6140
 ------
   7982
```

23.
```
    3
   250
 ×  70
 ------
 17,500
```

24.
```
      710
   5)3550
     35
     --
      05
       5
      --
      00
       0
      --
       0
```

25.
```
     14 R 12
  30)432
     30
     ---
     132
     120
     ---
      12
```

26.
```
     642 R 6
  9)5784
    54
    --
    38
    36
    --
     24
     18
     --
      6
```

27. 14 full sq. in. + about 5 full sq. in. = **19 sq. in.**

28. There is one flight that arrives in Los Angeles in the morning, and the airfare is **$412.00.**

29. The more economical round trip return flight is scheduled to land in Chicago on **July 29 at 12:29 a.m.**

30. When a flight from Chicago lands in Los Angeles, the time is about 2 hours later than the time of takeoff in Chicago.

2 hours + 2 hours = 4 hours

B. 4 hours

LESSON 117, WARM-UP

a. 7

b. 14

c. 21

d. 28

e. 42

f. 70

g. >

h. 79

Patterns

$$\frac{1}{4}, \frac{2}{8}, \frac{3}{12}, \frac{4}{16}, \frac{5}{20}, \frac{6}{24}, \frac{7}{28}, \cdots$$

LESSON 117, LESSON PRACTICE

a. Since 19,362 is greater than 15,000, it rounds up to **20,000.**

b. Since 31,289 is less than 35,000, it rounds down to **30,000.**

c. Since 868,367 is greater than 850,000, it rounds up to **900,000.**

d. Since 517,867 is less than 550,000, it rounds down to **500,000.**

e. Since 2,156,324 is less than 2,500,000, it rounds down to **2,000,000.**

f. Since 28,376,000 is greater than 25,000,000, it rounds up to **30,000,000.**

g. Since 412,500,000 is less than 450,000,000, it rounds down to **400,000,000.**

LESSON 117, MIXED PRACTICE

1.
$$\begin{array}{r} 11\text{ R}1 \\ 4\overline{)45} \\ \underline{4} \\ 05 \\ \underline{4} \\ 1 \end{array}$$

The largest group has $11 + 1$ merry men, which is **12 merry men.**

2. (a)
$$\begin{array}{r} \overset{1}{12}\text{ cm} \\ \times\ \ 8\text{ cm} \\ \hline \mathbf{96\text{ sq. cm}} \end{array}$$

(b) $12\text{ cm} + 8\text{ cm} + 12\text{ cm} + 8\text{ cm} = \mathbf{40\text{ cm}}$

3.

	90 questions
$\frac{5}{6}$ correct	15 questions
	15 questions
	15 questions
	15 questions
$\frac{1}{6}$ not correct	15 questions
	15 questions

$$\begin{array}{r} \overset{2}{15}\text{ questions} \\ \times\ \ 5 \\ \hline \mathbf{75\text{ questions}} \end{array}$$

4. (a) **Cylinder**

(b) **Sphere**

5. (a) $\dfrac{3}{6} \div \dfrac{3}{3} = \dfrac{3 \div 3}{6 \div 3} = \dfrac{1}{2}$

(b) $\dfrac{5}{15} \div \dfrac{5}{5} = \dfrac{5 \div 5}{15 \div 5} = \dfrac{1}{3}$

(c) $\dfrac{8}{12} \div \dfrac{4}{4} = \dfrac{8 \div 4}{12 \div 4} = \dfrac{2}{3}$

6. Multiples of 4: 4, 8, <u>12</u>
Multiples of 6: 6, <u>12</u>
$$\dfrac{3}{4} \times \dfrac{3}{3} = \dfrac{9}{12}$$
$$\dfrac{5}{6} \times \dfrac{2}{2} = \dfrac{10}{12}$$

7. **2** is in the ten-millions place.

8.
$$\begin{array}{r} 73\text{ miles} \\ -\ 23\text{ miles} \\ \hline \mathbf{50\text{ miles}} \end{array}$$

9. $100\% - 80\% = \mathbf{20\%}$

10. A nickel is one half of a dime. One half $= \mathbf{50\%}$

11.
$$
\begin{array}{r}
\overset{1}{4}.36 \\
12.7 \\
+\ 10.72 \\
\hline
\mathbf{27.78}
\end{array}
$$

12.
$$
\begin{array}{r}
4.\overset{1}{2}{}^1 0 \\
-\ 2.1\ 7 \\
\hline
2.0\ 3
\end{array}
$$

$$
\begin{array}{r}
8.54 \\
-\ 2.03 \\
\hline
\mathbf{6.51}
\end{array}
$$

13. $\dfrac{5}{9} + \dfrac{5}{9} = \dfrac{10}{9}$

$\dfrac{10}{9} = \mathbf{1\dfrac{1}{9}}$

14. $3\dfrac{2}{3} + 1\dfrac{2}{3} = 4\dfrac{4}{3}$

$4\dfrac{4}{3} = \mathbf{5\dfrac{1}{3}}$

15. $4\dfrac{5}{8} + 1 = \mathbf{5\dfrac{5}{8}}$

16. $7\dfrac{2}{3} + 1\dfrac{2}{3} = 8\dfrac{4}{3}$

$8\dfrac{4}{3} = \mathbf{9\dfrac{1}{3}}$

17. $4\dfrac{4}{9} + 1\dfrac{1}{9} = \mathbf{5\dfrac{5}{9}}$

18. Add only the numerators.

$\dfrac{12}{12} = \mathbf{1}$

19.
$$
\begin{array}{r}
\overset{4}{\overset{2}{570}} \\
\times\ \ \ 64 \\
\hline
2\ 280 \\
34\ 200 \\
\hline
\mathbf{36,480}
\end{array}
$$

20.
$$
\begin{array}{r}
\overset{2}{382} \\
\times\ \ \ 31 \\
\hline
382 \\
11\ 460 \\
\hline
\mathbf{11,842}
\end{array}
$$

21.
$$
\begin{array}{r}
\overset{3}{54} \\
\times\ \ 18 \\
\hline
432 \\
540 \\
\hline
\mathbf{972}
\end{array}
$$

22.
$$
\begin{array}{r}
\mathbf{533} \\
7\overline{)3731} \\
35\ \ \ \\
\hline
23\ \\
21\ \\
\hline
21 \\
21 \\
\hline
0
\end{array}
$$

23.
$$
\begin{array}{r}
\mathbf{603\ R\ 5} \\
9\overline{)5432} \\
54\ \ \ \\
\hline
03\ \\
0\ \\
\hline
32 \\
27 \\
\hline
5
\end{array}
$$

24.
$$
\begin{array}{r}
\mathbf{9\ R\ 8} \\
60\overline{)548} \\
540 \\
\hline
8
\end{array}
$$

25.
$$
\begin{aligned}
6 \times 6 &= \mathbf{36} \\
7 \times 7 &= \mathbf{49} \\
8 \times 8 &= \mathbf{64} \\
9 \times 9 &= \mathbf{81} \\
10 \times 10 &= \mathbf{100}
\end{aligned}
$$

26. Since 20,851,820 is greater than 20,500,000, it rounds up to **21,000,000.**

27. A trapezoid has exactly one pair of parallel lines.
D. trapezoid

28. 1 cm $=$ 10 mm
6 cm $=$ 60 mm
A square has four equal sides.

$$
\begin{array}{r}
\mathbf{15\ mm} \\
4\overline{)60} \\
4\ \\
\hline
20 \\
20 \\
\hline
0
\end{array}
$$

29. $2 \times 2 \times 2 = \mathbf{8\ smaller\ cubes}$

30. A cube has 8 vertices and this pyramid has 5 vertices.
8 vertices $-$ 5 vertices $= \mathbf{3\ more\ vertices}$

LESSON 118, WARM-UP

a. 17

b. $12\frac{1}{2}$

c. 25

d. 330

e. $0.87

f. 296

g. 260

h. 91

Problem Solving

100% − 30% = **70%** chance it will not rain.

70% is greater than 30%, so it is **more likely not to rain.**

An equal likelihood for an event to happen as not to happen means the chance for both possibilities is 100% ÷ 2, or **50%.**

LESSON 118, LESSON PRACTICE

a.
$$
\begin{array}{r}
4 \\
32)\overline{128} \\
\underline{128} \\
0
\end{array}
$$

b.
$$
\begin{array}{r}
4\ R\ 6 \\
21)\overline{90} \\
\underline{84} \\
6
\end{array}
$$

c.
$$
\begin{array}{r}
2\ R\ 18 \\
25)\overline{68} \\
\underline{50} \\
18
\end{array}
$$

d.
$$
\begin{array}{r}
5\ R\ 40 \\
42)\overline{250} \\
\underline{210} \\
40
\end{array}
$$

e.
$$
\begin{array}{r}
21\ R\ 19 \\
41)\overline{880} \\
\underline{82} \\
60 \\
\underline{41} \\
19
\end{array}
$$

f.
$$
\begin{array}{r}
50\ R\ 5 \\
11)\overline{555} \\
\underline{55} \\
05 \\
\underline{0} \\
5
\end{array}
$$

LESSON 118, MIXED PRACTICE

1. The temperature was the highest on **Wednesday.**

2. The high temperature on Tuesday was **57°F.**

3. 65°F − 50°F = **15°F**

4. (a)
$$
\begin{array}{r}
\overset{1}{2}4\ m \\
15\ m \\
24\ m \\
+\ 15\ m \\
\hline
78\ m
\end{array}
$$

(b)
$$
\begin{array}{r}
\overset{2}{2}4\ m \\
\times\ 15\ m \\
\hline
120 \\
240 \\
\hline
360\ \text{sq. m}
\end{array}
$$

5.
$$
\begin{array}{r}
\overset{2}{3}6 \\
49 \\
64 \\
\overset{2}{8}1 \\
+\ 100 \\
\hline
330
\end{array}
$$

$$
\begin{array}{r}
66 \\
5)\overline{330} \\
\underline{30} \\
30 \\
\underline{30} \\
0
\end{array}
$$

6. January, June, July begin with the letter J, so 3 out of 12 months begin with J.

$$\frac{3}{12} = \frac{1}{4} \qquad \frac{1}{4} = \textbf{25\%}$$

7. $\dfrac{4}{52} \div \dfrac{4}{4} = \dfrac{4 \div 4}{52 \div 4} = \dfrac{1}{13}$

8. (a) **Cylinder**

  (b) **Cone**

  (c) **Sphere**

9. (a) $\dfrac{6}{8} \div \dfrac{2}{2} = \dfrac{6 \div 2}{8 \div 2} = \dfrac{3}{4}$

  (b) The only number that divides 4 and 9 is 1.

  $\dfrac{4}{9}$

  (c) $\dfrac{4}{16} \div \dfrac{4}{4} = \dfrac{4 \div 4}{16 \div 4} = \dfrac{1}{4}$

10. Multiples of 3: 3, 6, 9, <u>12</u>

  Multiples of 4: 4, 8, <u>12</u>

  $\dfrac{2}{3} \times \dfrac{4}{4} = \dfrac{8}{12}$

  $\dfrac{3}{4} \times \dfrac{3}{3} = \dfrac{9}{12}$

11. 27,386,415

  Twenty-seven million, three hundred eighty-six thousand, four hundred fifteen

12. $4\dfrac{4}{5} + 3\dfrac{3}{5} = 7\dfrac{7}{5}$

  $7\dfrac{7}{5} = 8\dfrac{2}{5}$

13. $5\dfrac{1}{6} + 1\dfrac{2}{6} = 6\dfrac{3}{6}$

  $6\dfrac{3}{6} = 6\dfrac{1}{2}$

14. $7\dfrac{3}{4} + \dfrac{1}{4} = 7\dfrac{4}{4}$

  $7\dfrac{4}{4} = 8$

15.
```
    3 R 11
13)50
   39
   11
```

16.
```
  4
  28
  47
  74
  36
  91
  87
  21
  12
+ 14
 410
```

17.
```
   44
 +  N
   55
```
  $55 - 44 = 11$
  $N = 11$

18.
```
   4 R 9
72)297
   288
     9
```

19. $5\dfrac{3}{8} + 5\dfrac{1}{8} = 10\dfrac{4}{8}$

  $10\dfrac{4}{8} = 10\dfrac{1}{2}$

20. $4\dfrac{1}{6} + 2\dfrac{1}{6} = 6\dfrac{2}{6}$

  $6\dfrac{2}{6} = 6\dfrac{1}{3}$

21.
```
    1
   720
 ×  36
  4 320
 21 600
 25,920
```

22.
```
   2 3
   1 2
   147
 ×  54
   588
  7350
  7938
```

23.
```
    720 R 6
 8)5766
   56
   16
   16
    06
     0
     6
```

24.
$$\begin{array}{r} 21 \\ 21\overline{)441} \\ \underline{42} \\ 21 \\ \underline{21} \\ 0 \end{array}$$

25.
$$\begin{array}{r} {}^{1}_{1}4.75 \\ 16.14 \\ +\ \ 10.9 \\ \hline \mathbf{31.79} \end{array}$$

26.
$$\begin{array}{r} \overset{3}{\cancel{4}}.{}^{1}3\,2 \\ -\ 2.\,6 \\ \hline 1.\,7\,2 \end{array}$$

$$\begin{array}{r} 1\,\overset{7}{\cancel{8}}.\,\overset{13}{\cancel{4}}{}^{1}0 \\ -\ \ 1.\,7\,2 \\ \hline \mathbf{1\,6.\,6\,8} \end{array}$$

27. Since 18,976,457 is greater than 18,500,000, round up to **19,000,000.**

28. Since 297,576,320 is greater than 250,000,000, round up to **300,000,000.**

29. 75, 80, 85, 90, 90, 95, 95, 95, 100
Median: 90
$100 - 75 = 25$
Range: 25

30. The score that occurs most often is 95, so the mode is **95.**

LESSON 119, WARM-UP

a. 45

b. 9

c. 81

d. 420

e. $6.51

f. 18

g. 1667

h. <

Problem Solving

18 inches $\times$ 4 = 72 inches

72 inches $\div$ 12 = 6 feet

The wall is **about 6 feet** tall.

LESSON 119, LESSON PRACTICE

a.
$$\begin{array}{l} \dfrac{1}{2} \times \dfrac{3}{3} = \dfrac{3}{6} \\[2mm] +\ \dfrac{2}{6} \qquad = \dfrac{2}{6} \\[2mm] \hline \qquad\qquad\quad \dfrac{5}{6} \end{array}$$

b.
$$\begin{array}{l} \dfrac{1}{3} \times \dfrac{3}{3} = \dfrac{3}{9} \\[2mm] +\ \dfrac{1}{9} \qquad = \dfrac{1}{9} \\[2mm] \hline \qquad\qquad\quad \dfrac{4}{9} \end{array}$$

c.
$$\begin{array}{l} \dfrac{1}{8} \qquad\qquad = \dfrac{1}{8} \\[2mm] +\ \dfrac{1}{2} \times \dfrac{4}{4} = \dfrac{4}{8} \\[2mm] \hline \qquad\qquad\quad \dfrac{5}{8} \end{array}$$

d.
$$\begin{array}{l} \dfrac{3}{8} \qquad\qquad = \dfrac{3}{8} \\[2mm] -\ \dfrac{1}{4} \times \dfrac{2}{2} = \dfrac{2}{8} \\[2mm] \hline \qquad\qquad\quad \dfrac{1}{8} \end{array}$$

e.
$$\begin{array}{l} \dfrac{2}{3} \times \dfrac{3}{3} = \dfrac{6}{9} \\[2mm] -\ \dfrac{2}{9} \qquad = \dfrac{2}{9} \\[2mm] \hline \qquad\qquad\quad \dfrac{4}{9} \end{array}$$

f.
$$\begin{array}{l} \dfrac{7}{8} \qquad\qquad = \dfrac{7}{8} \\[2mm] -\ \dfrac{1}{2} \times \dfrac{4}{4} = \dfrac{4}{8} \\[2mm] \hline \qquad\qquad\quad \dfrac{3}{8} \end{array}$$

LESSON 119, MIXED PRACTICE

1.
$$\overset{1}{15} \text{ ft}$$
$$\times\ 12 \text{ ft}$$
$$\overline{\quad 30}$$
$$\underline{150\quad}$$
$$180 \text{ sq. ft } = \textbf{180 floor tiles}$$

2. (a)
$$\overset{1}{} 1.2 \text{ cm}$$
$$1.9 \text{ cm}$$
$$\underline{+\ 2.2 \text{ cm}}$$
$$\textbf{5.3 cm}$$

(b) None of the sides are equal, so the triangle is **scalene.**

3.

32 pencils

$\frac{3}{8}$ had no erasers.	4 pencils
	4 pencils
	4 pencils
$\frac{5}{8}$ had erasers.	4 pencils
	4 pencils
	4 pencils
	4 pencils
	4 pencils

$3 \times 4 \text{ pencils } = \textbf{12 pencils}$

4. One dozen $= 12$ eggs

$$\overset{\textbf{6 dozen eggs}}{12)\overline{72}}$$
$$\underline{72}$$
$$0$$

5. $50\% =$ one half
Half of 12 eggs is **6 eggs.**

6. $3 \text{ cm} \times 3 \text{ cm} \times 3 \text{ cm}$
$\qquad = \textbf{27 cubic centimeters}$

7.
$$\overset{1\ 4\ \ 3}{\$21.95}$$
$$\$21.95$$
$$\$14.99$$
$$\$14.99$$
$$\underline{+\ \$4.62}$$
$$\textbf{\$78.50}$$

8. Pattern:
Number of groups $\times$ number in each group
$\qquad\qquad\qquad\qquad\qquad = \text{total}$
Problem:
5 hours $\times$ N miles per hour $= 285$ miles

$$\overset{\textbf{57 miles per hour}}{5)\overline{285}}$$
$$\underline{25}$$
$$35$$
$$\underline{35}$$
$$0$$

9. A fraction is equivalent to one half when its denominator is exactly twice its numerator.
D. $\dfrac{12}{25}$

10. (a) $\dfrac{8}{10} \div \dfrac{2}{2} = \dfrac{8 \div 2}{10 \div 2} = \dfrac{4}{5}$

(b) $\dfrac{6}{15} \div \dfrac{3}{3} = \dfrac{6 \div 3}{15 \div 3} = \dfrac{2}{5}$

(c) $\dfrac{8}{16} \div \dfrac{8}{8} = \dfrac{8 \div 8}{16 \div 8} = \dfrac{1}{2}$

11. 123,415,720
One hundred twenty-three million, four hundred fifteen thousand, seven hundred twenty

12.
$$\overset{1}{8}.3$$
$$4.72$$
$$\overset{1}{}0.6$$
$$\underline{+\ 12.1}$$
$$\textbf{25.72}$$

13.
$$6.\overset{6}{\cancel{7}}{}^{1}0$$
$$\underline{-\ 1.2\ 3}$$
$$5.4\ 7$$

$$1\overset{6}{\cancel{7}}.\overset{1}{\cancel{4}}{}^{3}{}^{1}2$$
$$\underline{-\ \ 5.4\ 7}$$
$$\textbf{1 1.9 5}$$

14. $3\dfrac{3}{8} + 3\dfrac{3}{8} = 6\dfrac{6}{8}$

$6\dfrac{6}{8} = \mathbf{6\dfrac{3}{4}}$

15.
$$\dfrac{1}{4} \times \dfrac{2}{2} = \dfrac{2}{8}$$
$$+\ \dfrac{1}{8} \qquad = \dfrac{1}{8}$$
$$\overline{\qquad\qquad\quad \dfrac{3}{8}}$$

16. $\frac{1}{2} \times \frac{3}{3} = \frac{3}{6}$

$+\quad \frac{1}{6} \qquad = \frac{1}{6}$

$\qquad\qquad \frac{4}{6} = \mathbf{\frac{2}{3}}$

17. $5\frac{5}{6} - 1\frac{1}{6} = 4\frac{4}{6}$

$4\frac{4}{6} = \mathbf{4\frac{2}{3}}$

18. $\frac{1}{4} \times \frac{2}{2} = \frac{2}{8}$

$-\quad \frac{1}{8} \qquad = \frac{1}{8}$

$\qquad\qquad\quad \mathbf{\frac{1}{8}}$

19. $\frac{1}{2} \times \frac{3}{3} = \frac{3}{6}$

$-\quad \frac{1}{6} \qquad = \frac{1}{6}$

$\qquad\qquad \frac{2}{6} = \mathbf{\frac{1}{3}}$

20.
$$\begin{array}{r} \overset{4}{87} \\ \times\ 16 \\ \hline 522 \\ 870 \\ \hline \mathbf{1392} \end{array}$$

21.
$$\begin{array}{r} \overset{\overset{1}{3}}{340} \\ \times\ \ 49 \\ \hline 3\ 060 \\ 13\ 600 \\ \hline \mathbf{16,660} \end{array}$$

22.
$$\begin{array}{r} \overset{1}{504} \\ \times\ \ 30 \\ \hline \mathbf{15,120} \end{array}$$

23.
$$\begin{array}{r} \mathbf{\$5.90} \\ 6)\overline{\$35.40} \\ 30 \\ \hline 5\ 4 \\ 5\ 4 \\ \hline 00 \\ 0 \\ \hline 0 \end{array}$$

24.
$$\begin{array}{r} \mathbf{1446} \\ 4)\overline{5784} \\ 4 \\ \hline 17 \\ 16 \\ \hline 18 \\ 16 \\ \hline 24 \\ 24 \\ \hline 0 \end{array}$$

25.
$$\begin{array}{r} \mathbf{340\ R\ 5} \\ 7)\overline{2385} \\ 21 \\ \hline 28 \\ 28 \\ \hline 05 \\ 0 \\ \hline 5 \end{array}$$

26.
$$\begin{array}{r} \mathbf{15} \\ 30)\overline{450} \\ 30 \\ \hline 150 \\ 150 \\ \hline 0 \end{array}$$

27.
$$\begin{array}{r} \mathbf{14\ R\ 2} \\ 32)\overline{450} \\ 32 \\ \hline 130 \\ 128 \\ \hline 2 \end{array}$$

28.
$$\begin{array}{r} \mathbf{30} \\ 15)\overline{450} \\ 45 \\ \hline 00 \\ 0 \\ \hline 0 \end{array}$$

29. $\frac{13}{52} \div \frac{13}{13} = \frac{13 \div 13}{52 \div 13} = \mathbf{\frac{1}{4}}$

30. 30% = 3 tenths = 3 out of 10 square centimeters

5 cm

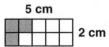
2 cm

LESSON 120, WARM-UP

a. 75

b. 70

c. 100

d. 460

e. $7.25

f. 162

g. 1900

h. 99

Patterns

The pattern is "count up by $\frac{1}{8}$'s."

$\frac{1}{8}, \frac{1}{4}, \frac{3}{8}, \frac{1}{2}, \frac{5}{8}, \frac{3}{4}, \frac{7}{8}, 1, \mathbf{1\frac{1}{8}}, \mathbf{1\frac{1}{4}}, \mathbf{1\frac{3}{8}}, \mathbf{1\frac{1}{2}},$

$\mathbf{1\frac{5}{8}}, \mathbf{1\frac{3}{4}}, \mathbf{1\frac{7}{8}}, \mathbf{2}, \dots$

LESSON 120, LESSON PRACTICE

a.
$$3\frac{1}{2} \times \frac{2}{2} = 3\frac{2}{4}$$
$$+ 1\frac{1}{4} \quad = 1\frac{1}{4}$$
$$\overline{\qquad \mathbf{4\frac{3}{4}}}$$

b.
$$4\frac{3}{4} \times \frac{2}{2} = 4\frac{6}{8}$$
$$+ 1\frac{1}{8} \quad = 1\frac{1}{8}$$
$$\overline{\qquad \mathbf{5\frac{7}{8}}}$$

c.
$$4\frac{1}{5} \times \frac{2}{2} = 4\frac{2}{10}$$
$$+ 1\frac{3}{10} \quad = 1\frac{3}{10}$$
$$\overline{\qquad 5\frac{5}{10} = \mathbf{5\frac{1}{2}}}$$

d.
$$6\frac{1}{6} \quad = 6\frac{1}{6}$$
$$+ 1\frac{1}{3} \times \frac{2}{2} = 1\frac{2}{6}$$
$$\overline{\qquad 7\frac{3}{6} = \mathbf{7\frac{1}{2}}}$$

e.
$$3\frac{7}{8} \quad = 3\frac{7}{8}$$
$$- 1\frac{1}{4} \times \frac{2}{2} = 1\frac{2}{8}$$
$$\overline{\qquad \mathbf{2\frac{5}{8}}}$$

f.
$$2\frac{3}{5} \times \frac{2}{2} = 2\frac{6}{10}$$
$$- 2\frac{1}{10} \quad = 2\frac{1}{10}$$
$$\overline{\qquad \frac{5}{10} = \mathbf{\frac{1}{2}}}$$

g.
$$6\frac{7}{12} \quad = 6\frac{7}{12}$$
$$- 1\frac{1}{6} \times \frac{2}{2} = 1\frac{2}{12}$$
$$\overline{\qquad \mathbf{5\frac{5}{12}}}$$

h.
$$4\frac{3}{4} \quad = 4\frac{3}{4}$$
$$- 1\frac{1}{2} \times \frac{2}{2} = 1\frac{2}{4}$$
$$\overline{\qquad \mathbf{3\frac{1}{4}}}$$

LESSON 120, MIXED PRACTICE

1. There are 4 quarts in 1 gallon.

11 gallons $\times$ 4 quarts per gallon = **44 quarts**

2.

$\frac{1}{4}$ perished.
$\frac{3}{4}$ survived.

60 fleas
15 fleas
15 fleas
15 fleas
15 fleas

$$\begin{array}{r} \overset{1}{1}5 \text{ fleas} \\ \times \quad 3 \\ \hline \mathbf{45 \text{ fleas}} \end{array}$$

3. (a) 10 mm $\times$ 10 mm = **100 sq. mm**

(b) 10 mm + 10 mm + 10 mm
+ 10 mm = **40 mm**

4. Jermaine: 61 in. − 8 in. = 53 in.

Jan: 53 in. − 5 in. = **48 in.**

5.
$$\overset{1}{61} \text{ in.}$$
$$53 \text{ in.}$$
$$+\ 48 \text{ in.}$$
$$162 \text{ in.}$$

$$\begin{array}{r} \mathbf{54 \text{ in.}} \\ 3\overline{)162} \\ \underline{15} \\ 12 \\ \underline{12} \\ 0 \end{array}$$

6.
$$\overset{1}{47} \text{ mi}$$
$$+\ 24 \text{ mi}$$
$$\mathbf{71 \text{ mi}}$$

7. 2 out of 6 numbers are greater than 4.
$$\frac{2}{6} = \mathbf{\frac{1}{3}}$$

8. Since $298,900 is greater than $250,000, it rounds up to **$300,000.**

9. (a) **Pyramid**

(b) **Rectangular solid (or rectangular prism)**

(c) **Cone**

10. (a) $\dfrac{9}{15} \div \dfrac{3}{3} = \dfrac{9 \div 3}{15 \div 3} = \mathbf{\dfrac{3}{5}}$

(b) $\dfrac{10}{12} \div \dfrac{2}{2} = \dfrac{10 \div 2}{12 \div 2} = \mathbf{\dfrac{5}{6}}$

(c) $\dfrac{12}{16} \div \dfrac{4}{4} = \dfrac{12 \div 4}{16 \div 4} = \mathbf{\dfrac{3}{4}}$

11. **119,247,984**

12.
$$\overset{7}{\cancel{8}}.\overset{1}{6}$$
$$-\ 4.\ 7$$
$$3.\ 9$$

$$14.94$$
$$-\ \ 3.90$$
$$\mathbf{11.04}$$

13.
$$1.37$$
$$+\ 2.2$$
$$3.57$$

$$6.\overset{7}{\cancel{8}}{}^{1}0$$
$$-\ 3.\ 5\ 7$$
$$\mathbf{3.\ 2\ 3}$$

14. $3\dfrac{2}{5} + 1\dfrac{4}{5} = 4\dfrac{6}{5}$

$4\dfrac{6}{5} = \mathbf{5\dfrac{1}{5}}$

15.
$$\frac{5}{8} = \frac{5}{8}$$
$$+\ \frac{1}{4} \times \frac{2}{2} = \frac{2}{8}$$
$$\mathbf{\frac{7}{8}}$$

16.
$$1\frac{1}{3} \times \frac{2}{2} = 1\frac{2}{6}$$
$$+\ 1\frac{1}{6} = 1\frac{1}{6}$$
$$2\frac{3}{6} = \mathbf{2\frac{1}{2}}$$

17.
$$5\frac{9}{10} = 5\frac{9}{10}$$
$$-\ 1\frac{1}{5} \times \frac{2}{2} = 1\frac{2}{10}$$
$$\mathbf{4\frac{7}{10}}$$

18.
$$\frac{5}{8} = \frac{5}{8}$$
$$-\ \frac{1}{4} \times \frac{2}{2} = \frac{2}{8}$$
$$\mathbf{\frac{3}{8}}$$

19.
$$\frac{1}{3} \times \frac{2}{2} = \frac{2}{6}$$
$$-\ \frac{1}{6} = \frac{1}{6}$$
$$\mathbf{\frac{1}{6}}$$

20.
$$\begin{array}{r} \overset{2}{\underset{}{1\,\overset{5}{}}} \\ 217 \\ \times\ \ \ 38 \\ \hline 1736 \\ 6510 \\ \hline \mathbf{8246} \end{array}$$

21.
$$\begin{array}{r} \overset{4\,1}{173} \\ \times\ \ \ \ 60 \\ \hline \mathbf{10,380} \end{array}$$

22. Multiply mentally: **45,000**

23.
$$\begin{array}{r} 420 \text{ R } 2 \\ 7\overline{)2942} \\ \underline{28} \\ 14 \\ \underline{14} \\ 02 \\ \underline{0} \\ 2 \end{array}$$

24.
$$\begin{array}{r} 45 \text{ R } 3 \\ 10\overline{)453} \\ \underline{40} \\ 53 \\ \underline{50} \\ 3 \end{array}$$

25.
$$\begin{array}{r} 41 \text{ R } 2 \\ 11\overline{)453} \\ \underline{44} \\ 13 \\ \underline{11} \\ 2 \end{array}$$

26.
$$3\frac{2}{5} \times \frac{2}{2} = 3\frac{4}{10}$$
$$+ \ 2\frac{1}{10} \qquad = 2\frac{1}{10}$$
$$5\frac{5}{10} = 5\frac{1}{2}$$

27. $3 \text{ ft} \times 2 \text{ ft} \times 2 \text{ ft} =$ **12 cubic feet**

28. $3\frac{1}{2}$ in. $- \ 1\frac{1}{2}$ in. $=$ **2 in.**

29. Since 493,782 is greater than 450,000, it rounds up to **500,000 people.**

30.
$$\frac{1}{2} \times \frac{2}{2} = \frac{2}{4}$$
$$+ \ \frac{1}{4} \qquad = \frac{1}{4}$$
$$\frac{3}{4}$$
$$\frac{3}{4} = \textbf{75\%}$$

INVESTIGATION 12

1. $N + 5 = 13; 8$

2. $11 = N + 6; 5$

3. $N + 5 = 11; 6$

4. $17 = N + 8; 9$

5. $2N = 16; 8$

6. $2N = 20; 10$

7. $3N = 12; 4$

8. $2N + 6 = 18; 6$

Solutions for

Appendix Topics

TOPIC A, LESSON PRACTICE

a. $ 3 4 5 . 2 3

b. $ 0 . 4 2

c. $ 5 . 2 0

d. $ 3 . 0 2

e. 3 . 4 2

f. 0 . 2 4

g. 1 2 . 0 3

h. 1 . 3

TOPIC B, LESSON PRACTICE

1. I

2. II

3. III

4. IV

5. V

6. VI

7. VII

8. VIII

9. IX

10. X

11. XI

12. XII

13. XIII

14. XIV

15. XV

16. XVI

17. XVII

18. XVIII

19. XIX

20. XX

21. XXI

22. XXII

23. XXIII

24. XXIV

25. XXV

26. XXVI

27. XXVII

28. XXVIII

29. XXIX

30. XXX

31. XXXI

32. **XXXII**

33. **XXXIII**

34. **XXXIV**

35. **XXXV**

36. **XXXVI**

37. **XXXVII**

38. **XXXVIII**

39. **XXXIX**

---

e. $38 = (1 \times 25) + (2 \times 5)$
  $+ (3 \times 1) = $ **123 (base 5)**

f. $86 = (3 \times 25) + (2 \times 5)$
  $+ (1 \times 1) = $ **321 (base 5)**

Topic C, Lesson Practice

a. $100 + 100 + 100 + 50$
  $+ 10 + 1 + 1 = $ **362**

b. $100 + 100 + 50 + 10$
  $+ 10 + 10 + 5 = $ **285**

c. $(500 - 100) = $ **400**

d. $(50 - 10) + 5 + 1 + 1 = $ **47**

e. $1000 + 1000 + 1000 + 100$
  $+ 100 + 50 + 5 + 1 = $ **3256**

f. $1000 + (1000 - 100) + (100 - 10)$
  $+ (10 - 1) = $ **1999**

Topic D, Lesson Practice

a. $31 = (1 \times 25) + (1 \times 5)$
  $+ (1 \times 1) = $ **111 (base 5)**

b. $51 = (2 \times 25) + (0 \times 5)$
  $+ (1 \times 1) = $ **201 (base 5)**

c. $10 = (2 \times 5) + (0 \times 1) = $ **20 (base 5)**

d. $100 = (4 \times 25) + (0 \times 5)$
  $+ (0 \times 1) = $ **400 (base 5)**

Solutions for

Supplemental Practice

SUPPLEMENTAL PRACTICE, LESSON 16

1. 19

2. 14

3. 7

4. 27

5. 7

6. 24

7. 17

8. 40

9. 28

10. 70

11. 37

12. 74

13. 18

14. 17

15. 17

16. 27

17. 57

18. 8

19. 74

20. 28

SUPPLEMENTAL PRACTICE, LESSON 17

1. 42

2. 84

3. 79

4. 138

5. 217

6. 202

7. 217

8. 222

9. 96

10. 543

11. 500

12. 292

13. 815

14. 616

15. 662

16. 600

17. 870

18. 710

19. 910

20. 483

SUPPLEMENTAL PRACTICE, LESSON 30

1. 116

2. 291

3. 184

4. 108

5. 139

6. 193

7. 226

8. 170

9. 237

10. 114

11. 28

12. 168

13. 36

14. 92

15. 483

16. 534

17. 191

18. 124

19. 280

20. 103

SUPPLEMENTAL PRACTICE, LESSON 34

1. three hundred sixty-three

2. one thousand, two hundred forty-six

3. twelve thousand, two hundred eighty

4. twenty-five thousand, three hundred sixty-two

5. one hundred twenty-three thousand, five hundred seventy

6. two hundred fifty-three thousand, five hundred

7. one hundred twelve thousand, sixty

8. two hundred twenty thousand, four hundred five

9. two hundred four thousand, fifty

10. five hundred forty-six thousand, three hundred twenty-five

11. 1278

12. 11,544

13. 22,430

14. 57,900

15. 171,230

16. 210,900

17. 563,058

18. 987,654

19. 105,070

20. 650,403

SUPPLEMENTAL PRACTICE, LESSON 37

A. $\frac{2}{3}$

B. $1\frac{1}{3}$

C. $2\frac{2}{3}$

D. $3\frac{1}{3}$

E. $\frac{3}{4}$

F. $2\frac{1}{4}$

G. $2\frac{3}{4}$

H. $3\frac{1}{4}$

I. $\frac{5}{6}$

J. $1\frac{5}{6}$

K. $2\frac{1}{6}$

L. $3\frac{1}{6}$

M. $\frac{3}{8}$

N. $1\frac{5}{8}$

O. $2\frac{7}{8}$

P. $3\frac{3}{8}$

Q. $\frac{7}{10}$

R. $1\frac{3}{10}$

S. $2\frac{1}{10}$

T. $3\frac{9}{10}$

SUPPLEMENTAL PRACTICE, LESSON 41

1. 164

2. 146

3. 57

4. 149

5. 351

6. 36

7. 64

8. 176

9. 51

10. 7

11. 264

12. 374

13. 21

14. 842

15. 251

16. 48

17. 752

18. 27

19. 283

20. 79

SUPPLEMENTAL PRACTICE, LESSON 43

1. $10.35

2. $1.84

3. $1.15

4. 34¢

5. $1.21

6. $1.17

7. $2.51

8. 8¢

9. $12.98

10. 5¢

11. $5.95

12. 43¢

13. $2.62

14. $0.63

15. $16.65

16. $0.63

17. $1.38

18. $0.37

19. $5.45

20. 9¢

SUPPLEMENTAL PRACTICE, LESSON 48

1. 72

2. 215

3. 171

4. 144

5. 152

6. 364

7. 135

8. 152

9. 168

10. 504

11. 329

12. 356

13. 512

14. 375

15. 567

16. 304

17. 582

18. 136

19. 378

20. 135

SUPPLEMENTAL PRACTICE, LESSON 50

1. 5.77

2. 17.68

3. 38.96

4. 22.5

5. 31.06

6. 48.21

7. 46.31

8. 10.01

9. 111.11

10. 99.99

11. 8.26

12. 17.2

13. 16.13

14. 10.68

15. 62.18

16. 9.37

17. 43.44

18. 23.72

19. 79.41

20. 70.33

21. 2.25

22. 20.9

23. 12.65

24. 10.9

25. 5.79

26. 24.13

27. 14.35

28. 7.79

29. 1.06

30. 2.14

31. 1.86

32. 11.16

33. 6.18

34. 1.98

35. 10.89

36. 27.09

37. 27.65

38. 55.86

39. 1.29

40. 16.91

SUPPLEMENTAL PRACTICE, LESSON 52

1. 1799

2. 829

3. 2093

4. 2928

5. 3151

6. 97

7. 1654

8. 1587

9. 1505

10. 3836

11. 185

12. 3169

13. 34,791

14. 7039

15. 23,577

16. 29,744

17. 28,689

18. 52,562

19. 339,968

20. 9028

SUPPLEMENTAL PRACTICE, LESSON 53

1. 7 R 1

2. 4 R 3

3. 8 R 1

4. 4 R 2

5. 7 R 3

6. 7 R 1

7. 7 R 2

8. 4 R 3

9. 7 R 1

10. 3 R 3

11. 6 R 3

12. 8 R 1

13. 6 R 1

14. 8 R 1

15. 6 R 6

16. 6 R 3

17. 1 R 7

18. 6 R 2

19. 6 R 2

20. 5 R 4

SUPPLEMENTAL PRACTICE, LESSON 58

1. 272

2. 1175

3. 1290

4. 1296

5. 1800

6. 4494

7. 1535

8. 3664

9. 4440

10. 2576

11. 3627

12. 3920

13. 1218

14. 4230

15. 1554

16. 592

17. 2832

18. 1290

19. 1375

20. 1232

SUPPLEMENTAL PRACTICE, LESSON 64

1. 16

2. 28

3. 18

4. 14

5. 16

6. 12

7. 15

8. 27

9. 13

10. 17

11. 38

12. 18

13. 18

14. 14

15. 26

16. 12

17. 21

18. 17

19. 12

20. 13

SUPPLEMENTAL PRACTICE, LESSON 65

1. 55

2. 14

3. 74

4. 54

5. 64

6. 65

7. 64

8. 96

9. 57

10. 68

11. 38

12. 25

13. 57

14. 73

15. 52

16. 27

17. 63

18. 95

19. 75

20. 49

SUPPLEMENTAL PRACTICE, LESSON 67

1. 640

2. $12.90

3. 2240

4. $13.60

5. 2820

6. $33.60

7. 1500

8. $14.70

9. 3480

10. $48.50

11. $104.30

12. 2280

13. 6080

14. $43.20

15. 1780

16. $112.50

17. $23.40

18. 2340

19. $26.80

20. 5880

SUPPLEMENTAL PRACTICE, LESSON 68

1. 23 R 1

2. 57 R 1

3. 36 R 2

4. 57 R 2

5. 48 R 1

6. 45 R 1

7. 23 R 3

8. 63 R 4

9. 27 R 2

10. 39 R 1

11. 24 R 4

12. 63 R 1

13. 94 R 1

14. 74 R 3

15. 36 R 1

16. 99 R 1

17. 54 R 2

18. 27 R 1

19. 46 R 2

20. 73 R 1

SUPPLEMENTAL PRACTICE, LESSON 76

1. 233 R 1

2. 123

3. 223

4. 124

5. 321 R 1

6. 114 R 3

7. 212 R 3

8. 214

9. 345

10. 534 R 1

11. 145 R 1

12. 277 R 3

13. 333 R 3

14. 125

15. 444 R 1

16. 253

17. 562 R 2

18. 166

19. 125

20. 748 R 3

SUPPLEMENTAL PRACTICE, LESSON 80

1. 240

2. 320

3. 402

4. 104

5. 520 R 2

6. 303 R 2

7. 400 R 1

8. 909 R 1

9. 680 R 4

10. 904

11. 500

12. 670 R 1

13. 807 R 1

14. 409 R 3

15. 150

16. 400 R 1

17. 190 R 1

18. 760

19. 409 R 5

20. 700 R 2

SUPPLEMENTAL PRACTICE, LESSON 90

1. 432

2. 690

3. 837

4. 1184

5. 2268

6. 1225

7. 882

8. 1716

9. 925

10. 4292

11. 3087

12. 1170

13. 3072

14. 3876

15. 912

16. 1786

17. 4214

18. 899

19. 2607

20. 2394

SUPPLEMENTAL PRACTICE, LESSON 104

1. $1\frac{1}{2}$

2. 3

3. $1\frac{1}{3}$

4. $1\frac{3}{4}$

5. $2\frac{2}{5}$

6. 2

7. $1\frac{1}{4}$

8. $1\frac{2}{5}$

9. 1

10. $1\frac{4}{5}$

11. $2\frac{1}{2}$

12. 2

13. 1

14. $1\frac{2}{3}$

15. $2\frac{1}{4}$

16. 3

17. 2

18. $3\frac{1}{3}$

19. $3\frac{1}{2}$

20. $2\frac{1}{3}$

SUPPLEMENTAL PRACTICE, LESSON 107

1. $4\frac{1}{2}$

2. $4\frac{2}{3}$

3. $1\frac{4}{5}$

4. $4\frac{1}{2}$

5. $7\frac{4}{5}$

6. $11\frac{5}{8}$

7. $5\frac{5}{7}$

8. $13\frac{1}{2}$

9. $3\frac{7}{9}$

10. $9\frac{9}{10}$

11. $4\frac{1}{3}$

12. $1\frac{3}{4}$

13. 6

14. 7

15. $\frac{3}{8}$

16. $4\frac{3}{5}$

17. $1\frac{4}{9}$

18. 1

19. 2

20. $3\frac{2}{7}$

SUPPLEMENTAL PRACTICE, LESSON 110

1. 23

2. 21

3. 13

4. 32

5. 12

6. 21

7. 21 R 20

8. 13 R 30

9. 34

10. 33 R 5

11. 24 R 10

12. 31 R 7

13. 25 R 1

14. 15 R 9

15. 46

16. 52 R 8

17. 44

18. 23 R 14

19. 56

20. 54 R 10

SUPPLEMENTAL PRACTICE, LESSON 112

1. $\frac{1}{2}$

2. $\frac{1}{2}$

3. $\frac{3}{4}$

4. $\frac{1}{3}$

5. $\frac{1}{3}$

6. $\frac{2}{5}$

7. $\frac{1}{2}$

8. $\frac{1}{6}$

9. $\frac{3}{4}$

10. $\frac{2}{3}$

11. $\frac{2}{3}$

12. $\frac{4}{5}$

13. $\frac{1}{4}$

14. $\frac{1}{4}$

15. $\frac{1}{5}$

16. $\frac{1}{2}$

17. $\frac{2}{3}$

18. $\frac{3}{5}$

19. $\frac{1}{3}$

20. $\frac{1}{2}$

SUPPLEMENTAL PRACTICE, LESSON 113

1. 3840

2. 2772

3. 3289

4. 4800

5. 3952

6. 7560

7. 6846

8. 3450

9. 10,050

10. 12,138

11. 43,680

12. 6902

13. 5488

14. 17,940

15. 7697

16. 4212

17. 25,826

18. 31,960

19. 10,353

20. 6882

SUPPLEMENTAL PRACTICE, LESSON 114

1. 1

2. 0

3. $\frac{1}{2}$

4. $\frac{1}{4}$

5. $\frac{1}{2}$

6. $\frac{2}{3}$

7. $1\frac{1}{3}$

8. $\frac{3}{4}$

9. $1\frac{2}{5}$

10. 2

11. 4

12. $2\frac{1}{2}$

13. $10\frac{1}{3}$

14. $2\frac{1}{3}$

15. $2\frac{1}{3}$

16. $2\frac{3}{5}$

17. $3\frac{1}{3}$

18. 10

19. $12\frac{3}{5}$

20. $4\frac{1}{2}$

SUPPLEMENTAL PRACTICE, LESSON 115

1. $\frac{3}{3}$

2. $\frac{5}{5}$

3. $\frac{6}{6}$

4. $\frac{3}{3}$

5. $\frac{2}{2}$

6. $\frac{3}{3}$

7. $\frac{2}{2}$

8. $\frac{4}{4}$

9. 2

10. 2

11. 2

12. 4

13. 6

14. 6

15. 4

16. 6

17. 10

18. 2

19. 9

20. 8

SUPPLEMENTAL PRACTICE, LESSON 118

1. 6

2. 4

3. 9 R 1

4. 3 R 2

5. 7

6. 5 R 2

7. 4 R 4

8. 5 R 5

9. 4 R 2

10. 6 R 3

11. 6 R 4

12. 5 R 10

13. 4 R 12

14. 7

15. 5 R 5

16. 7

17. 3 R 21

18. 8 R 8

19. 9 R 3

20. 5 R 5

21. 23

22. 32

23. 22

24. 42

25. 31

26. 23

27. 15

28. 42

29. 15 R 1

30. 31 R 2

31. 33 R 1

32. 23 R 2

33. 25

34. 22 R 5

35. 34 R 1

36. 44

37. 22 R 5

38. 50

39. 30 R 10

40. 32

SUPPLEMENTAL PRACTICE, LESSON 119

1. $\dfrac{3}{4}$

2. $\dfrac{3}{8}$

3. $\dfrac{5}{8}$

4. $\dfrac{1}{8}$

5. $\dfrac{3}{8}$

6. $\dfrac{1}{6}$

7. $\dfrac{5}{6}$

8. $\dfrac{7}{8}$

9. $\dfrac{5}{9}$

10. $\dfrac{1}{6}$

11. $\dfrac{1}{8}$

12. $\dfrac{1}{10}$

13. $\dfrac{11}{12}$

14. $\dfrac{7}{10}$

15. $\dfrac{5}{12}$

16. $\dfrac{7}{10}$

17. $\dfrac{2}{9}$

18. $\dfrac{5}{8}$

19. $\dfrac{9}{10}$

20. $\dfrac{1}{12}$

Solutions for

Facts Practice Tests

100 Addition Facts

Add.

4 + 4 **8**	7 + 5 **12**	0 + 1 **1**	8 + 7 **15**	3 + 4 **7**	3 + 2 **5**	8 + 3 **11**	2 + 1 **3**	5 + 6 **11**	2 + 9 **11**
0 + 9 **9**	8 + 9 **17**	7 + 6 **13**	1 + 3 **4**	6 + 8 **14**	7 + 3 **10**	1 + 6 **7**	4 + 7 **11**	0 + 3 **3**	6 + 4 **10**
9 + 3 **12**	2 + 6 **8**	3 + 0 **3**	6 + 1 **7**	3 + 6 **9**	4 + 0 **4**	5 + 7 **12**	1 + 1 **2**	5 + 4 **9**	2 + 8 **10**
4 + 3 **7**	9 + 9 **18**	0 + 7 **7**	9 + 4 **13**	7 + 7 **14**	8 + 6 **14**	0 + 4 **4**	5 + 8 **13**	7 + 4 **11**	1 + 7 **8**
9 + 5 **14**	1 + 5 **6**	9 + 0 **9**	3 + 8 **11**	1 + 9 **10**	9 + 1 **10**	8 + 8 **16**	2 + 2 **4**	4 + 5 **9**	6 + 2 **8**
7 + 9 **16**	1 + 2 **3**	6 + 7 **13**	0 + 8 **8**	9 + 2 **11**	4 + 8 **12**	8 + 0 **8**	3 + 9 **12**	1 + 0 **1**	6 + 3 **9**
2 + 0 **2**	8 + 4 **12**	3 + 5 **8**	9 + 8 **17**	5 + 0 **5**	5 + 5 **10**	3 + 1 **4**	7 + 2 **9**	8 + 5 **13**	2 + 5 **7**
5 + 2 **7**	0 + 5 **5**	6 + 9 **15**	1 + 8 **9**	9 + 6 **15**	7 + 1 **8**	4 + 6 **10**	0 + 2 **2**	6 + 5 **11**	4 + 9 **13**
1 + 4 **5**	3 + 7 **10**	7 + 0 **7**	2 + 3 **5**	5 + 1 **6**	6 + 6 **12**	4 + 1 **5**	8 + 2 **10**	2 + 4 **6**	6 + 0 **6**
5 + 3 **8**	4 + 2 **6**	9 + 7 **16**	0 + 6 **6**	7 + 8 **15**	0 + 0 **0**	5 + 9 **14**	3 + 3 **6**	8 + 1 **9**	2 + 7 **9**

B | 100 Subtraction Facts

Subtract.

7 − 0 **7**	10 − 8 **2**	6 − 3 **3**	14 − 5 **9**	3 − 1 **2**	16 − 9 **7**	7 − 1 **6**	18 − 9 **9**	11 − 3 **8**	13 − 7 **6**
13 − 8 **5**	7 − 4 **3**	10 − 7 **3**	0 − 0 **0**	12 − 8 **4**	10 − 9 **1**	6 − 2 **4**	13 − 4 **9**	4 − 0 **4**	10 − 5 **5**
5 − 3 **2**	7 − 5 **2**	2 − 1 **1**	6 − 6 **0**	8 − 4 **4**	7 − 2 **5**	14 − 7 **7**	8 − 1 **7**	11 − 6 **5**	3 − 3 **0**
1 − 1 **0**	11 − 9 **2**	10 − 4 **6**	9 − 2 **7**	14 − 6 **8**	17 − 8 **9**	6 − 0 **6**	10 − 6 **4**	4 − 1 **3**	9 − 5 **4**
7 − 7 **0**	14 − 8 **6**	12 − 9 **3**	9 − 8 **1**	12 − 7 **5**	12 − 3 **9**	16 − 8 **8**	9 − 1 **8**	15 − 6 **9**	11 − 4 **7**
8 − 6 **2**	15 − 9 **6**	11 − 8 **3**	3 − 2 **1**	4 − 4 **0**	8 − 2 **6**	11 − 5 **6**	5 − 0 **5**	17 − 9 **8**	6 − 1 **5**
5 − 5 **0**	4 − 3 **1**	8 − 7 **1**	7 − 3 **4**	7 − 6 **1**	5 − 1 **4**	10 − 3 **7**	12 − 6 **6**	10 − 1 **9**	6 − 4 **2**
2 − 2 **0**	13 − 6 **7**	15 − 8 **7**	2 − 0 **2**	13 − 9 **4**	16 − 7 **9**	5 − 2 **3**	12 − 4 **8**	3 − 0 **3**	11 − 7 **4**
8 − 0 **8**	9 − 4 **5**	10 − 2 **8**	6 − 5 **1**	8 − 3 **5**	9 − 0 **9**	5 − 4 **1**	12 − 5 **7**	4 − 2 **2**	9 − 3 **6**
9 − 9 **0**	15 − 7 **8**	8 − 8 **0**	14 − 9 **5**	9 − 7 **2**	13 − 5 **8**	1 − 0 **1**	8 − 5 **3**	9 − 6 **3**	11 − 2 **9**

C Multiplication Facts:
0's, 1's, 2's, 5's

Multiply.

$\begin{array}{r} 0 \\ \times\ 8 \\ \hline 0 \end{array}$	$\begin{array}{r} 3 \\ \times\ 2 \\ \hline 6 \end{array}$	$\begin{array}{r} 5 \\ \times\ 1 \\ \hline 5 \end{array}$	$\begin{array}{r} 4 \\ \times\ 5 \\ \hline 20 \end{array}$	$\begin{array}{r} 2 \\ \times\ 0 \\ \hline 0 \end{array}$	$\begin{array}{r} 1 \\ \times\ 8 \\ \hline 8 \end{array}$	$\begin{array}{r} 7 \\ \times\ 2 \\ \hline 14 \end{array}$	$\begin{array}{r} 1 \\ \times\ 1 \\ \hline 1 \end{array}$
$\begin{array}{r} 5 \\ \times\ 2 \\ \hline 10 \end{array}$	$\begin{array}{r} 4 \\ \times\ 0 \\ \hline 0 \end{array}$	$\begin{array}{r} 2 \\ \times\ 8 \\ \hline 16 \end{array}$	$\begin{array}{r} 1 \\ \times\ 3 \\ \hline 3 \end{array}$	$\begin{array}{r} 7 \\ \times\ 5 \\ \hline 35 \end{array}$	$\begin{array}{r} 7 \\ \times\ 0 \\ \hline 0 \end{array}$	$\begin{array}{r} 8 \\ \times\ 5 \\ \hline 40 \end{array}$	$\begin{array}{r} 0 \\ \times\ 5 \\ \hline 0 \end{array}$
$\begin{array}{r} 8 \\ \times\ 1 \\ \hline 8 \end{array}$	$\begin{array}{r} 6 \\ \times\ 5 \\ \hline 30 \end{array}$	$\begin{array}{r} 9 \\ \times\ 0 \\ \hline 0 \end{array}$	$\begin{array}{r} 2 \\ \times\ 6 \\ \hline 12 \end{array}$	$\begin{array}{r} 0 \\ \times\ 1 \\ \hline 0 \end{array}$	$\begin{array}{r} 4 \\ \times\ 2 \\ \hline 8 \end{array}$	$\begin{array}{r} 1 \\ \times\ 6 \\ \hline 6 \end{array}$	$\begin{array}{r} 9 \\ \times\ 2 \\ \hline 18 \end{array}$
$\begin{array}{r} 6 \\ \times\ 0 \\ \hline 0 \end{array}$	$\begin{array}{r} 3 \\ \times\ 5 \\ \hline 15 \end{array}$	$\begin{array}{r} 5 \\ \times\ 7 \\ \hline 35 \end{array}$	$\begin{array}{r} 4 \\ \times\ 1 \\ \hline 4 \end{array}$	$\begin{array}{r} 2 \\ \times\ 2 \\ \hline 4 \end{array}$	$\begin{array}{r} 8 \\ \times\ 0 \\ \hline 0 \end{array}$	$\begin{array}{r} 5 \\ \times\ 9 \\ \hline 45 \end{array}$	$\begin{array}{r} 1 \\ \times\ 2 \\ \hline 2 \end{array}$
$\begin{array}{r} 5 \\ \times\ 5 \\ \hline 25 \end{array}$	$\begin{array}{r} 1 \\ \times\ 7 \\ \hline 7 \end{array}$	$\begin{array}{r} 0 \\ \times\ 0 \\ \hline 0 \end{array}$	$\begin{array}{r} 8 \\ \times\ 2 \\ \hline 16 \end{array}$	$\begin{array}{r} 5 \\ \times\ 8 \\ \hline 40 \end{array}$	$\begin{array}{r} 5 \\ \times\ 6 \\ \hline 30 \end{array}$	$\begin{array}{r} 3 \\ \times\ 0 \\ \hline 0 \end{array}$	$\begin{array}{r} 9 \\ \times\ 1 \\ \hline 9 \end{array}$
$\begin{array}{r} 5 \\ \times\ 3 \\ \hline 15 \end{array}$	$\begin{array}{r} 0 \\ \times\ 4 \\ \hline 0 \end{array}$	$\begin{array}{r} 6 \\ \times\ 1 \\ \hline 6 \end{array}$	$\begin{array}{r} 9 \\ \times\ 5 \\ \hline 45 \end{array}$	$\begin{array}{r} 5 \\ \times\ 0 \\ \hline 0 \end{array}$	$\begin{array}{r} 7 \\ \times\ 1 \\ \hline 7 \end{array}$	$\begin{array}{r} 2 \\ \times\ 5 \\ \hline 10 \end{array}$	$\begin{array}{r} 0 \\ \times\ 9 \\ \hline 0 \end{array}$
$\begin{array}{r} 2 \\ \times\ 1 \\ \hline 2 \end{array}$	$\begin{array}{r} 6 \\ \times\ 2 \\ \hline 12 \end{array}$	$\begin{array}{r} 0 \\ \times\ 7 \\ \hline 0 \end{array}$	$\begin{array}{r} 2 \\ \times\ 3 \\ \hline 6 \end{array}$	$\begin{array}{r} 1 \\ \times\ 4 \\ \hline 4 \end{array}$	$\begin{array}{r} 2 \\ \times\ 9 \\ \hline 18 \end{array}$	$\begin{array}{r} 1 \\ \times\ 0 \\ \hline 0 \end{array}$	$\begin{array}{r} 5 \\ \times\ 4 \\ \hline 20 \end{array}$
$\begin{array}{r} 0 \\ \times\ 2 \\ \hline 0 \end{array}$	$\begin{array}{r} 1 \\ \times\ 9 \\ \hline 9 \end{array}$	$\begin{array}{r} 3 \\ \times\ 1 \\ \hline 3 \end{array}$	$\begin{array}{r} 2 \\ \times\ 7 \\ \hline 14 \end{array}$	$\begin{array}{r} 0 \\ \times\ 3 \\ \hline 0 \end{array}$	$\begin{array}{r} 1 \\ \times\ 5 \\ \hline 5 \end{array}$	$\begin{array}{r} 2 \\ \times\ 4 \\ \hline 8 \end{array}$	$\begin{array}{r} 0 \\ \times\ 6 \\ \hline 0 \end{array}$

Multiplication Facts:
2's, 5's, Squares

Multiply.

3 × 5 **15**	6 × 2 **12**	7 × 7 **49**	2 × 0 **0**	7 × 5 **35**	5 × 8 **40**	4 × 2 **8**
9 × 5 **45**	2 × 3 **6**	4 × 4 **16**	1 × 5 **5**	6 × 6 **36**	1 × 2 **2**	5 × 5 **25**
2 × 7 **14**	5 × 4 **20**	8 × 8 **64**	2 × 5 **10**	3 × 3 **9**	0 × 5 **0**	8 × 2 **16**
6 × 5 **30**	0 × 2 **0**	5 × 3 **15**	2 × 8 **16**	8 × 5 **40**	2 × 6 **12**	5 × 1 **5**
3 × 2 **6**	2 × 9 **18**	5 × 7 **35**	2 × 4 **8**	5 × 6 **30**	9 × 9 **81**	2 × 2 **4**
4 × 5 **20**	7 × 2 **14**	5 × 0 **0**	2 × 1 **2**	5 × 9 **45**	9 × 2 **18**	5 × 2 **10**

Saxon Math 5/4—Homeschool

E   Multiplication Facts: 2's, 5's, 9's, Squares

Multiply.

$\begin{array}{r} 9 \\ \times 9 \\ \hline 81 \end{array}$	$\begin{array}{r} 2 \\ \times 6 \\ \hline 12 \end{array}$	$\begin{array}{r} 3 \\ \times 5 \\ \hline 15 \end{array}$	$\begin{array}{r} 2 \\ \times 2 \\ \hline 4 \end{array}$	$\begin{array}{r} 8 \\ \times 2 \\ \hline 16 \end{array}$	$\begin{array}{r} 7 \\ \times 9 \\ \hline 63 \end{array}$	$\begin{array}{r} 2 \\ \times 3 \\ \hline 6 \end{array}$	$\begin{array}{r} 5 \\ \times 6 \\ \hline 30 \end{array}$
$\begin{array}{r} 4 \\ \times 5 \\ \hline 20 \end{array}$	$\begin{array}{r} 0 \\ \times 9 \\ \hline 0 \end{array}$	$\begin{array}{r} 9 \\ \times 8 \\ \hline 72 \end{array}$	$\begin{array}{r} 5 \\ \times 2 \\ \hline 10 \end{array}$	$\begin{array}{r} 6 \\ \times 9 \\ \hline 54 \end{array}$	$\begin{array}{r} 7 \\ \times 5 \\ \hline 35 \end{array}$	$\begin{array}{r} 9 \\ \times 3 \\ \hline 27 \end{array}$	$\begin{array}{r} 0 \\ \times 5 \\ \hline 0 \end{array}$
$\begin{array}{r} 6 \\ \times 6 \\ \hline 36 \end{array}$	$\begin{array}{r} 5 \\ \times 9 \\ \hline 45 \end{array}$	$\begin{array}{r} 5 \\ \times 8 \\ \hline 40 \end{array}$	$\begin{array}{r} 9 \\ \times 7 \\ \hline 63 \end{array}$	$\begin{array}{r} 2 \\ \times 0 \\ \hline 0 \end{array}$	$\begin{array}{r} 8 \\ \times 8 \\ \hline 64 \end{array}$	$\begin{array}{r} 7 \\ \times 2 \\ \hline 14 \end{array}$	$\begin{array}{r} 5 \\ \times 5 \\ \hline 25 \end{array}$
$\begin{array}{r} 4 \\ \times 2 \\ \hline 8 \end{array}$	$\begin{array}{r} 1 \\ \times 9 \\ \hline 9 \end{array}$	$\begin{array}{r} 3 \\ \times 3 \\ \hline 9 \end{array}$	$\begin{array}{r} 2 \\ \times 8 \\ \hline 16 \end{array}$	$\begin{array}{r} 4 \\ \times 9 \\ \hline 36 \end{array}$	$\begin{array}{r} 3 \\ \times 2 \\ \hline 6 \end{array}$	$\begin{array}{r} 8 \\ \times 9 \\ \hline 72 \end{array}$	$\begin{array}{r} 5 \\ \times 7 \\ \hline 35 \end{array}$
$\begin{array}{r} 2 \\ \times 1 \\ \hline 2 \end{array}$	$\begin{array}{r} 2 \\ \times 9 \\ \hline 18 \end{array}$	$\begin{array}{r} 7 \\ \times 7 \\ \hline 49 \end{array}$	$\begin{array}{r} 9 \\ \times 6 \\ \hline 54 \end{array}$	$\begin{array}{r} 5 \\ \times 0 \\ \hline 0 \end{array}$	$\begin{array}{r} 4 \\ \times 4 \\ \hline 16 \end{array}$	$\begin{array}{r} 9 \\ \times 1 \\ \hline 9 \end{array}$	$\begin{array}{r} 2 \\ \times 5 \\ \hline 10 \end{array}$
$\begin{array}{r} 1 \\ \times 5 \\ \hline 5 \end{array}$	$\begin{array}{r} 9 \\ \times 5 \\ \hline 45 \end{array}$	$\begin{array}{r} 2 \\ \times 7 \\ \hline 14 \end{array}$	$\begin{array}{r} 5 \\ \times 4 \\ \hline 20 \end{array}$	$\begin{array}{r} 9 \\ \times 2 \\ \hline 18 \end{array}$	$\begin{array}{r} 5 \\ \times 3 \\ \hline 15 \end{array}$	$\begin{array}{r} 0 \\ \times 2 \\ \hline 0 \end{array}$	$\begin{array}{r} 8 \\ \times 5 \\ \hline 40 \end{array}$
$\begin{array}{r} 3 \\ \times 9 \\ \hline 27 \end{array}$	$\begin{array}{r} 2 \\ \times 4 \\ \hline 8 \end{array}$	$\begin{array}{r} 6 \\ \times 5 \\ \hline 30 \end{array}$	$\begin{array}{r} 1 \\ \times 2 \\ \hline 2 \end{array}$	$\begin{array}{r} 9 \\ \times 0 \\ \hline 0 \end{array}$	$\begin{array}{r} 6 \\ \times 2 \\ \hline 12 \end{array}$	$\begin{array}{r} 9 \\ \times 4 \\ \hline 36 \end{array}$	$\begin{array}{r} 5 \\ \times 1 \\ \hline 5 \end{array}$

Multiplication Facts: Memory Group

Multiply.

6 × 4 **24**	7 × 6 **42**	8 × 7 **56**	8 × 3 **24**
4 × 8 **32**	4 × 3 **12**	3 × 7 **21**	7 × 4 **28**
3 × 6 **18**	8 × 6 **48**	4 × 6 **24**	7 × 8 **56**
6 × 8 **48**	3 × 8 **24**	8 × 4 **32**	3 × 4 **12**
7 × 3 **21**	4 × 7 **28**	6 × 3 **18**	6 × 7 **42**

G   64 Multiplication Facts

Multiply.

$\begin{array}{r} 4 \\ \times\ 6 \\ \hline 24 \end{array}$	$\begin{array}{r} 8 \\ \times\ 8 \\ \hline 64 \end{array}$	$\begin{array}{r} 5 \\ \times\ 7 \\ \hline 35 \end{array}$	$\begin{array}{r} 6 \\ \times\ 3 \\ \hline 18 \end{array}$	$\begin{array}{r} 5 \\ \times\ 6 \\ \hline 30 \end{array}$	$\begin{array}{r} 4 \\ \times\ 3 \\ \hline 12 \end{array}$	$\begin{array}{r} 9 \\ \times\ 8 \\ \hline 72 \end{array}$	$\begin{array}{r} 7 \\ \times\ 5 \\ \hline 35 \end{array}$
$\begin{array}{r} 2 \\ \times\ 6 \\ \hline 12 \end{array}$	$\begin{array}{r} 5 \\ \times\ 9 \\ \hline 45 \end{array}$	$\begin{array}{r} 3 \\ \times\ 3 \\ \hline 9 \end{array}$	$\begin{array}{r} 9 \\ \times\ 2 \\ \hline 18 \end{array}$	$\begin{array}{r} 9 \\ \times\ 4 \\ \hline 36 \end{array}$	$\begin{array}{r} 2 \\ \times\ 5 \\ \hline 10 \end{array}$	$\begin{array}{r} 7 \\ \times\ 6 \\ \hline 42 \end{array}$	$\begin{array}{r} 4 \\ \times\ 8 \\ \hline 32 \end{array}$
$\begin{array}{r} 5 \\ \times\ 2 \\ \hline 10 \end{array}$	$\begin{array}{r} 7 \\ \times\ 8 \\ \hline 56 \end{array}$	$\begin{array}{r} 2 \\ \times\ 3 \\ \hline 6 \end{array}$	$\begin{array}{r} 6 \\ \times\ 8 \\ \hline 48 \end{array}$	$\begin{array}{r} 3 \\ \times\ 7 \\ \hline 21 \end{array}$	$\begin{array}{r} 8 \\ \times\ 5 \\ \hline 40 \end{array}$	$\begin{array}{r} 6 \\ \times\ 2 \\ \hline 12 \end{array}$	$\begin{array}{r} 5 \\ \times\ 5 \\ \hline 25 \end{array}$
$\begin{array}{r} 3 \\ \times\ 4 \\ \hline 12 \end{array}$	$\begin{array}{r} 7 \\ \times\ 3 \\ \hline 21 \end{array}$	$\begin{array}{r} 5 \\ \times\ 8 \\ \hline 40 \end{array}$	$\begin{array}{r} 4 \\ \times\ 2 \\ \hline 8 \end{array}$	$\begin{array}{r} 6 \\ \times\ 4 \\ \hline 24 \end{array}$	$\begin{array}{r} 2 \\ \times\ 8 \\ \hline 16 \end{array}$	$\begin{array}{r} 4 \\ \times\ 4 \\ \hline 16 \end{array}$	$\begin{array}{r} 8 \\ \times\ 2 \\ \hline 16 \end{array}$
$\begin{array}{r} 2 \\ \times\ 2 \\ \hline 4 \end{array}$	$\begin{array}{r} 7 \\ \times\ 4 \\ \hline 28 \end{array}$	$\begin{array}{r} 3 \\ \times\ 8 \\ \hline 24 \end{array}$	$\begin{array}{r} 8 \\ \times\ 6 \\ \hline 48 \end{array}$	$\begin{array}{r} 2 \\ \times\ 9 \\ \hline 18 \end{array}$	$\begin{array}{r} 8 \\ \times\ 4 \\ \hline 32 \end{array}$	$\begin{array}{r} 9 \\ \times\ 3 \\ \hline 27 \end{array}$	$\begin{array}{r} 6 \\ \times\ 9 \\ \hline 54 \end{array}$
$\begin{array}{r} 6 \\ \times\ 7 \\ \hline 42 \end{array}$	$\begin{array}{r} 4 \\ \times\ 5 \\ \hline 20 \end{array}$	$\begin{array}{r} 7 \\ \times\ 2 \\ \hline 14 \end{array}$	$\begin{array}{r} 9 \\ \times\ 6 \\ \hline 54 \end{array}$	$\begin{array}{r} 7 \\ \times\ 9 \\ \hline 63 \end{array}$	$\begin{array}{r} 5 \\ \times\ 4 \\ \hline 20 \end{array}$	$\begin{array}{r} 3 \\ \times\ 2 \\ \hline 6 \end{array}$	$\begin{array}{r} 9 \\ \times\ 7 \\ \hline 63 \end{array}$
$\begin{array}{r} 4 \\ \times\ 7 \\ \hline 28 \end{array}$	$\begin{array}{r} 9 \\ \times\ 5 \\ \hline 45 \end{array}$	$\begin{array}{r} 3 \\ \times\ 6 \\ \hline 18 \end{array}$	$\begin{array}{r} 8 \\ \times\ 7 \\ \hline 56 \end{array}$	$\begin{array}{r} 3 \\ \times\ 5 \\ \hline 15 \end{array}$	$\begin{array}{r} 2 \\ \times\ 4 \\ \hline 8 \end{array}$	$\begin{array}{r} 7 \\ \times\ 7 \\ \hline 49 \end{array}$	$\begin{array}{r} 8 \\ \times\ 9 \\ \hline 72 \end{array}$
$\begin{array}{r} 8 \\ \times\ 3 \\ \hline 24 \end{array}$	$\begin{array}{r} 2 \\ \times\ 7 \\ \hline 14 \end{array}$	$\begin{array}{r} 6 \\ \times\ 5 \\ \hline 30 \end{array}$	$\begin{array}{r} 4 \\ \times\ 9 \\ \hline 36 \end{array}$	$\begin{array}{r} 3 \\ \times\ 9 \\ \hline 27 \end{array}$	$\begin{array}{r} 6 \\ \times\ 6 \\ \hline 36 \end{array}$	$\begin{array}{r} 9 \\ \times\ 9 \\ \hline 81 \end{array}$	$\begin{array}{r} 5 \\ \times\ 3 \\ \hline 15 \end{array}$

H   100 Multiplication Facts

Multiply.

9 × 1 **9**	2 × 2 **4**	5 × 1 **5**	4 × 3 **12**	0 × 0 **0**	9 × 9 **81**	3 × 5 **15**	8 × 5 **40**	2 × 6 **12**	4 × 7 **28**
5 × 6 **30**	7 × 5 **35**	3 × 0 **0**	8 × 8 **64**	1 × 3 **3**	3 × 4 **12**	5 × 9 **45**	0 × 2 **0**	7 × 3 **21**	4 × 1 **4**
2 × 3 **6**	8 × 6 **48**	0 × 5 **0**	6 × 1 **6**	3 × 8 **24**	1 × 1 **1**	9 × 0 **0**	2 × 8 **16**	6 × 4 **24**	0 × 7 **0**
7 × 7 **49**	1 × 4 **4**	6 × 2 **12**	4 × 5 **20**	2 × 4 **8**	4 × 9 **36**	7 × 0 **0**	1 × 2 **2**	8 × 4 **32**	6 × 5 **30**
3 × 2 **6**	4 × 6 **24**	1 × 9 **9**	5 × 7 **35**	8 × 2 **16**	0 × 8 **0**	4 × 2 **8**	9 × 8 **72**	3 × 6 **18**	5 × 5 **25**
8 × 9 **72**	3 × 7 **21**	9 × 7 **63**	1 × 7 **7**	6 × 0 **0**	0 × 3 **0**	7 × 2 **14**	1 × 5 **5**	7 × 8 **56**	4 × 0 **0**
8 × 3 **24**	5 × 2 **10**	0 × 4 **0**	9 × 5 **45**	6 × 7 **42**	2 × 7 **14**	6 × 3 **18**	5 × 4 **20**	1 × 0 **0**	9 × 2 **18**
7 × 6 **42**	1 × 8 **8**	9 × 6 **54**	4 × 4 **16**	5 × 3 **15**	8 × 1 **8**	3 × 3 **9**	4 × 8 **32**	9 × 3 **27**	2 × 0 **0**
8 × 0 **0**	3 × 1 **3**	6 × 8 **48**	0 × 9 **0**	8 × 7 **56**	2 × 9 **18**	9 × 4 **36**	0 × 1 **0**	7 × 4 **28**	5 × 8 **40**
0 × 6 **0**	7 × 1 **7**	2 × 5 **10**	6 × 9 **54**	3 × 9 **27**	1 × 6 **6**	5 × 0 **0**	6 × 6 **36**	2 × 1 **2**	7 × 9 **63**

I **90 Division Facts**

Divide.

$\frac{9}{2\overline{)18}}$	$\frac{1}{6\overline{)6}}$	$\frac{5}{3\overline{)15}}$	$\frac{9}{3\overline{)27}}$	$\frac{7}{2\overline{)14}}$	$\frac{5}{5\overline{)25}}$	$\frac{8}{6\overline{)48}}$	$\frac{3}{7\overline{)21}}$	$\frac{5}{2\overline{)10}}$	$\frac{7}{6\overline{)42}}$
$\frac{5}{4\overline{)20}}$	$\frac{7}{9\overline{)63}}$	$\frac{4}{1\overline{)4}}$	$\frac{2}{4\overline{)8}}$	$\frac{0}{7\overline{)0}}$	$\frac{2}{8\overline{)16}}$	$\frac{8}{3\overline{)24}}$	$\frac{8}{4\overline{)32}}$	$\frac{7}{8\overline{)56}}$	$\frac{0}{1\overline{)0}}$
$\frac{1}{5\overline{)5}}$	$\frac{8}{8\overline{)64}}$	$\frac{0}{3\overline{)0}}$	$\frac{1}{2\overline{)2}}$	$\frac{8}{5\overline{)40}}$	$\frac{3}{3\overline{)9}}$	$\frac{2}{9\overline{)18}}$	$\frac{0}{6\overline{)0}}$	$\frac{2}{5\overline{)10}}$	$\frac{1}{9\overline{)9}}$
$\frac{4}{8\overline{)32}}$	$\frac{1}{1\overline{)1}}$	$\frac{4}{9\overline{)36}}$	$\frac{5}{8\overline{)40}}$	$\frac{0}{2\overline{)0}}$	$\frac{4}{5\overline{)20}}$	$\frac{3}{9\overline{)27}}$	$\frac{3}{6\overline{)18}}$	$\frac{0}{4\overline{)0}}$	$\frac{6}{5\overline{)30}}$
$\frac{6}{2\overline{)12}}$	$\frac{9}{5\overline{)45}}$	$\frac{7}{1\overline{)7}}$	$\frac{2}{7\overline{)14}}$	$\frac{1}{3\overline{)3}}$	$\frac{3}{8\overline{)24}}$	$\frac{0}{5\overline{)0}}$	$\frac{4}{2\overline{)8}}$	$\frac{6}{7\overline{)42}}$	$\frac{6}{6\overline{)36}}$
$\frac{8}{7\overline{)56}}$	$\frac{0}{9\overline{)0}}$	$\frac{9}{8\overline{)72}}$	$\frac{7}{4\overline{)28}}$	$\frac{7}{7\overline{)49}}$	$\frac{2}{2\overline{)4}}$	$\frac{9}{9\overline{)81}}$	$\frac{2}{1\overline{)2}}$	$\frac{7}{5\overline{)35}}$	$\frac{7}{3\overline{)21}}$
$\frac{0}{8\overline{)0}}$	$\frac{4}{7\overline{)28}}$	$\frac{9}{4\overline{)36}}$	$\frac{3}{1\overline{)3}}$	$\frac{6}{4\overline{)24}}$	$\frac{2}{3\overline{)6}}$	$\frac{6}{9\overline{)54}}$	$\frac{8}{1\overline{)8}}$	$\frac{1}{4\overline{)4}}$	$\frac{5}{7\overline{)35}}$
$\frac{5}{9\overline{)45}}$	$\frac{9}{1\overline{)9}}$	$\frac{9}{6\overline{)54}}$	$\frac{2}{6\overline{)12}}$	$\frac{6}{3\overline{)18}}$	$\frac{8}{9\overline{)72}}$	$\frac{3}{5\overline{)15}}$	$\frac{4}{6\overline{)24}}$	$\frac{1}{8\overline{)8}}$	$\frac{8}{2\overline{)16}}$
$\frac{6}{1\overline{)6}}$	$\frac{3}{4\overline{)12}}$	$\frac{1}{7\overline{)7}}$	$\frac{3}{2\overline{)6}}$	$\frac{9}{7\overline{)63}}$	$\frac{4}{4\overline{)16}}$	$\frac{6}{8\overline{)48}}$	$\frac{4}{3\overline{)12}}$	$\frac{5}{6\overline{)30}}$	$\frac{5}{1\overline{)5}}$

90 Division Facts

Divide.

56 ÷ 7 = **8**	15 ÷ 3 = **5**	12 ÷ 6 = **2**	8 ÷ 2 = **4**	63 ÷ 7 = **9**	0 ÷ 4 = **0**
14 ÷ 2 = **7**	42 ÷ 6 = **7**	6 ÷ 1 = **6**	16 ÷ 8 = **2**	20 ÷ 5 = **4**	49 ÷ 7 = **7**
36 ÷ 4 = **9**	64 ÷ 8 = **8**	0 ÷ 3 = **0**	54 ÷ 9 = **6**	4 ÷ 2 = **2**	48 ÷ 8 = **6**
18 ÷ 9 = **2**	3 ÷ 1 = **3**	35 ÷ 5 = **7**	8 ÷ 4 = **2**	72 ÷ 8 = **9**	6 ÷ 6 = **1**
0 ÷ 5 = **0**	42 ÷ 7 = **6**	2 ÷ 2 = **1**	36 ÷ 9 = **4**	7 ÷ 1 = **7**	12 ÷ 3 = **4**
16 ÷ 2 = **8**	30 ÷ 5 = **6**	0 ÷ 1 = **0**	28 ÷ 7 = **4**	4 ÷ 4 = **1**	40 ÷ 8 = **5**
3 ÷ 3 = **1**	18 ÷ 6 = **3**	63 ÷ 9 = **7**	40 ÷ 5 = **8**	10 ÷ 2 = **5**	36 ÷ 6 = **6**
32 ÷ 8 = **4**	12 ÷ 4 = **3**	18 ÷ 3 = **6**	35 ÷ 7 = **5**	8 ÷ 8 = **1**	2 ÷ 1 = **2**
45 ÷ 5 = **9**	7 ÷ 7 = **1**	27 ÷ 9 = **3**	9 ÷ 1 = **9**	48 ÷ 6 = **8**	0 ÷ 7 = **0**
4 ÷ 1 = **4**	0 ÷ 9 = **0**	24 ÷ 3 = **8**	32 ÷ 4 = **8**	5 ÷ 5 = **1**	72 ÷ 9 = **8**
20 ÷ 4 = **5**	21 ÷ 7 = **3**	0 ÷ 2 = **0**	27 ÷ 3 = **9**	8 ÷ 1 = **8**	54 ÷ 6 = **9**
15 ÷ 5 = **3**	6 ÷ 3 = **2**	28 ÷ 4 = **7**	18 ÷ 2 = **9**	24 ÷ 6 = **4**	9 ÷ 9 = **1**
56 ÷ 8 = **7**	0 ÷ 6 = **0**	21 ÷ 3 = **7**	1 ÷ 1 = **1**	25 ÷ 5 = **5**	12 ÷ 2 = **6**
5 ÷ 1 = **5**	45 ÷ 9 = **5**	16 ÷ 4 = **4**	30 ÷ 6 = **5**	9 ÷ 3 = **3**	14 ÷ 7 = **2**
0 ÷ 8 = **0**	6 ÷ 2 = **3**	24 ÷ 8 = **3**	10 ÷ 5 = **2**	81 ÷ 9 = **9**	24 ÷ 4 = **6**

Solutions for

Tests

TEST 1

1. (a) **tens**   (b) **hundreds**   (c) **ones**

2. 3 hundreds + 4 tens + 8 ones = **348**

3. The rule is "Count up by sevens":
28, 35, 42, **49, 56, 63**

4. Counting by twos, we get:
2, 4, 6, 8, 10, 12, 14, **16 eyes**

5. Counting by tens, we get:
10, 20, 30, **40 cents**

6. The last digit is the one on the right, so the last digit is **3**.

7. 3 hundreds + 2 tens + 4 ones = **$324**

8. $341 = 3 hundreds + 4 tens + 1 one

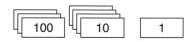

9. The rule is "Count down by fours":
40, 36, **32**, 28, …

10. The rule is "Count up by fives":
5, 10, 15, **20**, …

11. The rule is "Count down by threes":
27, **24**, 21, 18, …

12. 8 digits

13.

$$\begin{matrix} 3 \\ 7 \\ 6 \\ + 2 \end{matrix} \quad \begin{matrix} 10 \\ 10 \\ + \ 8 \\ \hline 18 \end{matrix}$$

14.

$$\begin{matrix} 9 \\ 4 \\ 1 \\ + 2 \end{matrix} \quad \begin{matrix} 10 \\ + \ 6 \\ \hline 16 \end{matrix}$$

15. 3 + 8 = 11   11 + N = 15
$N = 4$

16. 5 + 3 = 8 $N + 8 = 17$
$N = 9$

17. The rule is "Count up by threes":
3, 6, 9, **12**

18. **5 + 1 + 2 = 8**

19. Student may begin by drawing a diagram (not required).

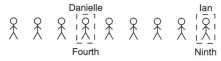

From the diagram, we count **4 people** between Danielle and Ian.

20. 8 tourists + 7 tourists = **15 tourists**

TEST 2

1. Student may begin by drawing a diagram (not required).

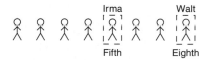

From the diagram, we count **2 people** between Irma and Walt.

2. $32 + $23 = **$55**

3. Two hundred forty-two = **242**

4. Nine hundred sixteen = **916**

5. 905 = **nine hundred five**

6. 521 = **five hundred twenty-one**

7. **2 + 6 = 8**   **8 − 2 = 6**
6 + 2 = 8   **8 − 6 = 2**

8. (a) **ones**   (b) **hundreds**   (c) **tens**

9. The hundreds place is the third place from the right, so the number is **3**.

10.
$$\begin{array}{r} \$55 \\ + \ \$23 \\ \hline \$78 \end{array}$$

11. $5 + 9 + 2 + 1 + 3 + 9$
$\qquad\quad 10$
$5 + 10 + 2 + 3 + 9$
$\qquad\qquad\quad 5$
$5 + 10 + 5 + 9$
$\qquad\quad 10$
$10 + 10 + 9 = $ **29**

12.
$$\begin{array}{r} 36 \\ + \ 25 \\ \hline 61 \end{array}$$

13.
$$\begin{array}{r} 14 \\ - \ 6 \\ \hline 8 \end{array}$$

14.
$$\begin{array}{r} 12 \\ - \ 5 \\ \hline 7 \end{array}$$

15. $7 + 3 = 10 \qquad 10 + N = 13$
$\qquad\qquad\qquad\qquad\quad N = $ **3**

16. $3 + 1 = 4 \qquad 4 + N = 11$
$\qquad\qquad\qquad\qquad\quad N = $ **7**

17. $6 + 4 + 1 + 2 = $ **13**

18. The rule is "Count up by fives":
$\qquad 5, 10, 15, $ **20, 25, 30**

19. The rule is "Count up by sixes":
$\qquad 36, 42, 48, $ **54, 60, 66**

20. **A. 330**

TEST 3

1. 42 children $+$ 25 children $= $ **67 children**

2. Seven hundred twenty-seven $= $ **727**

3. $391 = $ **three hundred ninety-one**

4. $6 > 3$

5. $5 + 6 = 11 \qquad 11 - 6 = 5$
$ 6 + 5 = 11 \qquad 11 - 5 = 6$

6. The last digit is even, so the number is **even.**

7. 17 coins $-$ 8 coins $= $ **9 coins**

8. Counting by fives fits the pattern. The number is **25.**

9. $594 > 495$

10. $321 > 213$

11. The rule is "Count up by fives":
$\qquad\qquad 35, $ **40,** $45, 50, \ldots$

12. The rule is "Count down by nines":
$\qquad\qquad 54, 45, $ **36,** $27, \ldots$

13. $13 - A = 8$
$ \quad A = $ **5**

14. $N - 4 = 6$
$ \quad N = $ **10**

15. $5 + 2 = 7 \qquad 7 + A = 15$
$\qquad\qquad\qquad\qquad\quad A = $ **8**

16.
$$\begin{array}{r} \$476 \\ + \ \$392 \\ \hline \$868 \end{array}$$

17. $38 - 17 = $ **21**

18. $42 - 27 = $ **15**

19. $6 + 3 + 1 + 9 + 4 + 4 + 5$
$\qquad\qquad\quad 10$
$6 + 3 + 10 + 4 + 4 + 5$
$\qquad\quad 10$
$10 + 3 + 10 + 4 + 5$
$\qquad\qquad\qquad\quad 9$
$10 + 3 + 10 + 9$
$\qquad\qquad 12$
$10 + 10 + 12 = $ **32**

20. **9 digits**

TEST 4

1. 98 pulsars – 47 pulsars = **51 pulsars**

2.
$$
\begin{array}{r}
400 \text{ cardinals} \\
200 \text{ cardinals} \\
+\ \ 50 \text{ cardinals} \\
\hline
\textbf{650 cardinals}
\end{array}
$$

3.
$$
\begin{array}{r}
\$359 \\
+\ \$241 \\
\hline
\textbf{\$600}
\end{array}
$$

4. $607 = \textbf{600} + \textbf{7}$

5. $506 < 516$

6. $313 > 285$

7. Count 53 minutes from the top of the clock to the long hand. Since it is morning, the time is **6:53 a.m.**

8. Counting by twos fits the pattern. The temperature is **74°F.**

9. 8 cm

10. The number 88 is closer to **90** than to 80.

11. $6.38 is closer to **$6** than to $7.

12. Student may begin by drawing a diagram (not required).

From the diagram, we count **6 people** between Feynman and Dirac.

13.
$$
\begin{array}{r}
31 \\
46 \\
12 \\
+\ 57 \\
\hline
\textbf{146}
\end{array}
$$

14.
$$
\begin{array}{r}
592 \\
+\ 336 \\
\hline
\textbf{928}
\end{array}
$$

15.
$$
\begin{array}{r}
\$81 \\
-\ \$53 \\
\hline
\textbf{\$28}
\end{array}
$$

16.
$$
\begin{array}{r}
C \\
-\ 24 \\
\hline
63
\end{array}
\qquad
\begin{array}{l}
(7) - 4 = 3 \\
(8) - 2 = 6 \\
C = \textbf{87}
\end{array}
$$

17.
$$
\begin{array}{r}
32 \\
+\ D \\
\hline
58
\end{array}
\qquad
\begin{array}{l}
2 + (6) = 8 \\
3 + (2) = 5 \\
D = \textbf{26}
\end{array}
$$

18.
$$
\begin{array}{r}
54 \\
-\ F \\
\hline
31
\end{array}
\qquad
\begin{array}{l}
4 - (3) = 1 \\
5 - (2) = 3 \\
F = \textbf{23}
\end{array}
$$

19.
$$
\begin{array}{r}
3 \\
43 \\
25 \\
+\ 10 \\
\hline
81
\end{array}
\qquad
\begin{array}{l}
81 + G = 100 \\
G = \textbf{19}
\end{array}
$$

20. Counting by fives, we get:

5, 10, 15, 20, 25, 30, 35, **40 dots**

TEST 5

1. 29 dollars – 16 dollars = **13 dollars**

2. Since the contestants could be divided into two equal groups, the total number of contestants must be even. The only number that is not even is 23, so the correct answer is **A. 23.**

3.
$$
\begin{array}{r}
46 \text{ people} \\
67 \text{ people} \\
+\ 73 \text{ people} \\
\hline
\textbf{186 people}
\end{array}
$$

4. Perimeter
= 14 mm + 14 mm + 14 mm + 14 mm
= **56 mm**

5. The rule is "Count up by sevens":

7, 14, **21, 28**, 35, **42,** …

6. Most doorways are about **2 meters** tall.

7. The number 76 is closer to **80** than to 70.

8. $13 - 7 \bigcirc 11 - 5$
    $6 = 6$

9. Count 23 minutes from the top of the clock to the
    long hand. Since it is afternoon, the time is
    2:23 p.m.

10. **Lindsey** is perpendicular to Berry.

11. 7 shaded parts $\dfrac{7}{10}$
    10 total parts

12. Counting by twos fits the pattern. The number
    is **288.**

13. **B.**

14. $5.95
    $+ \ 2.19$
    $8.14

15.   36
    $-\ 19$
    17

16. Q $(4) + 2 = 6$
    $+\ 52$ $(2) + 5 = 7$
      76       $Q = \textbf{24}$

17.   581
    $+\ 192$
    773

18.   96 $6 - (4) = 2$
    $-\ F$ $9 - (7) = 2$
      22        $F = \textbf{74}$

19.   647
    $-\ 415$
    232

20.   84
         21
         15
    $+\ 37$
    157

TEST 6

1.   83 people
        57 people
    $+\ 60$ people
    200 people

2.   40 crayons   40 crayons
    $-\ N$ crayons $-\ 17$ crayons
    17 crayons   **23 crayons**

3. N $46
    $+\ 28$ $-\ 28$
      $46   **$18**

4. $805 = \textbf{800} + \textbf{5}$
    $805 = $ **eight hundred five**

5. $\textbf{3} \times \textbf{0} = \textbf{2} \times \textbf{0}$

6. **10 cm**

7. (varies)

8. $14.64 is closer to **$15** than to $14.

9. The rule is "Count down by nines":
            54, **45, 36,** 27, 18, **9,** ...

10. Step 1: Count forward 30 minutes to 8:00 a.m.
    Step 2: Count forward 2 hours to **10:00 a.m.**

11. Perimeter $= 7\,\text{cm} + 7\,\text{cm} + 6\,\text{cm} = $ **20 cm**

12. **6 × 4** or   4
                   × 6

13. $1\,\text{m} = 100\,\text{cm}$
    $3\,\text{m} = $ **300 cm**

14. (a) $5 \times 5 = $ **25**
    (b) $9 \times 5 = $ **45**
    (c) $5 \times 7 = $ **35**

15.   590
    $-\ 320$
    270

16.
$$\begin{array}{r} 84 \\ -\ 37 \\ \hline \mathbf{47} \end{array}$$

17.
$$\begin{array}{r} 235 \\ +\ 679 \\ \hline \mathbf{914} \end{array}$$

18.
$$\begin{array}{r} 79 \\ -\ P \\ \hline 23 \end{array}$$
$9 - (6) = 3$
$7 - (5) = 2$
$P = \mathbf{56}$

19.
$$\begin{array}{r} 33 \\ +\ R \\ \hline 76 \end{array}$$
$3 + (3) = 6$
$3 + (4) = 7$
$R = \mathbf{43}$

20.
$$\begin{array}{r} 5 \\ 6 \\ 7 \\ 5 \\ 3 \\ 9 \\ 8 \\ 2 \\ 1 \\ 6 \\ 7 \\ +\ 4 \\ \hline \mathbf{63} \end{array}$$

TEST 7

1.
$$\begin{array}{r} N \text{ pages} \\ +\ 271 \text{ pages} \\ \hline 513 \text{ pages} \end{array} \qquad \begin{array}{r} 513 \text{ pages} \\ -\ 271 \text{ pages} \\ \hline \mathbf{242 \text{ pages}} \end{array}$$

2. To make a number greater than 600, the first digit must be a 7. To make an odd number, 5 must be the last digit since it is the only odd digit left. The number is **725.**

3.
```
x x x x x   or   x x x
x x x x x        x x x
x x x x x        x x x
                 x x x
                 x x x
```

4. (a) The first number is the month. The tenth month is **October.**

(b) The last number is the year. The 98 means **1998.**

5. ⟷ (varies)

6. The area of a rectangle is its length times its width. This rectangle's length is 5 cm, and its width is 2 cm, so
Area $= 5 \text{ cm} \times 2 \text{ cm} = \mathbf{10 \text{ sq. cm}}$

7. 7 shaded parts $\dfrac{\mathbf{7}}{\mathbf{10}}$
10 total parts

8. $\mathbf{4 \times 7}$ or $\begin{array}{r} 7 \\ \times\ 4 \\ \hline \end{array}$

9. The number 37 is closer to **40** than to 30. The number 44 is closer to **40** than to 50.
$40 + 40 = \mathbf{80}$

10. 2 nickels $= 10$ cents
4 dimes $= 40$ cents
$$\begin{array}{r} 10¢ \\ +\ 40¢ \\ \hline 50¢ \quad \mathbf{even} \end{array}$$

11. Counting by twenty-fives fits the pattern. The number is **125.**

12. (a) $4 \times 4 = \mathbf{16}$
(b) $8 \times 8 = \mathbf{64}$
(c) $5 \times 5 = \mathbf{25}$

13.
$$\begin{array}{r} 65 \\ -\ 56 \\ \hline \mathbf{9} \end{array}$$

14. $\sqrt{36} = \mathbf{6}$

15. $33 + 44 \bigcirc 22 + 54$
$77 \ > \ 76$

16.
$$\begin{array}{r} 23 \\ +\ W \\ \hline 79 \end{array}$$
$3 + (6) = 9$
$2 + (5) = 7$
$W = \mathbf{56}$

17.
$$\begin{array}{r} 636 \\ -\ X \\ \hline 214 \end{array}$$
$6 - (2) = 4$
$3 - (2) = 1$
$6 - (4) = 2$
$X = \mathbf{422}$

18. $4\dfrac{1}{3} = \mathbf{four\ and\ one\ third}$

19. four million $= \mathbf{4,000,000}$

20.
$$\begin{array}{r} 734 \\ -\ 368 \\ \hline \mathbf{366} \end{array}$$

TEST 8

1.
$$\begin{array}{r} \$0.40 \\ \$0.50 \\ + \ \$0.05 \\ \hline \$0.95 \end{array}$$

2. Student may draw and label a diagram (not required).

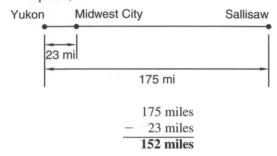

$$\begin{array}{r} 175 \text{ miles} \\ - \ \ 23 \text{ miles} \\ \hline \textbf{152 miles} \end{array}$$

3. There are 4 spaces between the whole numbers. Each space equals $\frac{1}{4}$. The arrow points to $\mathbf{2\frac{3}{4}}$.

4. $11 + 35 + 18 \bigcirc 8 \times 8$
$$64 = 64$$

5. **8 pints**

6. $876,482 = $ **eight hundred seventy-six thousand, four hundred eighty-two**

7. (varies)

8. $637 < 823$

9. A

10. Perimeter $= 4\,\text{cm} + 2\,\text{cm} + 4\,\text{cm} + 2\,\text{cm}$
$$= \textbf{12 cm}$$

11. The number 74 is closer to **70** than to 80.
The number 77 is closer to **80** than to 70.
$70 + 80 = \textbf{150}$

12. A circle's radius equals one half its diameter, so this circle's radius is two yards. A yard equals 3 feet, so this circle's radius is **6 feet.**

13. One full rectangle is shaded, plus three fourths of another, which equals $\mathbf{1\frac{3}{4}}$.

14.
$$\begin{array}{r} \$4.61 \\ - \ \$2.73 \\ \hline \$1.88 \end{array}$$

15. $\sqrt{49} = 7$

16.
$$\begin{array}{r} \$845 \\ \$753 \\ + \ \ \$29 \\ \hline \$1627 \end{array}$$

17. (a) $\begin{array}{r} 6 \\ \times \ 3 \\ \hline 18 \end{array}$ (b) $\begin{array}{r} 7 \\ \times \ 8 \\ \hline 56 \end{array}$ (c) $\begin{array}{r} 3 \\ \times \ 8 \\ \hline 24 \end{array}$

18.
$$\begin{array}{r} E \\ + \ 342 \\ \hline 621 \end{array} \qquad \begin{array}{r} 621 \\ - \ 342 \\ \hline \textbf{279} \end{array}$$

19.
$$\begin{array}{r} Y \\ - \ 232 \\ \hline 244 \end{array} \qquad \begin{array}{l} (6) - 2 = 4 \\ (7) = 3 - 4 \\ (4) - 2 = 2 \end{array}$$
$$Y = \textbf{476}$$

20.
$$\begin{array}{r} 25 \\ 51 \\ 84 \\ + \ 19 \\ \hline 179 \end{array} \qquad \begin{array}{r} 179 \\ + \ \ N \\ \hline 432 \end{array} \qquad \begin{array}{r} 432 \\ - \ 179 \\ \hline \textbf{253} \end{array}$$

TEST 9

1.
$$\begin{array}{r} 479 \text{ fish} \\ + \ \ N \text{ fish} \\ \hline 843 \text{ fish} \end{array} \qquad \begin{array}{r} 843 \text{ fish} \\ - \ 479 \text{ fish} \\ \hline \textbf{364 fish} \end{array}$$

2.
$$\begin{array}{r} 648 \text{ bags} \\ - \ 533 \text{ bags} \\ \hline \textbf{115 bags} \end{array}$$

3.
$$\begin{array}{r} \$5.00 \\ - \ \$3.27 \\ \hline \$1.73 \end{array}$$

4. The number 845 is closer to **800** than to 900.

5. $2\frac{5}{6} = $ (varies)

6. $16.3 = $ **sixteen and three tenths**

7. Count backward 40 minutes to **9:15 p.m.**

8. $\mathbf{1\frac{1}{4}}$ **in.**

9. There are 6 spaces between the whole numbers. Each space equals $\frac{1}{6}$. The arrow points to $10\frac{5}{6}$.

10.
$$\begin{array}{r} N \\ +\ 356 \\ \hline 497 \end{array} \qquad \begin{array}{r} 497 \\ -\ 356 \\ \hline \mathbf{141} \end{array}$$

11.
$$\begin{array}{r} 597 \\ -\ S \\ \hline 356 \end{array} \qquad \begin{array}{r} 7 - (1) = 6 \\ 9 - (4) = 5 \\ 5 - (2) = 3 \\ S = \mathbf{241} \end{array}$$

12. $9N = 63$
$N = \mathbf{7}$

13.
$$\begin{array}{r} \$9.06 \\ -\ \$3.48 \\ \hline \mathbf{\$5.58} \end{array}$$

14.
$$\begin{array}{r} 31 \\ \times\ \ 9 \\ \hline \mathbf{279} \end{array}$$

15.
$$\begin{array}{r} 40 \\ \times\ \ 3 \\ \hline \mathbf{120} \end{array}$$

16. $7 \times 8 = 56$
$$\begin{array}{r} 84 \\ +\ 56 \\ \hline \mathbf{140} \end{array}$$

17. $4 + 3 = 7$
$6 \times 7 = \mathbf{42}$

18.
$$\begin{array}{r} \$6.54 \\ \$0.68 \\ +\ \$3.00 \\ \hline \mathbf{\$10.22} \end{array}$$

19. $\sqrt{36} + \sqrt{81} = 6 + 9 = \mathbf{15}$

20. $0.65 - 0.30 = \mathbf{0.35}$

TEST 10

1.
$$\begin{array}{r} 12 \text{ players on each team} \\ \times\ \ 5 \text{ teams} \\ \hline \mathbf{60 \text{ players}} \end{array}$$

2. D. $1\frac{1}{2}$ ft

3.
$$\begin{array}{r} 394 \text{ monkeys} \\ +\ 273 \text{ monkeys} \\ \hline \mathbf{667 \text{ monkeys}} \end{array}$$

4.
$$\begin{array}{r} 837 \text{ shells} \\ -\ 326 \text{ shells} \\ \hline \mathbf{511 \text{ shells}} \end{array}$$

5. **3**

6. A number that is between 6500 and 6900 must begin with 6. To be even, the number must end with an even digit. Since 8 is the only even digit left, the last digit is 8. The hundreds digit must be 7 for the number to be between 6500 and 6900. The remaining digit 9 goes in the tens place. Thus, the number is **6798.**

7. Three full circles are shaded, plus three eighths of another, which equals $3\frac{3}{8}$.

8. Perimeter
$= 3\text{ ft} + 5\text{ ft} + 3\text{ ft} + 2\text{ ft} + 4\text{ ft} + 4\text{ ft}$
$= \mathbf{21\text{ ft}}$

9. The number 281 is closer to **300** than to 200.

10.
$$\begin{array}{r} 84{,}048 \\ +\ 15{,}569 \\ \hline \mathbf{99{,}617} \end{array}$$

11.
$$\begin{array}{r} \$5.50 \\ -\ \$2.69 \\ \hline \mathbf{\$2.81} \end{array}$$

12.
$$\begin{array}{r} N \\ +\ 192 \\ \hline 671 \end{array} \qquad \begin{array}{r} 671 \\ -\ 192 \\ \hline \mathbf{479} \end{array}$$

13.
$$\begin{array}{r} 41 \\ \times\ \ 8 \\ \hline \mathbf{328} \end{array}$$

14.
$$\begin{array}{r} 27 \\ \times\ \ 8 \\ \hline \mathbf{216} \end{array}$$

15.
$$\begin{array}{r} Z \\ -\ 546 \\ \hline 312 \end{array} \qquad \begin{array}{r} (8) - 6 = 2 \\ (5) - 4 = 1 \\ (8) - 5 = 3 \\ Z = \mathbf{858} \end{array}$$

16.
$$\begin{array}{r} 6.43 \\ -\ 3.8 \\ \hline \mathbf{2.63} \end{array}$$

17.
$$\begin{array}{r} 5.1 \\ +\ 3.72 \\ \hline \mathbf{8.82} \end{array}$$

18.
$$\begin{array}{r} \$5.32 \\ \$3.00 \\ \$0.57 \\ +\ \$0.08 \\ \hline \mathbf{\$8.97} \end{array}$$

19. $(5 \times 6) + 7 \bigcirc 5 \times (6 + 7)$

$\quad\quad 30 + 7 \bigcirc 5 \times 13$

$\quad\quad\quad\quad 37 < 65$

20. (a) $9\overline{)54}$ → $\mathbf{6}$

(b) $32 \div 8 = \mathbf{4}$

(c) $\dfrac{63}{9} = \mathbf{7}$

TEST 11

1. $7\overline{)56}$ → **8 weeks**

2. Factors of 18: **1, 2, 3, 6, 9, 18**

3. $5\overline{)45}$ → **9 books**

4. 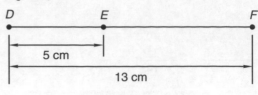 (varies)

5. 1 decade = 10 years
5 decades = **50 years**

6. $8 \times 6 = 48$
$$\begin{array}{r} 48 \\ -\ 17 \\ \hline \mathbf{31} \end{array}$$

7. $\dfrac{1}{2} \bigcirc 75\%$

$\dfrac{1}{2} < \dfrac{3}{4}$

8. Student may draw and label a diagram (not required).

$$13 \text{ cm } - 5 \text{ cm } = \mathbf{8 \text{ cm}}$$

9. Perimeter = 6 cm + 4 cm + 6 cm + 4 cm
$\quad\quad\quad\quad = \mathbf{20 \text{ cm}}$

10. The number 1760 is closer to **2000** than to 1000.

11.
$$\begin{array}{r} 42{,}092 \\ +\ 8{,}768 \\ \hline \mathbf{50{,}860} \end{array}$$

12.
$$\begin{array}{r} \$17.00 \\ -\ \$9.27 \\ \hline \mathbf{\$7.73} \end{array}$$

13.
$$\begin{array}{r} 35{,}456 \\ -\ 17{,}468 \\ \hline \mathbf{17{,}988} \end{array}$$

14.
$$\begin{array}{r} 83 \\ \times\ 8 \\ \hline \mathbf{664} \end{array}$$

15.
$$\begin{array}{r} 1.54 \\ 3.8 \\ +\ 14.2 \\ \hline \mathbf{19.54} \end{array}$$

16. $24 \div 3 = 8$
$\sqrt{36} = 6$
$6 + 8 = \mathbf{14}$

17. $6\overline{)55}$ → **9 R 1**
$$\begin{array}{r} 54 \\ \hline 1 \end{array}$$

18.
$$\begin{array}{r} 307 \\ -\ 49 \\ \hline \mathbf{258} \end{array}$$

19.
$$\begin{array}{r} Z \\ +\ 937 \\ \hline 1284 \end{array} \quad\quad \begin{array}{r} 1284 \\ -\ 937 \\ \hline \mathbf{347} \end{array}$$

20.
$$\begin{array}{r} N \\ -\ 472 \\ \hline 500 \end{array}$$
$$(2) - 2 = 0$$
$$(7) - 7 = 0$$
$$(9) - 4 = 5$$
$$N = \textbf{972}$$

TEST 12

1.
$$\begin{array}{r} 2282 \\ -\ 326 \\ \hline \textbf{1956} \end{array}$$

2.
$$\begin{array}{r} 16\text{ cars} \\ \times\ \ 9\text{ clowns in each car} \\ \hline \textbf{144 clowns} \end{array}$$

3. $\overset{\textbf{5 minutes}}{8\overline{)40}}$

4. $\overset{\textbf{8 quarts}}{6\overline{)48}}$

5.
$$\begin{array}{r} 52\text{ miles per hour} \\ \times\ \ 5\text{ hours} \\ \hline \textbf{260 miles} \end{array}$$

6. 9 shaded parts $\quad \dfrac{\textbf{9}}{\textbf{14}}$
14 total parts

7. 30×10 years $= 300$ years $= \textbf{3 centuries}$

8. (varies)

$$\dfrac{1}{3} < \dfrac{5}{8}$$

9. $582 + 321 \approx 600 + 300 = \textbf{900}$

10. $\overline{AC}$ or $\overline{CA}$

11. $6 \times 9 = \textbf{54}$
$$\begin{array}{r} 54 \\ -\ 23 \\ \hline \textbf{31} \end{array}$$

12.
$$\begin{array}{r} \$65.98 \\ +\ \$11.45 \\ \hline \textbf{\$77.43} \end{array}$$

13.
$$\begin{array}{r} 645{,}972 \\ -\ 208{,}394 \\ \hline \textbf{437,578} \end{array}$$

14.
$$\begin{array}{r} 2.50 \\ +\ 1.53 \\ \hline 4.03 \end{array} \qquad \begin{array}{r} 7.89 \\ -\ 4.03 \\ \hline \textbf{3.86} \end{array}$$

15.
$$\begin{array}{r} 14 \\ \times\ 6 \\ \hline 84 \end{array} \qquad \begin{array}{r} 2000 \\ -\ 84 \\ \hline \textbf{1916} \end{array}$$

16.
$$\begin{array}{r} 720 \\ \times\ 6 \\ \hline \textbf{4320} \end{array}$$

17.
$$\begin{array}{r} 549 \\ \times\ 7 \\ \hline \textbf{3843} \end{array}$$

18. $\sqrt{49} = 7$
$$7 \div 7 = \textbf{1}$$

19. $\overset{\textbf{8 R 3}}{8\overline{)67}}$
$$\begin{array}{r} 64 \\ \hline 3 \end{array}$$

20. $\overset{\textbf{9}}{8\overline{)72}}$
$$\begin{array}{r} 72 \\ \hline 0 \end{array}$$

TEST 13

1.
$$\begin{array}{r} 3\text{ feet per yard} \\ \times\ 8\text{ yards} \\ \hline \textbf{24 feet} \end{array}$$

2. $\overset{\textbf{12 necklaces}}{8\overline{)96}}$
$$\begin{array}{r} 8 \\ \hline 16 \\ 16 \\ \hline 0 \end{array}$$

3. $\dfrac{5}{5} - \dfrac{4}{5} = \dfrac{\textbf{1}}{\textbf{5}}$

4. 50% of 24 $\bigcirc$ 2×5
$$12 > 10$$

5. Area $= 6$ in. $\times 6$ in. $= \textbf{36 sq. in.}$

6. 87 rounds to 90.
$$90 \times 6 = \textbf{540}$$

7.
$$\begin{array}{r} 26\text{ miles per gallon} \\ \times\ \ 9\text{ gallons} \\ \hline \textbf{234 miles} \end{array}$$

8. The shape has 6 sides, so it is a **hexagon**.

9. Perimeter = 6 × 18 in. = **108 in.**

10.

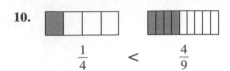

$$\frac{1}{4} \quad < \quad \frac{4}{9}$$

11.
$$\begin{array}{r} 100\% \\ - \quad 60\% \\ \hline \textbf{40\%} \text{ were girls} \end{array}$$

12.
$$\begin{array}{r} 32{,}624 \\ + \ 109{,}876 \\ \hline \textbf{142{,}500} \end{array}$$

13.
$$\begin{array}{r} \$12.00 \\ - \quad \$0.18 \\ \hline \textbf{\$11.82} \end{array}$$

14. 6 × 8 = 48
$$\begin{array}{r} 48 \\ \times \quad 3 \\ \hline \textbf{144} \end{array}$$

15. $6^2 = 36$
$8^2 = 64$
$$\begin{array}{r} 36 \\ + \ 64 \\ \hline \textbf{100} \end{array}$$

16.
$$\begin{array}{r} 276 \\ \times \quad 8 \\ \hline \textbf{2208} \end{array}$$

17.
$$\begin{array}{r} 47 \\ 8\overline{)376} \\ 32 \\ \hline 56 \\ 56 \\ \hline 0 \end{array}$$

18. 20 ÷ 4 = 5
60 ÷ 5 = **12**

19.
$$\begin{array}{r} 6.9 \\ 4.83 \\ + \ 15.2 \\ \hline \textbf{26.93} \end{array}$$

20.
$$\begin{array}{r} 13 \\ 5\overline{)65} \\ 5 \\ \hline 15 \\ 15 \\ \hline 0 \end{array}$$

TEST 14

1.
$$\begin{array}{r} \textbf{21 heads} \\ 4\overline{)84} \text{ heads} \\ 8 \\ \hline 04 \\ 4 \\ \hline 0 \end{array}$$

Student may draw a diagram (not required).

84 heads

21 heads	} $\frac{1}{4}$ eaten
21 heads	
21 heads	} $\frac{3}{4}$ not eaten
21 heads	

2.
$$\begin{array}{r} 100\% \\ - \quad 65\% \\ \hline \textbf{35\%} \text{ of the lights were off} \end{array}$$

3.
$$\begin{array}{r} 98 \text{ cookies} \\ \times \quad 7 \text{ chocolate chips in each cookie} \\ \hline \textbf{686 chocolate chips} \end{array}$$

4. The number 6417 is closer to 6000 than to 7000.
The number 8692 is closer to 9000 than to 8000.
$$\begin{array}{r} 6{,}000 \\ + \ 9{,}000 \\ \hline \textbf{15{,}000} \end{array}$$

5.
$$\begin{array}{r} \$0.08 \\ \$0.30 \\ \$0.50 \\ + \ \$0.15 \\ \hline \textbf{\$1.03} \end{array}$$

6.
$$\begin{array}{r} \textbf{16 swimmers} \\ 5\overline{)80} \text{ swimmers} \\ 5 \\ \hline 30 \\ 30 \\ \hline 0 \end{array}$$

Student may draw a diagram (not required).

80 swimmers

16 swimmers	} $\frac{1}{5}$ earned medals.
16 swimmers	
16 swimmers	} $\frac{4}{5}$ did not earn medals.
16 swimmers	
16 swimmers	

7. **−4**

8.
$$\begin{array}{r} 5 \text{ sides on a pentagon} \\ + \ 8 \text{ sides on an octagon} \\ \hline \textbf{13 sides} \end{array}$$

9. (a) **5 cm**   (b) **50 mm**

10. Perimeter = 6 cm + 3 cm + 6 cm + 3 cm
= **18 cm**

11. To find area, the student may draw a diagram and count the squares (not required).

6 cm

3 cm

There are 3 rows of 6 squares. The area is **18 sq. cm.**

Alternately,
Area = 6 cm × 3 cm = 18 sq. cm

12.
$$\begin{array}{r} \$254.26 \\ + \ \$336.50 \\ \hline \mathbf{\$590.76} \end{array}$$

13.
$$\begin{array}{r} \$30.00 \\ - \ \$29.74 \\ \hline \mathbf{\$0.26} \end{array}$$

14.
$$\begin{array}{r} \$2.06 \\ \times \quad 8 \\ \hline \mathbf{\$16.48} \end{array}$$

15.
$$\begin{array}{r} 34 \\ \times \ 20 \\ \hline \mathbf{680} \end{array}$$

16. $5^2 - \sqrt{16} = 25 - 4 = \mathbf{21}$

17.
$$\begin{array}{r} \mathbf{56\ R\ 3} \\ 5\overline{)283} \\ \underline{25} \\ 33 \\ \underline{30} \\ 3 \end{array}$$

18. $9 \times 7 = 63$
$$\begin{array}{r} 63 \\ \times \ 3 \\ \hline \mathbf{189} \end{array}$$

19.
$$\begin{array}{r} 4.43 \\ - \ 2.6 \\ \hline \mathbf{1.83} \end{array}$$

20.
$$\begin{array}{r} \mathbf{16} \\ 7\overline{)112} \\ \underline{7} \\ 42 \\ \underline{42} \\ 0 \end{array}$$

TEST 15

1.
$$\begin{array}{r} \mathbf{44}\ \textbf{pages} \\ 5\overline{)220} \\ \underline{20} \\ 20 \\ \underline{20} \\ 0 \end{array}$$

2. $\dfrac{5}{5} - \dfrac{3}{5} = \dfrac{\mathbf{2}}{\mathbf{5}}$

3.
$$\begin{array}{r} 32 \\ - \quad 6 \\ \hline \mathbf{26}\ \textbf{children} \end{array}$$

4.
$$\begin{array}{r} 32 \\ + \ 26 \\ \hline 58 \end{array} \qquad \begin{array}{r} 64 \\ - \ 58 \\ \hline \mathbf{6}\ \textbf{children} \end{array}$$

5. pizza

6. 1 week = 7 days
7 weeks = **49 days**

7. 2 letter P's $\dfrac{\mathbf{2}}{\mathbf{11}}$
11 letters in all

8. 1 km = 1000 m
5 km = **5000 m**

9.
$$\begin{array}{r} \mathbf{3\ km} \\ 5\overline{)15\ km} \\ \underline{15} \\ 0 \end{array}$$
Student may draw a diagram (not required).

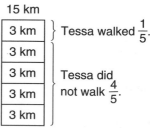

15 km

3 km	} Tessa walked $\frac{1}{5}$.
3 km	
3 km	Tessa did
3 km	not walk $\frac{4}{5}$.
3 km	

10. Perimeter = 18 mm + 22 mm + 30 mm
= **70 mm**

SOLUTIONS

11. Student may draw and label a diagram (not required).

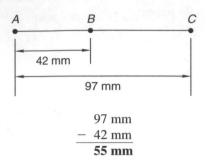

```
    97 mm
 -  42 mm
    55 mm
```

12.
```
   $18.61          $34.00
 +  $0.95        - $19.56
   $19.56          $14.44
```

13.
```
   44,317
 -    726
   43,591
```

14.
```
   4.6
   3.57
   0.34
 + 1.0
   9.51
```

15.
```
   $3.87
 ×     7
   $27.09
```

16.
```
      50 R 1
   9)451
     45
     ―――
     01
      0
     ――
      1
```

17.
```
      48 R 1
   4)193
     16
     ――
     33
     32
     ――
      1
```

18.
```
      108
   4)432
     4
     ――
     03
      0
     ――
     32
     32
     ――
      0
```

19.
```
    374        589        756
 +  215      +   N      - 589
    589        756        167
```

20. $2 \times 4^2 \bigcirc \sqrt{16} \times \sqrt{36}$

$2 \times 16 \bigcirc 4 \times 6$

$32 > 24$

TEST 16

1. **A. north**

2.
```
    83 cents per pound
 ×   3 pounds
    249 cents = $2.49
```

3.
```
     D          113
 +  52        -  52
    113          61
```

4.   **24 diners**
```
   3)72
     6
     ――
     12
     12
     ――
      0
```

Student may draw a diagram (not required).

72 diners

| 24 diners |
| 24 diners | $\}$ $\frac{1}{3}$ in each room
| 24 diners |

5.

6. 5 letters are not E's $\dfrac{5}{9}$
9 letters in all

7. 4×2000 pounds = **8000 pounds**

8. **250 g**

9. **A.**

10.
```
    $67,342
 -  $48,796
    $18,546
```

11. 347 rounds to 300.

623 rounds to 600.

300 + 600 = **900**

12.
$$\begin{array}{r} \textbf{41} \textbf{ packages} \\ 3\overline{)123} \\ \underline{12} \\ 03 \\ \underline{3} \\ 0 \end{array}$$

13.
$$\begin{array}{r} 429 \\ \times \quad 7 \\ \hline 3003 \end{array} \qquad \begin{array}{r} 6002 \\ - 3003 \\ \hline \textbf{2999} \end{array}$$

14.
$$\begin{array}{r} \$8.58 \\ \times \quad 6 \\ \hline \textbf{\$51.48} \end{array}$$

15.
$$\begin{array}{r} 53.7 \\ + \quad 6.41 \\ \hline \textbf{60.11} \end{array}$$

16.
$$\begin{array}{r} \textbf{506} \\ 8\overline{)4048} \\ \underline{40} \\ 04 \\ \underline{0} \\ 48 \\ \underline{48} \\ 0 \end{array}$$

17.
$$\begin{array}{r} \textbf{946 R 2} \\ 6\overline{)5678} \\ \underline{54} \\ 27 \\ \underline{24} \\ 38 \\ \underline{36} \\ 2 \end{array}$$

18.
$$\begin{array}{r} \textbf{68} \\ 4\overline{)272} \\ \underline{24} \\ 32 \\ \underline{32} \\ 0 \end{array}$$

19.
$$\begin{array}{r} 2367 \\ + 1825 \\ \hline 4192 \end{array} \qquad \begin{array}{r} 4192 \\ + \quad N \\ \hline 5000 \end{array} \qquad \begin{array}{r} 5000 \\ - 4192 \\ \hline \textbf{808} \end{array}$$

20. 2 × 1000 grams = 2000 grams

2 × 1000 milliliters = 2000 milliliters

2000 = 2000

TEST 17

1. Count backward 25 minutes from 8:05 a.m. to **7:40 a.m.**

2.
$$\begin{array}{r} 4 \\ 15\overline{)60} \end{array}$$

In one hour (60 minutes) there are four groups of 15 minutes. So Josie can walk **4 miles** in an hour at the given rate.

3. (a)
$$\begin{array}{r} \$1.29 \\ \times \quad 4 \\ \hline \$5.16 \end{array} \qquad \begin{array}{r} \$5.16 \\ + \$0.23 \\ \hline \textbf{\$5.39} \end{array} \qquad \text{(b)} \quad \begin{array}{r} \$10.00 \\ - \$5.39 \\ \hline \textbf{\$4.61} \end{array}$$

4. 7 households have dogs; 9 households have cats

$$\begin{array}{r} 7 \\ + \quad 9 \\ \hline \textbf{16 households} \end{array}$$

5. To find area, we can count squares.

4 miles

 3 miles

The area of the land is 12 square miles. Half of 12 square miles is **6 square miles.**

6. **C.**

7. 5243 rounds down to **5000.**

8. 18 × 2000 pounds = **36,000 pounds**

9.
$$\begin{array}{r} \textbf{2 roses} \\ 7\overline{)14} \text{ flowers} \\ \underline{14} \\ 0 \end{array}$$

Student may draw a diagram (not required).

14 flowers

2 flowers	} $\frac{1}{7}$ roses
2 flowers	
2 flowers	
2 flowers	
2 flowers	$\frac{6}{7}$ not roses
2 flowers	
2 flowers	

295

10. $\dfrac{345}{1000} = 0.345$

11. $0.543 = \dfrac{543}{1000}$

12.
$$\begin{array}{r} \$68.24 \\ + \ \$11.98 \\ \hline \mathbf{\$80.22} \end{array}$$

13.
$$\begin{array}{r} 31.428 \\ + \ 16.888 \\ \hline \mathbf{48.316} \end{array}$$

14.
$$\begin{array}{r} 487 \\ \times \quad 8 \\ \hline \mathbf{3896} \end{array}$$

15.
$$\begin{array}{r} 32 \\ \times \quad 100 \\ \hline \mathbf{3200} \end{array}$$

16.
$$\begin{array}{r} \mathbf{\$0.93} \\ 8\overline{)\$7.44} \\ 7\ 2 \\ \hline 24 \\ 24 \\ \hline 0 \end{array}$$

17.
$$6\overline{)24} \quad \begin{array}{r} 4 \\ \end{array} \qquad 4\overline{)2400} \begin{array}{r} \mathbf{600} \\ 24 \\ \hline 00 \\ 0 \\ \hline 00 \\ 0 \\ \hline 0 \end{array}$$

18.
$$6\overline{)540} \begin{array}{r} \mathbf{90} \\ 54 \\ \hline 00 \\ 0 \\ \hline 0 \end{array}$$

19.
$$\begin{array}{r} 19 \\ 6 \\ 12 \\ 23 \\ + \ 108 \\ \hline 168 \end{array} \qquad \begin{array}{l} 168 + N = 198 \\ \qquad N = \mathbf{30} \end{array}$$

20.
$$\begin{array}{r} 27 \\ \times \quad 50 \\ \hline \mathbf{1350} \end{array}$$

TEST 18

1.
$$\begin{array}{ccc} \$0.25 & \$0.10 & \$0.05 \\ \times \quad 2 & \times \quad 7 & \times \quad 1 \\ \hline \$0.50 & \$0.70 & \$0.05 \end{array}$$

$$\begin{array}{r} \$0.50 \\ \$0.70 \\ \$0.05 \\ + \ \$0.08 \\ \hline \mathbf{\$1.33} \end{array}$$

2. First divide.
$$7\overline{)46} \begin{array}{r} 6\,\text{R}\,4 \\ 42 \\ \hline 4 \end{array}$$

The remainder gives the number of books left over after each stack had six books, so four stacks received an extra book. This means that 4 stacks had 7 books. Since there were 7 stacks in all, 3 stacks had exactly 6 books.

(a) **3 stacks**   (b) **4 stacks**

3.
$$\begin{array}{r} \$1.00 \\ - \ \$0.41 \\ \hline \mathbf{\$0.59} \end{array}$$

4.
$$\begin{array}{r} 15 \quad \text{words} \\ \times \quad 10 \text{ times each word is written} \\ \hline \mathbf{150}\ \textbf{words} \end{array}$$

5. The number 5438 rounds to **5000**.
The number 2263 rounds to **2000**.
$5000 + 2000 = \mathbf{7000}$

6. $\overline{AB}$ (or $\overline{BA}$) **and** $\overline{ED}$ (or $\overline{DE}$)

7.
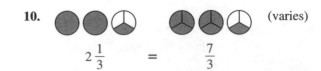

8. A square has 4 sides of equal length.
$$4\overline{)60} \begin{array}{r} \mathbf{15}\ \textbf{millimeters} \\ \end{array}$$

9. $19 + 15 + N = 65$

$$\begin{array}{ccc} 19 & 34 & 65 \\ + \ 15 & + \ N & - \ 34 \\ \hline 34 & 65 & \mathbf{31\ mm} \end{array}$$

10.
(varies)

$$2\tfrac{1}{3} \quad = \quad \tfrac{7}{3}$$

11. $72.75
  + $186.75
  $259.50

12. 26,345
  − 6,721
  19,624

13. $4.53
  × 10
  $45.30

14. 47
  × 36
  282
  141
  1692

15.
$$\begin{array}{r}\textbf{\$8.40}\\6\overline{)\$50.40}\\48\\\hline 2\,4\\2\,4\\\hline 00\\0\\\hline 0\end{array}$$

16. $7^2 - \sqrt{49} = 49 - 7 = \textbf{42}$

17. 5.732
  + 2.18
  7.912

18. $800 \times 30 = \textbf{24,000}$

19.
$$\begin{array}{r}N\\+\ 977\\\hline 5368\end{array}\qquad\begin{array}{r}5368\\-\ \ 977\\\hline \textbf{4391}\end{array}$$

20.
$$\begin{array}{r}5\\6\\8\\12\\+\ 4\\\hline 35\end{array}\qquad\begin{array}{r}35\\+\ N\\\hline 52\end{array}\qquad\begin{array}{r}52\\-\ 35\\\hline \textbf{17}\end{array}$$

TEST 19

1. First, find the age of Ruth.

$$\begin{aligned}\text{Ruth} &= \text{twice as old as James}\\ &= 2 \times 7 \text{ years old}\\ &= 14 \text{ years old}\end{aligned}$$

Now, find the age of Jan.

$$\begin{aligned}\text{Jan} &= \text{six years older than Ruth}\\ &= 6 \text{ years } + 14 \text{ years old}\\ &= \textbf{20 years old}\end{aligned}$$

2.  (varies)

3. First, find out how much is on the package.
  37¢
  × 5
  185¢ = $1.85

Now, find out how much more is needed.
  $2.37
  − $1.85
  $0.55

4. First, do the division.

$$\begin{array}{r}6\ \text{R}\ 3\\7\overline{)45}\\42\\\hline 3\end{array}$$

The remainder gives the cans left over after all shelves have six, so three shelves have an extra can. This means that 3 shelves have 7 cans. Since there are 7 shelves, 4 shelves have 6 cans.
(a) **4 shelves**
(b) **3 shelves**

5.  (varies)

6. (a) 17 parts of 100 are shaded: **0.17**
  (b) $100 - 17 = 83$
  83 parts of 100 are not shaded: **0.83**

7. **3**

8. 42 rounds to 40.   42
  68 rounds to 70.   × 68
  $40 \times 70 = \textbf{2800}$ 336
      252
      2856

9. 1 liter = 1000 milliliters

$$\begin{array}{r} \textbf{125 milliliters} \\ 8\overline{)1000 \text{ milliliters}} \\ \underline{80} \\ 20 \\ \underline{16} \\ 40 \\ \underline{40} \\ 0 \end{array}$$

10. Area = 50 ft × 60 ft = **3000 sq. ft**

11. $\begin{array}{r} 12 \\ 3\overline{)36} \end{array}$

2 × 12 children = **24 children**

Student may draw a diagram (not required).

36 children

12 children
12 children
12 children

$\frac{2}{3}$ reading books

$\frac{1}{3}$ not reading books

12. $\begin{array}{r} \textbf{68 miles} \\ 7\overline{)476} \\ \underline{42} \\ 56 \\ \underline{56} \\ 0 \end{array}$

13. $\frac{82}{100}$ = **0.82**

14. $\begin{array}{r} 5.6 \\ -\ 0.56 \\ \hline \textbf{5.04} \end{array}$

15. 60 × 800 = **48,000**

16. $\begin{array}{r} \$7.43 \\ \times\ \ \ \ 8 \\ \hline \$59.44 \end{array}$

17. $\begin{array}{r} 32 \\ \times\ 23 \\ \hline 96 \\ 640 \\ \hline \textbf{736} \end{array}$

18. $\begin{array}{r} \textbf{250} \\ 8\overline{)2000} \\ \underline{16} \\ 40 \\ \underline{40} \\ 00 \\ \underline{0} \\ 0 \end{array}$

19. $\begin{array}{r} \textbf{185} \\ 4\overline{)740} \\ \underline{4} \\ 34 \\ \underline{32} \\ 20 \\ \underline{20} \\ 0 \end{array}$

20. $\begin{array}{r} 2 \\ 6 \\ 7 \\ 4 \\ 5 \\ 3 \\ 3 \\ +\ 6 \\ \hline 36 \end{array}$ $\quad\begin{array}{r} 36 \\ +\ N \\ \hline 52 \end{array}$ $\quad\begin{array}{r} 52 \\ -\ 36 \\ \hline \textbf{16} \end{array}$

TEST 20

1. First, find the total.

$$\begin{array}{r} 4 \text{ orange wedges} \\ 7 \text{ orange wedges} \\ 8 \text{ orange wedges} \\ 6 \text{ orange wedges} \\ +\ \ 5 \text{ orange wedges} \\ \hline 30 \text{ orange wedges} \end{array}$$

Then, divide into equal groups.

$$\begin{array}{r} \textbf{6 orange wedges} \\ 5\overline{)30} \end{array}$$

2. $7\frac{4}{10}$ = **7.4**

3. The 3 in 27.43 is in the hundredths place. The digit **8** is in the hundredths place in 97.685.

4. $\begin{array}{r} 9 \\ 7\overline{)63} \\ \underline{63} \\ 0 \end{array}$ 3 × 9 marchers = **27 marchers**

Student may draw a diagram (not required).

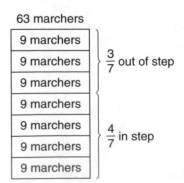

63 marchers

9 marchers
9 marchers
9 marchers
9 marchers
9 marchers
9 marchers
9 marchers

$\frac{3}{7}$ out of step

$\frac{4}{7}$ in step

5.

Sodas
75¢
a can

Sodas
$0.75
a can

6. (a) Radius = **15 millimeters**
(b) Diameter = **3 centimeters**

7. 3.62 = **three and sixty-two hundredths**

8. 81 rounds to 80.
56 rounds to 60.
80 × 60 = **4800**

9. 43 cents per pound
× 6 pounds
‾‾‾‾‾‾‾‾‾‾‾‾‾‾
258 cents = **$2.58**

10. Perimeter = 7 in. + 4 in. + 7 in. + 4 in.
= **22 in.**
Area = 7 in. × 4 in. = **28 sq. in.**

11. 378,422
+ 8,635
‾‾‾‾‾‾‾‾‾
387,057

12. First, find how many cookies Sarah has.
Sarah = 6 fewer cookies than Nathan
= 10 cookies − 6 cookies
= 4 cookies

Now, find how many cookies Will has.
Will = twice as many cookies as Sarah
= 2 × 4 cookies
= **8 cookies**

13.

$$3 \quad = \quad \frac{6}{2}$$

14. The scores in order are 75, 80, 85, 85, 95, 100, and 100. The middle, or median score is **85.**

15. A. cone

16. 70 × 70 = **4900**

17. 63
× 52
‾‾‾‾‾
126
315
‾‾‾‾‾
3276

18.
$$
\begin{array}{r}
\$6.07 \\
3)\overline{\$18.21} \\
\underline{18} \\
0\,2 \\
\underline{0} \\
21 \\
\underline{21} \\
0
\end{array}
$$

19. 31.4
− 2.71
‾‾‾‾‾‾‾
28.69

20. $(5^2 + 15) \div \sqrt{16} \bigcirc 3 + \sqrt{16}$
$(25 + 15) \div 4 \bigcirc 3 + 4$
$40 \div 4 \bigcirc 7$
$10 > 7$

TEST 21

1. 7 + 9 + 2 + 2 + 2 + 1 + 1 = **24 hours**

2. 1 hour spent eating $\dfrac{1}{24}$
24 total hours

3. Gabby sleeps for 9 hours, so count forward 9 hours from 8:30 p.m. to **5:30 a.m.**

4. 7 hours
+ 1 hour
‾‾‾‾‾‾‾‾
8 hours

5. 9 cats × $\frac{1}{3}$ can for each cat = **3 cans**

6. 52¢
× 3
‾‾‾‾‾
156¢ = **$1.56**

7. A square has 4 sides of equal length.

$$
\begin{array}{r}
\textbf{18 inches} \\
4)\overline{72} \\
\underline{4} \\
32 \\
\underline{32} \\
0
\end{array}
$$

8. (a) **35 mm**
(b) **3.5 cm**

9. $5\overline{)8}$ $\dfrac{8}{5} = 1\dfrac{3}{5}$

     $\dfrac{5}{3}$

10. $1 = \dfrac{\mathbf{18}}{\mathbf{18}}$

11.

$8.49	$20.00
$8.49	− $18.14
+ $1.16	**$1.86**
$18.14	

12. (a)   **45 miles per hour**

$5\overline{)225}$

   $\dfrac{20}{25}$

     $\dfrac{25}{0}$

(b)

   45 miles per hour

$\times$ 7 hours

   315 miles

13.

$47.00

− $21.68

$25.32

14.

$5.21

$\times$ 8

$41.68

15. $60 \times 60 = \mathbf{3600}$

16.

    67

$\times$ 92

  134

603

6164

17. $840 \div 2^3 = 840 \div 8$ $\dfrac{105}{8\overline{)840}}$

                   $\dfrac{8}{04}$

                     $\dfrac{0}{40}$

                     $\dfrac{40}{0}$

18. $\dfrac{63\,\mathbf{R}\,\mathbf{5}}{10\overline{)635}}$

   $\dfrac{60}{35}$

    $\dfrac{30}{5}$

19.

3.1	4.30
+ 0.48	− 3.58
3.58	**0.72**

20.

    647     $7 - (2) = 5$

  − N $4 - (3) = 1$

    315     $6 - (3) = 3$

            $N = \mathbf{332}$

TEST 22

1. Eight pounds falls in the third row down. Zone 3 is the last column on the right. The price is **$3.14** to ship the package.

2.

   $3.50

− $2.56

$0.94

3. $\dfrac{11\ \text{tadpoles}}{7\overline{)77\ \text{tadpoles}}}$ $3 \times 11\ \text{tadpoles} = \mathbf{33\ tadpoles}$

    $\dfrac{7}{07}$ Student may draw a diagram

     $\dfrac{7}{0}$ (not required).

77 tadpoles

11 tadpoles
11 tadpoles
11 tadpoles

$\dfrac{3}{7}$ had back legs.

11 tadpoles
11 tadpoles
11 tadpoles
11 tadpoles

$\dfrac{4}{7}$ did not have back legs.

4. (a) **48 mm**

(b) **4.8 cm**

5. $\dfrac{2}{3} \times \dfrac{2}{2} = \dfrac{\mathbf{4}}{\mathbf{6}}$

$\dfrac{2}{3} \times \dfrac{3}{3} = \dfrac{\mathbf{6}}{\mathbf{9}}$

$\dfrac{2}{3} \times \dfrac{4}{4} = \dfrac{\mathbf{8}}{\mathbf{12}}$

6. 266 rounds to 300.

92 rounds to 90.

$300 \times 90 = \mathbf{27{,}000}$

7. $215.6 = $ **two hundred fifteen and six tenths**

8. $3\overline{)4}$ $\quad \frac{4}{3} = 1\frac{1}{3}$
   $\quad \frac{3}{1}$

9.   14 marbles in each pouch
   $\underline{\times\ 12 \text{ pouches}}$
   $\quad\quad 28$
   $\quad\underline{140}$
   168 marbles

10. First, find the cost of one golf ball.
   $\quad\quad\quad \$0.80$
   $\quad\quad 6\overline{)\$4.80}$

   Now, find the cost of 30 golf balls.
   $\quad\quad \$0.80 \times 30 = \textbf{\$24.00}$

   Alternately, 30 golf balls is 5 packs of 6.
   $\quad\quad 5 \times \$4.80 = \$24.00$

11. $\quad \$30.00$
   $\underline{-\ \$19.35}$
   $\quad \textbf{\$10.65}$

12. $\quad\quad 6.72$
   $\quad\quad 17.5$
   $\underline{+\quad 6.3}$
   $\quad\ \textbf{30.52}$

13. $\quad\quad 15.8$
   $\underline{-\quad 4.72}$
   $\quad\ \textbf{11.08}$

14. $2^3 + \sqrt{36} = 8 + 6 = \textbf{14}$

15. $\quad\quad 33$
   $\underline{\times\quad 80}$
   $\quad\ \textbf{2640}$

16. $\quad\ \ \overset{68}{7\overline{)476}}$
   $\quad\quad \underline{42}$
   $\quad\quad 56$
   $\quad\quad \underline{56}$
   $\quad\quad\ \ 0$

17. $\overset{\textbf{2 R 6}}{70\overline{)146}}$
   $\quad \underline{140}$
   $\quad\quad 6$

18. $\overset{\textbf{52 R 7}}{10\overline{)527}}$
   $\quad \underline{50}$
   $\quad 27$
   $\quad \underline{20}$
   $\quad\ \ 7$

19. $bh = 9 \times 4 = \textbf{36}$

20. There are two equally possible outcomes, but "heads" is only one of them. So the probability is $\frac{1}{2}$.

TEST 23

1. $\quad \$8.25$
   $\quad \$8.25$
   $\quad \$4.15$
   $\quad \$4.15$
   $\underline{+\ \$4.15}$
   $\quad \textbf{\$28.95}$

2. Number of full squares $=\ 8$
   Squares made from pieces $=\ 7$
   Total Area $=\ 8 + 7 = \textbf{15 sq. cm}$

3. A dollar is worth 100 cents.
   A nickel is worth 5 cents.
   $\frac{5}{100} = \textbf{5\%}$

4. Area of rectangle:
   $6\,\text{cm} \times 4\,\text{cm} = 24\,\text{sq. cm}$

   The shaded portion is half this area:
   $\frac{1}{2} \times 24\,\text{sq. cm} = \textbf{12 sq. cm}$

5. $\frac{3}{4} \times \frac{4}{4} = \frac{\textbf{12}}{\textbf{16}}$

6. $\frac{\textbf{8}}{\textbf{8}}$

7.   (varies)
   $\frac{2}{3} = \frac{4}{6}$

8. First, find how many miles each hour.

$$\begin{array}{r} 23 \text{ miles} \\ 2\overline{)46 \text{ miles}} \\ \underline{4} \\ 06 \\ \underline{6} \\ 0 \end{array}$$

Now, find how many miles in 5 hours.

$$\begin{array}{r} 23 \text{ miles per hour} \\ \times \quad 5 \text{ hours} \\ \hline \textbf{115 miles} \end{array}$$

9. First, find out how much for the sleeping bags and skewers.

$4 \times \$19.99 = \$79.96$ for sleeping bags
$4 \times \$1.29 = \$5.16$ for skewers

Now, find the total and the change.

$$\begin{array}{rr} \$79.96 & \$100.00 \\ \$5.16 & -\quad \$91.29 \\ +\quad \$6.17 & \quad\textbf{\$8.71} \\ \hline \$91.29 \end{array}$$

10. First, find how many can ride.
7×8 people $= 56$ people

Now, find how many people cannot ride.

$$\begin{array}{r} 80 \text{ people} \\ -\ 56 \text{ people} \\ \hline \textbf{24 people} \end{array}$$

11.
$$\begin{array}{r} 32.61 \\ 3.51 \\ +\ 11.6 \\ \hline \textbf{47.72} \end{array}$$

12.
$$\begin{array}{r} 128.62 \\ -\ 41.9 \\ \hline \textbf{86.72} \end{array}$$

13. $\dfrac{2}{7} + \dfrac{2}{7} = \dfrac{4}{7}$

14. $\dfrac{8}{9} - \dfrac{4}{9} = \dfrac{4}{9}$

15. $2\dfrac{3}{4} + 1\dfrac{1}{4} = 3\dfrac{4}{4} = \textbf{4}$

16.
$$\begin{array}{r} \$1.25 \\ \times \quad 32 \\ \hline 250 \\ 375 \\ \hline \textbf{\$40.00} \end{array}$$

17.
$$\begin{array}{r} \$9.06 \\ 4\overline{)\$36.24} \\ \underline{36} \\ 02 \\ \underline{0} \\ 24 \\ \underline{24} \\ 0 \end{array}$$

18.
$$\begin{array}{r} 84 \\ 7\overline{)588} \\ \underline{56} \\ 28 \\ \underline{28} \\ 0 \end{array}$$

19.
$$\begin{array}{r} 16 \\ 30\overline{)480} \\ \underline{30} \\ 180 \\ \underline{180} \\ 0 \end{array}$$

20. (a) $\dfrac{4 \div 4}{12 \div 4} = \dfrac{1}{3}$

(b) $\dfrac{2 \div 2}{8 \div 2} = \dfrac{1}{4}$

(c) $\dfrac{10 \div 5}{15 \div 5} = \dfrac{2}{3}$